Asterisk™
The Future of Telephony

Other resources from O'Reilly

Related titles
Ethernet: The Definitive Guide

TCP/IP Network Administration

Switching to VoIP

T1: A Survival Guide

oreilly.com
oreilly.com is more than a complete catalog of O'Reilly books. You'll also find links to news, events, articles, weblogs, sample chapters, and code examples.

oreillynet.com is the essential portal for developers interested in open and emerging technologies, including new platforms, programming languages, and operating systems.

Conferences
O'Reilly brings diverse innovators together to nurture the ideas that spark revolutionary industries. We specialize in documenting the latest tools and systems, translating the innovator's knowledge into useful skills for those in the trenches. Visit *conferences.oreilly.com* for our upcoming events.

Safari Bookshelf (*safari.oreilly.com*) is the premier online reference library for programmers and IT professionals. Conduct searches across more than 1,000 books. Subscribers can zero in on answers to time-critical questions in a matter of seconds. Read the books on your Bookshelf from cover to cover or simply flip to the page you need. Try it today with a free trial.

Asterisk™
The Future of Telephony

Jim Van Meggelen, Jared Smith, and Leif Madsen

O'REILLY®

Beijing · Cambridge · Farnham · Köln · Paris · Sebastopol · Taipei · Tokyo

Asterisk™: The Future of Telephony
by Jim Van Meggelen, Jared Smith, and Leif Madsen

Published by O'Reilly Media, Inc., 1005 Gravenstein Highway North, Sebastopol, CA 95472.

O'Reilly books may be purchased for educational, business, or sales promotional use. Online editions are also available for most titles (*safari.oreilly.com*). For more information, contact our corporate/institutional sales department: (800) 998-9938 or *corporate@oreilly.com*.

Editor:	Mike Loukides
Production Editor:	Colleen Gorman
Cover Designer:	Ellie Volckhausen
Interior Designer:	David Futato

Printing History:

September 2005:	First Edition.

 This book uses RepKover™, a durable and flexible lay-flat binding.

ISBN: 0-596-00962-3
[M]

Table of Contents

Foreword . . . . **ix**

Preface . . . . **xiii**

1. A Telephony Revolution **1**
VoIP: Bridging the Gap Between Traditional Telephony and Network
Telephony   2
Massive Change Requires Flexible Technology   3
Asterisk: The Hacker's PBX   5
Asterisk: The Professional's PBX   5
The Asterisk Community   5
The Business Case   7
This Book   8

2. Preparing a System for Asterisk **9**
Server Hardware Selection   10
Environment   18
Telephony Hardware   22
Types of Phone   25
Linux Considerations   30
Conclusion   30

3. Installing Asterisk **31**
What Packages Do I Need?   31
Obtaining the Source Code   32
Compiling Zaptel   35
Compiling libpri   41

Compiling Asterisk   41
Installing Additional Prompts   46
Updating Your Source Code   46
Common Compiling Issues   47
Loading Zaptel Modules   50
Loading libpri   52
Loading Asterisk   52
Directories Used by Asterisk   54
Conclusion   57

4. Initial Configuration of Asterisk **58**
What Do I Really Need?   58
Working with Interface Configuration Files   59
FXO and FXS Channels   60
Configuring an FXO Channel   61
Configuring an FXS Channel   65
Configuring SIP   67
Configuring Inbound IAX Connections   72
Configuring Outbound IAX Connections   74
Debugging   75
Conclusion   76

5. Dialplan Basics **77**
Dialplan Syntax   77
A Simple Dialplan   81
Adding Logic to the Dialplan   84
Conclusion   98

6. More Dialplan Concepts **99**
Expressions and Variable Manipulation   99
Dialplan Functions   102
Conditional Branching   103
Voicemail   106
Macros   110
Using the Asterisk Database (AstDB)   112
Handy Asterisk Features   115
Conclusion   118

7. Understanding Telephony **119**

Analog Telephony   119
Digital Telephony   122
The Digital Circuit-Switched Telephone Network   130
Packet-Switched Networks   135
Conclusion   136

8. Protocols for VoIP **137**

The Need for VoIP Protocols   138
VoIP Protocols   139
Codecs   144
Quality of Service   148
Echo   151
Asterisk and VoIP   152
Conclusion   155

9. The Asterisk Gateway Interface (AGI) **156**

Fundamentals of AGI Communication   156
Writing AGI Scripts in Perl   159
Creating AGI Scripts in PHP   163
Writing AGI Scripts in Python   169
Debugging in AGI   172
Conclusion   174

10. Asterisk for the Über-Geek **175**

Festival   175
Call Detail Recording   178
Customizing System Prompts   179
Manager   180
Call Files   182
DUNDi   184
Conclusion   189

11. Asterisk: The Future of Telephony **190**

The Problems with Traditional Telephony   190
Paradigm Shift   193
The Promise of Open Source Telephony   193
The Future of Asterisk   200

A. VoIP Channels 209

B. Application Reference 229

C. AGI Reference 292

D. Configuration Files 301

E. Asterisk Command-Line Interface Reference 337

Index 359

Foreword

Once upon a time, there was a boy.

...with a computer

...and a phone.

This simple beginning begat much trouble!

It wasn't that long ago that telecommunications, both voice and data, as well as software, were all proprietary products and services, controlled by one select club of companies that created the technologies, and another select club of companies who used the products to provide services. By the late 1990s, data telecommunications had been opened by the expansion of the Internet. Prices plummeted. New and innovative technologies, services, and companies emerged. Meanwhile, the work of free software pioneers like Richard Stallman, Linus Torvalds, and countless others were culminating in the creation of a truly open software platform called Linux (or GNU/Linux). However, voice communications, ubiquitous as they were, remained proprietary. Why? Perhaps it was because voice on the old public telephone network lacked the glamor and promise of the shiny new World Wide Web. Or, perhaps it's because a telephone just isn't as effective at supplying adult entertainment. Whatever the reason, one thing was clear. Open source voice communications was about as widespread as open source copy protection software.

Necessity (and in some cases simply being cheap) is truly the mother of invention. In 1999, having started Linux Support Services to offer free and commercial technical support for Linux, I found myself in need (or at least in perceived need) of a phone system to assist me in providing 24-hour technical support. The idea was that people would be able to call in, enter their customer identity, and leave a message. The system would in turn page a technician to respond to the customer's request in short order. Since I had started the company with about $4000 of capital, I was in no position to be able to afford a phone system of the sort that I needed to implement this scenario. Having already been a Linux user since 1994, and having already gotten my feet wet in Open Source software development by starting l2tpd, gaim, and cheops, and in the complete absence of anyone having explained the complexity of such a task, I decided that I would simply make my own phone system using hardware borrowed from Adtran, where I had worked as a co-op student. Once I got a call into a PC, I fantasized, I could do *anything* with it. In fact, it is from this conjecture that the official Asterisk motto (which any sizable, effective project must have) is derived:

It's only software!

For better or worse, I rarely think small. Right from the start, it was my intent that Asterisk would do *everything* related to telephony. The name "Asterisk" was chosen because it was both a key on a standard telephone and also the wildcard symbol in Linux (e.g., `rm -rf *`).

So, in 1999, I have a free telephony platform I've put out on the web and I go about my business trying to eke out a living at providing Linux technical support. However, by 2001, as the economy was tanking, it became apparent that Linux Support Services might do better by pursuing Asterisk than general purpose Linux technical support. That year, we would make contact with Jim "Dude" Dixon of the Zapata Telephony project. Dude's exciting work was a fantastic companion to Asterisk, and provided a business model for us to start pursuing Asterisk with more focus. After creating our first PCI telephony interface card in conjunction with Dude, it became clear that "Linux Support Services" was not the best name for a telephony company, and so we changed the name to "Digium," which is a whole other story that cannot be effectively conveyed in writing. Enter the expansion of Voice over IP ("VoIP") with its disruptive transition of voice from the old, circuit-switched networks to new IP-based networks and things really started to take hold.

Now, as we've already covered, clearly most people don't get very excited about telephones. Certainly, few people could share my excitement the moment I heard dialtone coming from a phone connected to my PC. However, those who *do* get excited about telephones get *really* excited about telephones. And facilitated by the Internet, this small group of people were now able to unite and apply our bizarre passions to a common, practical project for the betterment of many.

To say that telecom was ripe for an open source solution would be an immeasurable understatement. Telecom is an enormous market due to the ubiquity of telephones in work and personal life. The direct market for telecom products has a highly technical audience that is willing and able to contribute. People demand their telecom solutions be infinitely customizable. Proprietary telecom is very expensive. Creating Asterisk was simply the spark in this fuel rich backdrop.

Asterisk sits at the apex of a variety of transitions (Proprietary → Open Source, Circuit Switched → VoIP, Voice only → Voice, Video, and Data, Digital Signal Processing → Host Media Processing, Centralized Directory → Peer to Peer) while easing those transitions by providing bridges back to the older ways of doing things. Asterisk can talk to anything from a 1960s era pulse dial phone to the latest wireless VoIP devices, and provide features from simple tandem switching all the way to bluetooth presence and DUNDi.

Most important of all, though, Asterisk demonstrates how a community of motivated people and companies can work together to create a project with a scope so significant that no one person or company could have possibly created it on its own. In making Asterisk possible, I particularly would like to thank Linus Torvalds, Richard Stallman, the entire Asterisk community and whoever invented Red Bull.

So where is Asterisk going from here? Think about the history of the PC. When it was first introduced in 1980, it had fairly limited capabilities. Maybe you could do a spreadsheet, maybe do some word processing, but in the end, not much. Over time, however, its open architecture led to price reductions and new products allowing it to slowly expand its applications, eventually displacing the mini computer, then the mainframe. Now, even Cray supercomputers are built using Linux-based x86 architectures. I anticipate that Asterisk's future will look very similar. Today, there is a large subset of telephony that is served by Asterisk. Tomorrow, who knows what the limit might be.

So, what are you waiting for? Read, learn, and participate in the future of open telecommunications by joining the Asterisk revolution!

—Mark Spencer

Preface

This is a book for anyone who is new to Asterisk™.

Asterisk is an open source, converged telephony platform, which is designed primarily to run on Linux. Asterisk combines over 100 years of telephony knowledge into a robust suite of tightly integrated telecommunications applications. The power of Asterisk lies in its customizable nature, complemented by unmatched standards-compliance. No other PBX can be deployed in so many creative ways.

Applications such as voicemail, hosted conferencing, call queuing and agents, music on hold, and call parking are all standard features built right into the software. Moreover, Asterisk can integrate with other business technologies in ways that closed, proprietary PBXs can scarcely dream of.

Asterisk can appear quite daunting and complex to a new user, which is why documentation is so important to its growth. Documentation lowers the barrier to entry and helps people contemplate the possibilities.

Produced with the generous support of O'Reilly Media, *Asterisk: The Future of Telephony* was inspired by the work started by the Asterisk Documentation Project. We have come a long way, and this book is the realization of a desire to deliver documentation which introduces the most fundamental elements of Asterisk-the things someone new to Asterisk needs to know. It is the first volume in what we are certain will become a huge library of knowledge relating to Asterisk.

This book was written for, and by, the Asterisk community.

Audience

This book is for those new to Asterisk, but we assume that you're familiar with basic Linux administration, networking, and other IT disciplines. If not, we encourage you to explore the vast and wonderful library of books O'Reilly publishes on these subjects. We also assume you're fairly new to telecommunications, both traditional switched telephony and the new world of voice over IP.

Organization

The book is organized into these chapters:

Chapter 1, *A Telephony Revolution*
> This is where we chop up the kindling, and light the fire. Asterisk is going to change the world of telecom, and this is where we discuss our reasons for that belief.

Chapter 2, *Preparing a System for Asterisk*
> Covers some of the engineering considerations you should have in mind when designing a telecommunications system. Much of this material can be skipped if you want to get right to installing, but these are important concepts to understand, should you ever plan on putting an Asterisk system into production.

Chapter 3, *Installing Asterisk*
> Covers the obtaining, compiling and installation of Asterisk.

Chapter 4, *Initial Configuration of Asterisk*
> Describes the initial configuration of Asterisk. Here we will cover the important configuration files that must exist to define the channels and features available to your system.

Chapter 5, *Dialplan Basics*
> Introduces the heart of Asterisk, the dialplan.

Chapter 6, *More Dialplan Concepts*
> Goes over some more advanced dialplan concepts.

Chapter 7, *Understanding Telephony*
> Taking a break from Asterisk, this chapter discusses some of the more important technologies in use in the Public Telephone Network.

Chapter 8, *Protocols for VoIP*
> Following the discussion of legacy telephony, this chapter discusses Voice over IP.

Chapter 9, *The Asterisk Gateway Interface (AGI)*
> Introduces one of the more amazing components, the Asterisk Gateway Interface. Using Perl, PHP, and Python, we demonstrate how external programs can be used to add nearly limitless functionality to your PBX.

Chapter 10, *Asterisk for the Über-Geek*
> Briefly covers what is, in fact, a rich and varied cornucopia of incredible features and functions; all part of the Asterisk phenomenon.

Chapter 11, *Asterisk: The Future of Telephony*
> Predicts a future where open source telephony completely transforms an industry desperately in need of a revolution.

Software

This book is focused on documenting Asterisk Version 1.2, however many of the conventions and information in this book are version-agnostic. Linux is the operating system we have run and tested Asterisk on, with a leaning towards Red Hat syntax. We decided that while Red Hat-based distributions may not be the preferred choice of everyone; its layout and utilities are nevertheless familiar to many experienced Linux administrators.

Conventions Used in This Book

The following typographical conventions are used in this book:

Italic

> Indicates new terms, URLs, email addresses, filenames, file extensions, pathnames, directories, and Unix utilities.

`Constant width`

> Indicates commands, options, parameters, and arguments that must be substituted into commands.

`Constant width bold`

> Shows commands or other text that should be typed literally by the user. Also used for emphasis in code.

`Constant width italic`

> Shows text that should be replaced with user-supplied values.

`[ Keywords and other stuff ]`

> Indicates optional keywords and arguments.

`{ choice-1 | choice-2 }`

> Signifies either choice-1 or choice-2.

 This icon signifies a tip, suggestion, or general note.

 This icon indicates a warning or caution.

Using Code Examples

This book is here to help you get your job done. In general, you may use the code in this book in your programs and documentation. You do not need to contact us for permission unless you're reproducing a significant portion of the code. For example, writing a program that uses several chunks of code from this book does not require permission. Selling or distributing a CD-ROM of examples from O'Reilly books does require permission. Answering a question by citing this book and quoting example code does not require permission. Incorporating a significant amount of example code from this book into your product's documentation does require permission.

We appreciate, but do not require, attribution. An attribution usually includes the title, author, publisher, and ISBN. For example: "*Asterisk: The Future of Telephony*, by Jim Van Meggelen, Jared Smith, and Leif Madsen. Copyright 2005 O'Reilly Media, Inc., 0-596-00962-3."

If you feel your use of code examples falls outside fair use or the permission given above, feel free to contact us at *permissions@oreilly.com*.

Safari® Enabled

 When you see a Safari® enabled icon on the cover of your favorite technology book, that means the book is available online through the O'Reilly Network Safari Bookshelf.

Safari offers a solution that's better than e-books. It's a virtual library that lets you easily search thousands of top tech books, cut and paste code samples, download chapters, and find quick answers when you need the most accurate, current information. Try it for free at *http://safari.oreilly.com*.

How to Contact Us

Please address comments and questions concerning this book to the publisher:

O'Reilly Media, Inc.
1005 Gravenstein Highway North
Sebastopol, CA 95472
(800) 998-9938 (in the United States or Canada)
(707) 829-0515 (international or local)
(707) 829-0104 (fax)

We have a web page for this book, where we list errata, examples, and any additional information. You can access this page at:

http://www.oreilly.com/catalog/asterisk

To comment or ask technical questions about this book, send email to:

bookquestions@oreilly.com

For more information about our books, conferences, Resource Centers, and the O'Reilly Network, see our web site at:

http://www.oreilly.com

Acknowledgments

Firstly, we have to thank our fantastic editor Michael Loukides, who offered invaluable feedback and found incredibly tactful ways to tell us to re-write a section (or chapter) when it was needed, and have us think it was our idea. Mike built us up when we were down, and brought us back to earth when we got uppity. You are a master, Mike, and seeing how many books have received your editorial oversight contributes to an understanding of why O'Reilly Media is the success it is.

Thanks also to Rachel Wheeler, our copy editor, Colleen Gorman, our production editor, and the rest of the unsung heroes in O'Reilly's production department. These are the folks that take our book and make it an *O'Reilly book*.

Everyone in the Asterisk community needs to thank Jim Dixon for creating the first open-source telephony hardware interfaces, starting the revolution, and giving his creations to the community at large.

Thanks to Tim O'Reilly, for giving us a chance to write this book.

To our most generous and merciless review team:

- Rich Adamson, President of Network Partners Inc., for your encyclopedic knowledge of the PSTN, and your tireless willingness to share your experience. Your generosity, even in the face of daunting challenge, is inspiring to us all.
- Dr. Edward Guy, Chief Scientist, Pulver Innovations, for your comprehensive and razor-sharp evaluation of each and every chapter, and for your championing of Asterisk.
- Kristian Kielhofner, President, KrisCompanies and creator of AstLinux, for the most excellent AstLinux distribution.
- Joel Sisko, Systems Integrator, for braving the fire.
- Travis Smith, for your valuable and timely feedback.
- Ted Wallingford, for leading the way with O'Reilly's: Switching to VoIP.
- Brian K. West, for your commitment to the community, Asterisk, our book, and open-source telephony.
- Joshua Colp, for putting up with, and answering, the numerous questions posed by Leif.
- Robert M. Zigweid, not only for your thorough evaluation of our book (especially for slogging through the appendices), but also for having the coolest name in the universe.

Anthony Minessale (a.k.a. anthm) is one of the unsung heroes of Asterisk development. The number of people who have contributed to Asterisk development are many; the number who can claim to have matched Anthony's efforts are few.

Finally, and most importantly, thanks go to Mark Spencer for GAIM, Asterisk and DUNDi, and for contributing his creations to the open source community.

Leif Madsen

The road to this book is a long one—nearly three years in the making. Back when I started using Asterisk, possibly much like you, I didn't know anything about Asterisk, very little about traditional telephony and even less about voice over IP. I delved right into this new and very exciting world and took in all I could. For two months during a co-op term, for which I couldn't immediately find work, I absorbed as much as I could, asking questions, trying things and seeing what the system could do. Unfortunately very little to no documentation existed for Asterisk aside from some dialplan examples I was able to find by John Todd and having questions answered by Brian K. West on IRC. Of course, this method wasn't going to scale.

Not being much of a coder, I wanted to contribute something back to the community, and what do coders hate doing more than anything? Documentation! So I started The Asterisk Documentation Assignment (TADA), a basic outline with some information for the beginnings of a book.

Shortly after releasing it on my website, an intelligent fellow calling himself Jared Smith introduced himself. He had similar aspirations for creating a "dead-tree" format book for the community, and we humbly started the Asterisk Documentation Project. Jared setup a simple web site at *http://www.asteriskdocs.org*, a CVS server and the very first DocBook formatted version of a book for Asterisk. From there we started filling in information, and soon had information submitted by a number of members of the community.

In June of 2004, an animated chap by the name of Jim Van Meggelen started showing up on the mailing lists, and contributing lots of information and documentation - this was definitely a guy we wanted on our team! Jim had the vision and the drive to really get Jared and my butts in gear and to work on something grander. Jim brought us years of experience and a writing flair which we could hardly have imagined.

With the core documentation team established, we embarked on a plan for the creation of volumes of Asterisk knowledge, eventually to lead to a complete library and wealth of information. This book is essentially the beginning of that dream.

Firstly and mostly, I have to thank my parents, Rick and Carol for always supporting my efforts, allowing me to realize my dreams, and always putting my needs ahead of theirs. Without their vision, understanding and insight into the future, it would have been impossible to have accomplished what I have. I love you both very much!

I'd like to thank Felix Carapaica and Bill Farkas of the Sheridan Institute of Technology for their dedication to the advancement of knowledge. Their teaching has complemented my prior learning, and has allowed me to expand my understanding of routing and telecommunications exponentially.

There are far too many people to thank individually, but of particular importance, the following people were, and are, the most influential to my understanding of Asterisk:, Olle Johansson, Steven Sokol, Joshua Colp, Brian K. West, John Todd—and William Suffill for my very first VoIP phone. And for those who I said I'd mention in the book, thanks!

And of course, I must thank Jared Smith and Jim Van Meggelen for having the vision and understanding of how important documentation really is—all of this would have been impossible without you.

Jared Smith

I first started working with Asterisk in the spring of 2002. I had recently started a new job with a market research company, and ended up taking a long road trip to a remote call center with the CIO. On the long drive home we talked about innovation in telephony, and he mentioned a little open-source telephony project he had heard of called Asterisk. Over the next few months, I was able to talk the company into buying a developers kit from Digium and start playing with Asterisk on company time.

Over the next few months, I became more and more involved with the Asterisk community. I read the mailing lists. I scoured the archives. I hung out in the IRC channel, just hoping to find nuggets of Asterisk knowledge. As time went on, I was finally able to figure out enough to get Asterisk up and running.

That's when the real fun began.

With the help of the CIO and the approval of the CEO, we moved forward with plans to move our entire telecom infrastructure to Asterisk, including our corporate office and all of our remote call centers. Along the way, we ran into a lot of uncharted territory, and I began thinking about creating a good repository of Asterisk knowledge. Over the course of the project, we were able to do some really innovative things, such as invent IAX trunking!

When all was said and done, we ended up with around forty Asterisk servers spread across many different geographical locations, all communicating with each other to provide a cohesive enterprise-class VoIP phone system. It currently handles approximately one million minutes of calls per month, serves several hundred employees, connects to 27 voice T1s, and saves the company around $20,000 (USD) per month on their telecom costs. In short, our Asterisk project was a resounding success!

While in the middle of implementing this project, I met Leif in one of the Asterisk IRC channels. We talked about ways we could help out new Asterisk users and lower the barrier to entry, and we decided to push ahead with plans to more fully document Asterisk. I really wanted some good documentation in "dead-tree" format — basically a book that a new user could pick up and learn the basics of Asterisk. About that same time, the number of new users on the Asterisk mailing lists and in the IRC channels grew tremendously, and we felt that writing an Asterisk book would greatly improve the signal-to-noise ratio. The Asterisk Documentation Project was born! The rest, they say, is history.

Since then, we've been writing Asterisk documentation. I never thought it would be this arduous, yet rewarding. (I joked with Leif and Jim that it might be easier and less controversial to write an in-depth tome called "Religion, Gun Control, and Sushi" than cover everything that Asterisk has to offer in sufficient detail!) What you see here is a direct result of a lot of late nights and long weekends spent helping the Asterisk community—after all, it's the least we could do, considering what Asterisk has given to us. We hope it will inspire other members of the Asterisk community to help document changes and new features, for the benefit of all involved.

Now to thank some people:

First of all, I'd like to thank my beautiful wife. She's put up with a lot of lonely nights while I've been slaving away at the keyboard, and I'd like her to know how much I appreciate her and her endless support. I'd also like to thank my kids for doing their best to remind me of the important things in life. I love you!

To my parents: thanks for everything you've done to help me stretch and grow and learn over the years. You're the best parents a person could ask for.

To Dave Carr and Michael Lundberg: thanks for letting me learn Asterisk on company time. Working with both of you was truly a pleasure. May God smile upon you and grant you success and joy in all you do.

To Leif and Jim: thanks for putting up with my stupid jokes, my insistence that we do things "the right way," and my crazy schedule. Thanks for pushing me along, and making me a better writer. I've really enjoyed working with you two, and hope to collaborate with you on future projects!

To Mark Spencer: thank you for your continued support and dedication and friendship. You've been an invaluable resource to our effort, and I truly believe that you've started a revolution in the world of telephony. You're always welcome in my home and at my dinner table!

To the other great people at Digium: thank you for your help and support. We're especially thankful for you willingness to give us more insight into the Asterisk code, and for donating hardware so that we can better document the Asterisk Developer's Kit.

To Steven Sokol, Steven Critchfield, Olle E. Johansson, and all the others who have contributed to the Asterisk Documentation Project and to this book: thank you! We couldn't have done it without your help and suggestions.

Jim Van Meggelen

For me, it all started in the spring of 2004, sitting at my desk in the technical support department of the telecom company I'd worked at for nearly fifteen years. With no challenges worthy of my skills, I spent my time trying to figure out what I had achieved in the last fifteen years. I was stuck in an industry that had squandered far too many opportunities, and had as a result caused itself a spectacular and embarrassing fall from being the darling of investors to a joke known to even the most uneducated. I was supposed to feel fortunate to be one of the few who still had work, but what thankless, purposeless work it was. We knew why our industry had collapsed: the products we sold could not hope to deliver the solutions our customers required—even though the industry promised that they could. They lacked flexibility, and were priced totally out of step with the functionality they were delivering (or, more to the point, were failing to deliver). Nowhere in the industry were there any signs this was going to change any time soon.

I had been dreaming of an open-source PBX for many long years, but I really didn't know how such a thing could ever come to be—I'd given up on the idea several years before. I knew that to be successful, an open source PBX would need to effectively bridge the worlds of legacy and network-based telecom. I always failed to find anything that seemed ready.

Then, one fine day in spring, I half-heartedly seeded a Google search with the phrase "open source telephony," and discovered a bright new future for telecom: Asterisk, the Open Source Linux PBX.

There it was: the very thing I'd been dreaming of for so many years. The clouds parted, the sun shone through; adventure lay ahead. I had no idea how I was going to contribute, but I knew this: open-source telephony was going to cause a necessary and beneficial revolution in the telecom industry; and one way or another, I was going to be a part of it.

For me, more of a systems integrator than developer, I needed a way to contribute to the community. There didn't seem to be a shortage of developers, but there sure was a shortage of documentation. This sounded like something I could do. I knew how to write, I knew a thing or two about PBXs, and I desperately needed to talk about this phenomenon that suddenly made telecom fun again.

If I contribute only one thing to this book, I hope you will catch some of my enthusiasm for the subject of open-source telephony. This is an incredible gift we have been given, but also an incredible responsibility. What a wonderful challenge. What a cosmic opportunity. What delicious fun!

First of all, I need to thank Leif and Jared for inviting me to join the Asterisk Documentation Project. I have immensely enjoyed working with both of you, and I am constantly amazed at how well our personalities and skills complement each other. A truly balanced team, are we.

To my wife Killi, and my children Kaara, Joonas, and Joosep (who always remember to visit me when I disappear into my underground lair for too long): you are a source of inspiration to me. Your love is the fuel that feeds my fire, and I thank you.

Obviously, I need to thank my parents Jack and Martiny, for always believing in me, no matter how many rules I broke. In a few years, I'll have my own teenagers, and it'll be your turn to laugh!

To Mark Spencer: thanks for all the things that everybody else thanks you for, but also, personally, thanks for giving generously of your time to the Asterisk community. The Toronto Asterisk Users' Group (*http://www.taug.ca*) made a quantum leap forward as a result of your taking the time to speak to us, and that event will forever form a part of our history. Oh yeah, and thanks for the beers, too. :-)

Finally, thanks to the Asterisk Community. This book is our gift to you. We hope you enjoy reading it as much as we've enjoyed writing it.

## CHAPTER 1

A Telephony Revolution

It does not require a majority to prevail,
but rather an irate, tireless minority
keen to set brush fires in people's minds.
—Samuel Adams

An incredible revolution is under way. It has been a long time in coming, but now that it has started, there will be no stopping it. It is taking place in an area of technology that has lapsed embarrassingly far behind every other industry that calls itself high-tech. The industry is telecommunications, and the revolution is being fueled by an open source Private Branch eXchange (PBX) called *Asterisk™*.

Telecommunications is arguably the last major electronics industry that has (until now) remained untouched by the open source revolution. Major telecommunications manufacturers still build ridiculously expensive, incompatible systems, running complicated, ancient code on impressively engineered yet obsolete hardware.

As an example, Nortel's Business Communications Manager kludges together a Windows NT 4.0 server, a 15-year-old VXWorks-based Key Telephone Switch, and a 700-MHz PC. All this can be yours for between 5 and 15 thousand dollars, not including telephones. If you want it to actually do anything interesting, you'll have to pay extra licensing fees for closed, limited-functionality, shrink-wrapped applications. Customization? Forget it—it's not in the plan. Future technology and standards compliance? Give them a year or two—they're working on it.

All of the major telecommunications manufacturers offer similar-minded products. They don't want you to have flexibility or choice; they want you to be locked in to their product cycles.

Asterisk changes all that. With Asterisk, no one is telling you how your phone system works, or what technology you are limited to. If you want it, you can have it. Asterisk lovingly embraces the concept of standards compliance, while also enjoying the freedom to develop its own innovations. What you choose to implement is up to you-Asterisk imposes no limits.

Naturally, this incredible flexibility comes with a price: Asterisk is not a simple system to configure. This is not because it's illogical, confusing, or cryptic; to the contrary, it is very sensible and practical. People's eyes light up when they first see an Asterisk dialplan and begin to contemplate the possibilities. But when there are literally thousands of ways to achieve a result, the process naturally requires extra effort. Perhaps it can be compared to building a house: the components are relatively easy to understand, but a person contemplating such a task must either a) enlist competent help or b) develop the required skills through instruction, practice, and a good book on the subject.

VoIP: Bridging the Gap Between Traditional Telephony and Network Telephony

While Voice over IP (VoIP) is often thought of as little more than a method of obtaining free long-distance calling, the real value (and—let's be honest—challenge as well) of VoIP is that it allows voice to become nothing more than another application in the data network.

It sometimes seems that we've forgotten that the purpose of the telephone is to allow people to communicate. It is a simple goal, really, and it should be possible for us to make it happen in far more flexible and creative ways than are currently available to us. Since the industry has demonstrated an unwillingness to pursue this goal, a large community of passionate people have taken on the task.

The challenge comes from the fact that an industry that has changed very little in the last century shows little interest in starting now.

The Zapata Telephony Project

The Zapata Telephony Project was conceived of by Jim Dixon, a telecommunications consulting engineer who was inspired by the incredible advances in CPU speeds that the computer industry has now come to take for granted. Dixon's belief was that far more economical telephony systems could be created if a card existed that had nothing more on it than the basic electronic components required to interface with a telephone circuit. Rather than having expensive components on the card, Digital Signal Processing (DSP)* would be handled in the CPU by software. While this would impose a tremendous load on the CPU, Dixon was certain that the low cost of CPUs relative to their performance made them far more attractive than

---

* The term DSP also means Digital Signal Processor, which is a device (usually a chip) that is capable of interpreting and modifying signals of various sorts. In a voice network, DSPs are primarily responsible for encoding, decoding, and transcoding audio information. This can require a lot of computational effort.

expensive DSPs, and, more importantly, that this price/performance ratio would continue to improve as CPUs continued to increase in power.

Like so many visionaries, Dixon believed that many others would see this opportunity, and that he merely had to wait for someone else to create what to him was an obvious improvement. After a few years, he noticed that not only had no one created these cards, but it seemed unlikely that anyone was ever going to. At that point it was clear that if he wanted a revolution, he was going to have to start it himself. And so the Zapata Telephony Project was born.

> Since this concept was so revolutionary, and was certain to make a lot of waves in the industry, I decided on the Mexican revolutionary motif, and named the technology and organization after the famous Mexican revolutionary Emiliano Zapata. I decided to call the card the 'tormenta' which, in Spanish, means 'storm,' but contextually is usually used to imply a big storm, like a hurricane or such.[*]

Perhaps we should be calling ourselves Asteristas. Regardless, we owe Jim Dixon a debt of thanks, partly for thinking this up and partly for seeing it through, but mostly for giving the results of his efforts to the open source community. As a result of Jim's contribution, Asterisk's Public Switched Telephone Network (PSTN) engine came to be.

Massive Change Requires Flexible Technology

The most successful key telephone system in the world has a design limitation that has survived 15 years of users begging for what appears to be a simple change: when you determine the number of times your phone will ring before it forwards to voicemail, you can choose from 2, 3, 4, 6, or 10 ring cycles. Have you any idea how many times people ask for five rings? Yet the manufacturers absolutely cannot get their heads around the idea that this is a problem. That's the way it works, they say, and users need to get over it.

That's just one example—the industry is rife with them.

Another example from the same system is that the name you program on your set can only be seven characters in length. Back in the late 1980s, when this particular system was built, RAM was pretty dear, and storing those seven characters for dozens of sets represented a huge hardware expense. So what's the excuse today? None. Are there any plans to change it? Hardly—the issue is not even officially acknowledged as a problem.

Now, it's all very well and good to pick on one system, but the reality is that every PBX in existence suffers shortcomings. No matter how fully featured it is, something will always be left out, because even the most feature-rich PBX will always fail to anticipate

---

[*] Jim Dixon, "The History of Zapata Telephony and How It Relates to the Asterisk PBX" (*http://www.asteriskdocs.org/modules/tinycontent/index.php?id=10*).

the creativity of the customer. A small group of users will desire an odd little feature that the design team either did not think of or could not justify the cost of building, and, since the system is closed, the users will not be able to build it themselves.

If the Internet had been thusly hampered by regulation and commercial interests, it is doubtful that it would have developed the wide acceptance it currently enjoys. The openness of the Internet meant that anyone could afford to get involved. So, everyone did. The tens of thousands of minds that collaborated on the creation of the Internet delivered something that no corporation ever could have.

As with many other open source projects, such as Linux and the Internet, the explosion of Asterisk was fueled by the dreams of folks who knew that there had to be something more than what the industry was producing. The strength of the community is that it is composed not of employees assigned to specific tasks, but rather of folks from all sorts of industries, with all sorts of experiences, and all sorts of ideas about what flexibility means, and what openness means. These people knew that if one could take the best parts of various PBXs and separate them into interconnecting components—akin to a boxful of LEGO bricks—one could begin to conceive of things that would not survive a traditional corporate risk-analysis process. While no one can seriously claim to have a complete picture of what this thing should look like, there is no shortage of opinions and ideas.

Many people new to Asterisk see it as unfinished. Perhaps these people can be likened to visitors to an art studio, looking to obtain a signed, numbered print. They often leave disappointed, because they discover that Asterisk is the blank canvas, the tubes of paint, the unused brushes waiting.

Even at this early stage in its success, Asterisk is nurtured by a greater number of artists than any other PBX. Most manufacturers dedicate no more than a few developers to any one product; Asterisk has scores. Most proprietary PBXs have a worldwide support team comprised of a few dozen real experts; Asterisk has hundreds.

The depth and breadth of expertise that surrounds this product is unmatched in the telecom industry. Asterisk enjoys the loving attention of old Telco guys who remember when rotary dial mattered, enterprise telecom people who recall when voicemail was the hottest new technology, and data communications geeks and coders who helped build the Internet. These people all share a common belief: that the telecommunications industry needs a *proper* revolution.[*]

Asterisk is the catalyst.

---

[*] The telecom industry has been predicting a revolution since before the crash; time will tell how well they respond to the *open source* revolution.

Asterisk: The Hacker's PBX

Telecommunications companies who choose to ignore Asterisk do so at their peril. The flexibility it delivers creates possibilities that the best proprietary systems can scarcely dream of. This is because Asterisk is the ultimate hacker's PBX.

If someone asks you not to use the term hacker, refuse. That term does not belong to the mass media. They stole it and corrupted it to mean "malicious cracker." It's time we took it back. Hackers built the networking engine that is the Internet. Hackers built the Apple Macintosh and the Unix operating system. Hackers are also building your next telecom system. Do not fear; these are the good guys, and they'll be able to build a system that's far more secure than anything that exists today, because rather than being constricted by the dubious and easily cracked security of closed systems, they will be able to quickly respond to changing trends in security and fine-tune the telephone system in response to both corporate policy and industry best practices.

Like other open source systems, Asterisk will be able to evolve into a far more secure platform than any proprietary system, not in spite of its hacker roots, but rather because of them.

Asterisk: The Professional's PBX

Never in the history of telecommunications has a system so suited to the needs of business been available, at any price. Asterisk is an enabling technology, and, as with Linux, it will become increasingly rare to find an enterprise that is not running some version of Asterisk, in some capacity, somewhere in the network, solving a problem as only Asterisk can.

This acceptance is likely to happen much faster than it did with Linux, though, for several reasons:

1. Linux has already blazed the trail that led to open source acceptance, so Asterisk can follow that lead.
2. The telecom industry is crippled, with no leadership being provided by the giant industry players. Asterisk has a compelling, realistic, and exciting vision.
3. End users are fed up with incompatible, limited functionality, and horrible support. Asterisk solves the first two problems; the community has shown a passion for the latter.

The Asterisk Community

One of the compelling strengths of Asterisk is the passionate community that developed and supports it. This community, led by Mark Spencer of Digium, is keenly aware of the cultural significance of Asterisk, and they are giddy about the future.

One of the more powerful side effects caused by the energy of the Asterisk community is the cooperation it has spawned among the telecommunications professionals, networking professionals, and information technology professionals who share a love for this phenomenon. While these professions have traditionally been at odds with each other, in the Asterisk community they delight in each other's skills. The significance of this cooperation cannot be underestimated.

Still, if the dream of Asterisk is to be realized, the community must grow—yet one of the key challenges the community currently faces is a rapid influx of new users. The members of the existing community, having birthed this thing called Asterisk, are generally welcoming of new users, but they've grown impatient with being asked the kinds of questions whose answers can often be obtained independently, if one is willing to put forth the time needed to research and experiment.

Obviously, new users do not fit any particular kind of mold. While some will happily spend hours experimenting and reading various blogs describing the trials and tribulations of others, many people who have become enthusiastic about this technology are completely uninterested in such pursuits. They want a simple, straightforward, step-by-step guide that'll get them up and running, followed by some sensible examples describing the best methods of implementing common functionality (such as voicemail, auto attendants, and the like).

To the members of the expert community, who (correctly) perceive that Asterisk is like a programming language, this approach doesn't make any sense. To them, it's clear that you have to immerse yourself in Asterisk to appreciate its subtleties. Would one ask for a step-by-step guide to programming and expect to learn from it all that a language has to offer?

Clearly, there's no one approach that's right for everyone. Asterisk is a different animal altogether, and it requires a totally different mindset. As you explore the community, though, be aware that there are people with many different skill sets and attitudes here. Some of these folks do not display much patience with new users, but that's often due to their passion for the subject, not because they don't welcome your participation.

The Asterisk Mailing Lists

As with any community, there are places where members of the Asterisk community meet to discuss matters of mutual interest. Of the mailing lists you will find at *http:// lists.digium.com*, these three are currently the most important:

Asterisk-Biz

> Anything commercial with respect to Asterisk belongs in this list. If you're selling something Asterisk-related, sell it here. If you want to buy an Asterisk service or product, post here.

Asterisk-Dev

> The Asterisk developers hang out here. The purpose of this list is the discussion of the development of the software that is Asterisk, and its participants

vigorously defend that purpose. Expect a lot of heat if you post anything to this list not relating to programming or development.

Asterisk-Users

This is where most Asterisk users hang out. This list generates several hundred messages per day and has over ten thousand subscribers. While you can go here for help, you are expected to have done some reading on your own before you post a query.

The Asterisk Wiki

The Asterisk Wiki is a source of much enlightenment and confusion. A community-maintained repository of VoIP knowledge, *http://www.voip-info.org* contains a truly inspiring mess of fascinating, informative, and frequently contradictory information about many subjects, just one of which is Asterisk.

Since Asterisk documentation forms by far the bulk of the information on this web site, and it probably contains more Asterisk knowledge than all other sources put together (with the exception of the mailing-list archives), it is commonly referred to as the place to go for Asterisk knowledge.

The IRC Channels

The Asterisk community maintains Internet Relay Chat channels on *irc.freenode.net*. The two most active are *#Asterisk* and *#Asterisk-Dev*. To cut down on spam-bot intrusions, both of these channels now require registration to join.

The Asterisk Documentation Project

The Asterisk Documentation Project was started by Leif Madsen and Jared Smith. Many people in the community have contributed.

The goal of the documentation project is to provide a structured repository of written work on Asterisk. In contrast with the flexible and ad hoc nature of the Wiki, the Docs project is passionate about building a more focused approach to various Asterisk-related subjects.

As part of the efforts of the Asterisk Docs project to make documentation available online, this book is available at the *http://www.asteriskdocs.org* web site, under a Creative Commons license.

The Business Case

It is very rare to find businesses these days that do not have to reinvent themselves every few years. It is equally rare to find a business that can afford to replace its communications infrastructure each time it goes in a new direction. Today's businesses need extreme flexibility in all of their technology, including telecom.

In his book *Crossing the Chasm* (HarperBusiness), Geoffrey Moore opines, "The idea that the value of the system will be discovered rather than known at the time of installation implies, in turn, that product flexibility and adaptability, as well as ongoing account service, should be critical components of any buyer's evaluation checklist." What this means, in part, is that the true value of a technology is often not known until it has been deployed.

How compelling, then, to have a system that holds at its very heart the concept of openness and the value of continuous innovation.

This Book

So where to begin? Well, when it comes to Asterisk, there is far more to talk about than we can fit into one book. For now, we're not going to take you down all the roads that the über-geeks follow—we're just going to give you the basics.

In Chapter 2, we cover some of the engineering considerations you should have in mind when designing a telecommunications system. You can skip much of this material if you want to get right to installing, but these are important concepts to understand, should you ever plan on putting an Asterisk system into production.

Chapter 3 covers obtaining, compiling, and installing Asterisk, and Chapter 4 deals with the initial configuration of Asterisk. Here we cover the important configuration files that must exist to define the channels and features available to your system. This will prepare you for Chapter 5, where we introduce the heart of Asterisk, the dialplan. Having covered dialplan basics, Chapter 6 introduces some more advanced dialplan concepts.

We will take a break from Asterisk in Chapter 7, and discuss some of the more important technologies in use in the PSTN. Naturally, following the discussion of legacy telephony, Chapter 8 discusses Voice over IP.

Chapter 9 introduces one of the more amazing components, the Asterisk Gateway Interface (AGI). Using Perl, PHP, and Python, we demonstrate how external programs can be used to add nearly limitless functionality to your PBX. In Chapter 10, we briefly cover what is, in fact, a rich and varied cornucopia of incredible features and functions, all of which are part of the Asterisk phenomenon. To conclude, Chapter 11 looks forward, predicting a future where open source telephony completely transforms an industry desperately in need of a revolution. You'll also find a wealth of reference information in the book's five appendixes.

This book can only lay down the basics, but from this foundation, you will be able to come to an understanding of the concept of Asterisk—and from that, who knows what you will build?

## CHAPTER 2

Preparing a System for Asterisk

Very early on, I knew that someday in some "perfect"
future out there over the horizon, it would be
commonplace for computers to handle all of the
necessary processing functionality internally,
making the necessary external hardware to connect up
to telecom interfaces VERY inexpensive
and in some cases trivial.

—Jim Dixon, "The History of Zapata Telephony and
How It Relates to the Asterisk PBX"

By this point, you must be anxious to get your Asterisk system up and running. If you are building a hobby system, you can probably jump right to the next chapter and begin the installation. For a mission-critical deployment, however, some thought must be given to the environment in which the Asterisk system will run. Make no mistake: Asterisk, being a very flexible piece of software, will happily and success-fully install on nearly any Linux platform you can conceive of, and several non-Linux platforms as well.* However, to arm you with an understanding of the type of operat-ing environment Asterisk will really thrive in, this chapter will discuss issues you need to be aware of in order to deliver a reliable, well-designed system.

In terms of its resource requirements, Asterisk's needs are similar to those of an embedded, real-time application. This is due in large part to its need to have priority access to the processor and system buses. It is therefore imperative that any func-tions on the system not directly related to the call-processing tasks of Asterisk be run at a low priority, if at all. On smaller systems and hobby systems, this might not be as much of an issue. However, on high-capacity systems, performance shortcomings will manifest as audio quality problems for users, often experienced as echo, static,

---

* People have successfully compiled and run Asterisk on WRAP boards, Linksys WRT54G routers, Soekris systems, Pentium 100s, PDAs, Apple Macs, Sun SPARCs, laptops, and more. Of course, whether you would *want* to put such a system into production is another matter entirely. (Actually, the AstLinux distribution, by Kristian Kielhofner, runs very well indeed on the Soekris 4801 board. Once you've grasped the basics of Asterisk, this is something worth looking into further. Check out *http://www.astlinux.org*.)

and the like. The symptoms will resemble those experienced on a cell phone when going out of range, although the underlying causes will be different. As loads increase, the system will have increasing difficulty maintaining connections. For a PBX, such a situation is nothing short of disastrous, so careful attention to performance requirements is a critical consideration during the platform selection process.

Table 2-1 lists some very basic guidelines that you'll want to keep in mind when planning your system. The next section takes a close look at the various design and implementation issues that will affect its performance.

Table 2-1. System requirement guidelines

Purpose	Number of channels	Minimum recommended
Hobby system	No more than 5	400-MHz x86, 256 MB RAM
SOHOª system	5 to 10	1-GHz x86, 512 MB RAM
Small business system	Up to 15	3-GHz x86, 1 GB RAM
Medium to large system	More than 15	Dual CPUs, possibly also multiple servers in a distributed architecture

ª Small Office/home Office—less than three lines and five sets.

With large Asterisk installations, it is common to deploy functionality across several servers. One or more central units will be dedicated to call processing; these will be complemented by one or more ancillary servers handling peripherals (such as a database, voicemail, conferencing, management, a web interface, a firewall, and so on). As is true in most Linux environments, Asterisk is well suited to growing with your needs: a small system that used to be able to handle all your call-processing and peripheral tasks can be distributed between several servers when increased demands exceed its abilities. Flexibility is a key reason why Asterisk is extremely cost-effective for rapidly growing businesses—there is no effective maximum or minimum size to consider when budgeting the initial purchase. While some scalability is possible with most telephone systems, we have yet to hear of one that can scale as inexpensively as Asterisk. Having said that, distributed Asterisk systems are not simple to design—this is not a task for someone new to Asterisk.*

Server Hardware Selection

The selection of a server is both simple and complicated: simple because, really, any x86-based platform will suffice; but complicated because the reliable performance of your system will depend on the care that is put into the platform design. When

---

* If you are sure that you need to set up a distributed Asterisk system, you will want to study the DUNDi protocol. You should probably get the interest of the *Asterisk-Users* mailing list as well, but be sure to wear your flame-retardant suit; for some reason, this subject can spur a heated (but generally very educational) debate.

selecting your hardware, you must carefully consider the overall design of your system and what functionality you need to support. This will help you determine your requirements for the CPU, motherboard, and power supply. If you are simply setting up your first Asterisk system for the purpose of learning, you can safely ignore the information in this section. If, however, you are building a mission-critical system suitable for deployment, these are issues that require some thought.

Performance Issues

Among other considerations, when selecting the hardware for an Asterisk installation you must bear in mind this critical question: how powerful must the system be? This is not an easy question to answer, because the manner in which the system is to be used will play a big role in the resources it will consume. There is no such thing as an Asterisk performance-engineering matrix, so you will need to understand how Asterisk uses the system in order to make intelligent decisions about what kinds of resources will be required. You will need to consider several factors, including:

The maximum number of concurrent connections the system will be expected to support
Each connection will increase the workload on the system.

The percentage of traffic that will require processor-intensive Digital Signal Processing (DSP) of compressed codecs (such as G.729 and GSM)[*]
The DSP work that Asterisk performs in software can have a staggering impact on the number of concurrent calls it will support. A system that can happily handle 50 concurrent G.711 calls can be brought to its knees by a request to conference together 10 G.729 compressed channels.

Whether conferencing will be provided, and what level of conferencing activity is expected
Will the system be used heavily? Conferencing requires the system to transcode and mix each individual incoming audio stream into multiple outgoing streams. Mixing multiple audio streams in near-real-time can place an enormous load on the CPU.

Echo cancellation[†]
Echo cancellation may be required on any calls where a Public Switched Telephone Network (PSTN) interface is involved. Since echo cancellation is a mathematical function—the more of it the system has to perform, the higher the load on the CPU will be.

Dialplan scripting logic
Whenever Asterisk has to pass call control to an external program, there is a performance penalty. As much logic as possible should be built into the dialplan. If external scripts are used, they should be designed with performance and efficiency as important goals.

---

[*] We'll talk more about G.729, GSM, G.711, and many other codecs in Chapter 8.

[†] Do not fear. Echo cancellation is another topic for Chapter 8.

As for the exact performance impact of these factors, the jury's still out. The effect of each is known in general terms, but an accurate performance calculator has not yet been successfully defined. This is partly because the effect of each component of the system is dependent on numerous variables, such as CPU power, motherboard chipset and overall quality, total traffic load on the system, Linux kernel optimizations, network traffic, number and type of PSTN interfaces, and PSTN traffic—not to mention any non-Asterisk services the system is performing concurrently. Let's take a look at the effects of several key factors:

Codecs and transcoding
> Simply put, a *codec* (short for coder/decoder or compression/decompression) is a set of mathematical rules that define how an analog waveform will be digitized. The differences between the various codecs are due in large part to the levels of compression and quality that they offer. Generally speaking, the more compression that's required, the more work the DSP must do to code or decode the signal. Uncompressed codecs, therefore, put far less strain on the CPU (but require more network bandwidth). Codec selection must strike a balance between bandwidth and processor usage.

Central Processing Unit (and Floating Point Unit)
> A CPU is comprised of several components, one of which is the Floating Point Unit (FPU). The speed of the CPU, coupled with the efficiency of its FPU, will play a significant role in the number of users a system can effectively support. The next section, "Choosing a Processor," offers guidelines for choosing a CPU that will meet the needs of your system.

Other processes running concurrently on the system
> Being Unix-like, Linux is designed to be able to multitask several different processes. A problem arises when one of those processes (such as Asterisk) demands a very high level of responsiveness from the system. By default, Linux will equally distribute resources amongst every application that requests them. If you install a system with many different server applications, those applications will each be allowed their fair use of the CPU. Since Asterisk requires frequent high-priority access to the CPU, it does not get along well with other applications, and if Asterisk must coexist with other apps, the system may require special optimizations. This primarily involves the assignment of priorities to various applications in the system, and, during installation, careful attention to which applications are installed as services.

Kernel optimizations
> A kernel optimized for the performance of one specific application is something that very few Linux distributions offer by default, and thus it requires some thought. At the very minimum—whichever distribution you choose—a fresh copy of the Linux kernel (available from *http://www.kernel.org*) should be downloaded and compiled on your platform. You may also be able to acquire patches

that will yield performance improvements, but these are considered hacks to the officially supported kernel.

IRQ latency

Interrupt request (IRQ) latency is basically the delay between the moment a peripheral card (such as a telephone interface card) requests that the CPU stop what it's doing and the moment when the CPU actually responds and is ready to handle the task. Asterisk's peripherals (especially the Zaptel cards) are extremely intolerant of IRQ latency.

 Linux has historically had problems with its ability to service IRQs quickly; this problem has caused enough trouble for audio developers that several patches have been created to address this shortcoming. So far, there has been some mild controversy over how to incorporate these patches into the Linux kernel.

Because the Digium cards require so much, it is generally recommended that only one Digium card be run in a system. If you require more connectivity than a single card can provide, either replace your existing card with one of higher density, or add another server to your environment.[*]

Kernel version

Asterisk is officially supported on Linux Version 2.6.

Linux distribution

Linux distributions are many and varied. In the next chapter, we will discuss the challenge of selecting a Linux distribution, and how to obtain and install both Linux and Asterisk.

Choosing a Processor

Since the performance demands of Asterisk will generally involve a large number of math calculations, it is essential that you select a processor with a powerful FPU. The signal processing that Asterisk performs can quickly demand a staggering quantity of complex mathematical computations from the CPU. The efficiency with which these tasks are carried out will be determined by the power of the FPU within that processor.

To actually name a best processor for Asterisk would fly in the face of the rapid advances in the computer industry. Even in the time between the authoring and publishing of this book, processor speeds will undergo rapid improvements, as will Asterisk's support for various architectures. Obviously, this is a good thing, but it also makes the giving of advice on the topic a thankless task. Naturally, the more

---

[*] Many people report that Sangoma cards are more robust when it comes to dealing with unpredictable motherboard chipsets, and thus can handle sharing motherboard IRQ resources. Regardless, it is still worth considering using multiple servers, as the redundancy that can be gained from this strategy can quickly offset the cost.

powerful the FPU is, the more concurrent DSP tasks Asterisk will be able to handle, so that is the ultimate consideration. When you are selecting a processor, the raw clock speed is only part of the equation. How well it handles floating-point operations will be a key differentiator, as DSP operations in Asterisk will place a large demand on it.

Both Intel and AMD CPUs have powerful FPUs. As of this writing, the Intel chips are commonly preferred for 32-bit systems, while AMD gets the nod if you're going to 64-bit. When you read this book, that may no longer be true.*

The obvious conclusion is that you should get the most powerful CPU your budget will allow. However, don't be too quick to buy the most expensive CPU out there. You'll need to keep the requirements of your system in mind—after all, a Formula 1 Ferrari is ill-suited to the rigors of rush-hour traffic.

In order to attempt to provide a frame of reference from which we can contemplate our platform decision, we have chosen to define three sizes of Asterisk systems: small, medium, and large.

Small systems

Small systems (up to 10 phones) are not immune to the performance requirements of Asterisk, but the typical load that will be placed on a smaller system will generally fall within the capabilities of a modern processor.

If you are building a small system from older components you have lying around, be aware that the resulting system cannot be expected to perform at the same level as a more powerful machine, and will run into performance degradation under a much lighter load. Hobby systems can be run successfully on very low-powered hardware,† although this is by no means recommended for anyone who is not a whiz at Linux performance tuning.

If you are setting up an Asterisk system for the purposes of learning, you will be able to build a fully featured platform using a relatively low-powered CPU. The authors run several Asterisk lab systems with 433-MHz to 700-MHz Celeron processors; the workload of these systems is typically minimal.

---

* If you want to be completely up to the minute on which CPUs are leading the performance race, surf on over to Tom's Hardware (*http://www.tomshardware.com*) or AnandTech (*http://www.anandtech.com*), where you will find a wealth of information about both current and out-of-date CPUs, motherboards, and chipsets.

† A 133-MHz Pentium system is known to be running Asterisk, but performance problems are likely, and properly configuring such a system requires an expert knowledge of Linux. We do not recommend running Asterisk on anything less than a 500-MHz system (for a production system, 2 GHz might be a sensible minimum), but we think the fact that Asterisk is so flexible is remarkable.

Medium systems

Medium-sized systems (from 10 to 50 phones) are where performance consider-ations will be the most challenging to resolve. Generally, these systems will be deployed on one or two servers only, and thus each machine will be required to han-dle more than one specific task. As loads increase, the limits of the platform will become increasingly stressed. Users may begin to perceive quality problems without realizing that the system is not faulty in any way, but simply exceeding its capacity. These problems will get progressively worse as more and more load is placed on the system, with the user experience degrading accordingly. It is critical that perfor-mance problems be identified and addressed before they are noticed by users.

Monitoring performance on these systems and quickly acting on any developing trends is a key to ensuring that a quality telephony platform is provided.

Large systems

Large systems (over 50 users) can be distributed across multiple cores, and thus per-formance concerns can be managed through the addition of machines. Very large Asterisk systems—from 500 to over 1,000 users—have been created in this way. Building a large system requires an advanced level of knowledge in many different disciplines. We will not discuss it in detail in this book, other than to say that the issues you'll encounter will be similar to those encountered during any deployment of multiple servers handling a single, distributed task.

Choosing a Motherboard

Just to get any anticipation out of the way, we also cannot recommend specific motherboards in this book. With new motherboards coming out on a weekly basis, any recommendations we made would be rendered moot by obsolescence before the published copy hit the shelves. Not only that, but motherboards are like automo-biles: while they are all very similar in principle, the difference is in the details. And as Asterisk is a performance application, the details matter.

What we will do, therefore, is give you some idea of the kinds of motherboards that can be expected to work well with Asterisk, and the features that will make for a good motherboard. The key is to have both stability and high performance. Here are some guidelines to follow:

- The various system buses must provide the minimum possible latency. If you are planning a PSTN connection using analog or PRI interfaces (discussed later in this chapter), the Digium Zaptel cards will generate 1,000 interrupt requests per second. Having devices on the bus that interfere with this process will result in degradation of call quality. Chipsets from Intel (for Intel CPUs) and nVidia nForce (for AMD CPUs) seem to score the best marks in this area. Review the specific chipset of any motherboard you are evaluating to ensure that it does not have known problems with IRQ latency.

- If you are running Zaptel cards in your system, you will want to ensure that your BIOS allows you maximum control over IRQ assignment. As a rule, high-end motherboards will offer far greater flexibility with respect to BIOS tweaking; value-priced boards will generally offer very little control. This may be a moot point, however, as APIC-enabled motherboards turn IRQ control over to the operating system.

- Server-class motherboards generally implement a different PCI standard than workstation-class motherboards. While there are many differences, the most obvious and well known is that the two versions have different voltages. Depending on which cards you purchase, you will need to know if you require 3.3V or 5V PCI slots. Figure 2-1 shows the difference between 3.3V and 5V slots. Most server motherboards will have both types, but workstations will typically have only the 5V version.

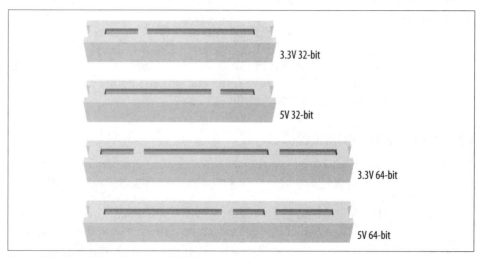

Figure 2-1. Visual identification of PCI slots

- Consider using multiple processors. This will provide an improvement in the system's ability to handle multiple tasks. For Asterisk, this will be of special benefit in the area of floating-point operations.

 It should be noted that evidence now suggests that connecting together two completely separate, single-CPU systems may provide far more benefits than simply using two processors in the same machine. You not only double your CPU power, but you also achieve a much better level of redundancy at a similar cost to a single-chassis, dual-CPU machine. Keep in mind, though, that a dual-server Asterisk solution will be more complex to design than a single-machine solution.

- Avoid motherboards that include built-in audio and video components. If you want a sound card, install one. As for a video card, you may not need one at all—certainly Asterisk does not require one. It has traditionally been the more value-priced motherboards that have had these components on-board, and these board designs often make compromises to keep down the costs.

- If possible, install an external modem. If you must have an internal modem, you will need to ensure that it is not a so-called "Win-modem"—it must be a completely self-sufficient unit (note that these are very difficult, if not impossible, to find).

- Consider that with built-in networking, if you have a network component failure, the entire motherboard will need to be replaced. On the other hand, if you install a peripheral Network Interface Card (NIC), there may be an increased chance of failure due to the extra mechanical connections involved. Some value can probably be gained from having both primary and backup cards installed in the system.

- The stability and quality of your Asterisk system will be dependent on the components you select for its architecture. Asterisk is a beast, and it expects to be fed the best. As with just about anything, high cost is not always synonymous with quality, but you will want to become a connoisseur of computer components.

Having said all that, we need to get back to the original point: Asterisk can and will happily install on pretty nearly any PC platform. The lab systems used to write this book, for example, included everything from a 433-MHz Pentium III on an Intel chipset to an Athlon XP 2000 on a VIA-based motherboard. We have not experienced any performance or stability problems running less than five concurrent telephone connections. For the purposes of learning, do not be afraid to install Asterisk on whatever system you can scrounge up. When you are ready to put your system into production, however, you will need to understand the ramifications of the choices you make with respect to your hardware.

Power Supply Requirements

One often-overlooked component in a PC is the power supply (and the supply of power). For a telecommunications system, these components can play a significant role in the quality of the user experience.

Computer power supplies

The power supply you select for your system will play a vital role in the stability of the entire platform. Asterisk is not a particularly power-hungry application, but anything relating to multimedia (whether it be telephony, professional audio, video, or the like) is generally sensitive to power *quality*.

This oft-neglected component can turn an otherwise top-quality system into a poor performer. By the same token, a top-notch power supply might enable an otherwise cheap PC to perform like a champ.

The power supplied to a system must provide not only the energy the system needs to perform its tasks, but also stable, clean signal lines for all of the voltages your system expects from it.

Redundant power supplies

In a carrier-grade or high-availability environment, it is common to deploy servers that use a redundant power supply. Essentially, this involves two completely independent power supplies, either one of which is capable of supplying the power requirements of the system.

If this is important to you, keep in mind that best practices suggest that to be properly redundant, these power supplies should be connected to completely independent Uninterruptible Power Supplies (UPSs) that are in turn fed by totally isolated electrical circuits.

Environment

Your system's environment consists of all those factors that are not actually part of the server itself, but nevertheless play a crucial role in the reliability and quality that can be expected from the system. Electrical supplies, room temperature and humidity, sources of interference, and security are all factors that should be contemplated.

Power Conditioning and Uninterruptible Power Supplies

When selecting the power sources for your system, consideration should be given not only to the amount of power the system will use, but also to the manner in which that power is delivered.

Power is not as simple as voltage coming from the outlet in the wall, and you should never just plug a production system into whatever electrical source is near at hand.* Giving some consideration to the supply of power to your system can provide a far more stable power environment, leading to a far more stable system.

---

* Okay, look, you *can* plug it in wherever you'd like, and it'll probably work, but if your system has strange stability problems, please give this section another read. Deal?

Properly grounded, conditioned power feeding a premium-quality power supply will ensure a clean *logic ground* (a.k.a. 0-volt) reference* for the system and keep electrical noise on the motherboard to a minimum. These are industry-standard best practices for this type of equipment, which should not be neglected. A relatively simple way to achieve this is through the use of a *power-conditioned* UPS.†

Power-conditioned UPSs

The UPS is well known for its role as a battery backup, but the power-conditioning benefits that high-end UPS units also provide are less well understood.

Power conditioning can provide a valuable level of protection from the electrical environment by regenerating clean power through an isolation transformer. A quality power conditioner in your UPS will eliminate most electrical noise from the power feed and help to ensure a rock-steady supply of power to your system.

Unfortunately, not all UPS units are created equal; many of the less expensive units do not provide clean power. What's worse, manufacturers of these devices will often promise all kinds of protection from surges, spikes, overvoltages, and transients. While such devices may protect your system from getting fried in an electrical storm, they will not clean up the power being fed to your system, and thus will do nothing to contribute to stability.

Make sure your UPS is power conditioned. If it doesn't say exactly that, it isn't.

Grounding

Voltage is defined as the difference in electrical potential between two points. When considering a *ground* (which is basically nothing more than an electrical path to earth), the common assumption is that it represents 0 volts. But if we do not define that 0V in *relation* to something, we are in danger of assuming things that may not be so. If you measure the voltage between two grounding references, you'll often find that there is a voltage potential between them. This voltage potential between grounding points can be significant enough to cause logic errors—or even damage—in a system where more than one path to ground is present.

---

* In electronic devices, a binary zero (0) is generally related to a 0-volt signal, while a binary one (1) can be represented by many different voltages (commonly between 2.5 and 5 volts). The grounding reference that the system will consider 0 volts is often referred to as the "logic ground." A poorly grounded system might have electrical potential on the logic ground to such a degree that the electronics mistake a binary zero for a binary one. This can wreak havoc with the system's ability to process instructions.

† It is a commonly misunderstood belief that all UPSs provide clean power. This is not at all true.

One of the authors recalls once frying a sound card he was trying to connect to a friend's stereo system—even though both the computer and the stereo were in the same room, more than 6 volts of difference was measured between the ground conductors of the two electrical outlets they were plugged into! The wire between the stereo and the PC (by way of the sound card) provided a path that the voltage eagerly followed, thus frying a sound card that was not expecting an electrical current on its signal leads. Connecting both the PC and the stereo to the same outlet fixed the problem.

When considering electrical regulations, the purpose of a ground is primarily human safety. In a computer, the ground is used as a 0V logic reference. An electrical system that provides proper safety will not always provide a proper logic reference—in fact, the goals of safety and power quality are sometimes in disagreement. Naturally, when a choice must be made, safety has to take precedence.

Since the difference between a binary zero and a binary one is represented in computers by voltage differences of sometimes less than 3V, it is entirely possible for unstable power conditions caused by poor grounding or electrical noise to cause all manner of intermittent system problems. Some power and grounding advocates estimate that more than 80% of unexplained computer glitches can be traced to power quality.

Modern switching power supplies are somewhat isolated from power quality issues, but any high-performance system will always benefit from a well-designed power environment. In mainframes, proprietary PBXs, and other expensive computing platforms, the grounding of the system is never left to chance. The electronics and frames of these systems are always provided with a dedicated ground that does not depend on the safety grounds supplied with the electrical feed.

Regardless of how much you are willing to invest in grounding, when you specify the electrical supply to any PBX, ensure that the electrical circuit is completely dedicated to your system (as discussed in the next section) and that an insulated, isolated grounding conductor is provided. This can be expensive to provision, but it will contribute greatly to a quality power environment for your system.*

It is also vital that each and every peripheral you connect to your system be connected to the same electrical receptacle (or, more specifically, the same ground reference). This will cut down on the occurrence of ground loops, which can cause anything from buzzing and humming noises to damaged or destroyed equipment.

---

* On a hobby system, this is probably too much to ask, but if you are planning on using Asterisk for anything important, at least be sure to give it a fighting chance—don't put anything like air conditioners, photocopiers, laser printers, or motors on the same circuit.

Electrical Circuits

If you've ever seen the lights dim when an electrical appliance kicks in, you've seen the effect that a high-energy device can have on an electrical circuit. If you were to look at the effects of a multitude of such devices, each drawing power in its own way, you would see that the harmonically perfect 50- or 60-Hz sine wave you may think you're getting with your power is anything but. Harmonic noise is extremely common on electrical circuits, and it can wreak havoc on sensitive electronic equipment. For a PBX, these problems can manifest as audio problems, logic errors, and system instability.

Never install a server on an electrical circuit that is shared with any other devices. There should be only one outlet on the circuit, and you should connect only your telephone system (and associated peripherals) to it. The wire (including the ground) should be run unbroken directly back to the electrical panel. The grounding conductor should be insulated, and isolated. There are far too many stories of photocopiers, air conditioners, and vacuum cleaners wreaking havoc with sensitive electronics to ignore this rule of thumb.

The electrical regulations in your area must always take precedence over any ideas presented here. If in doubt, consult a power quality expert in your area on how to ensure that you adhere to electrical regulations. Remember, electrical regulations take into account the fact that human safety is far more important than the safety of the equipment.

The Equipment Room

Environmental conditions can wreak havoc on systems, and yet it is quite common to see critical systems deployed with little or no attention given to these matters. If one looks at the statistics, it becomes obvious that attention to environmental factors can play a significant role in the stability and reliability of systems.

Humidity

Simply put, humidity is water in the air. Water is a disaster for electronics, for two main reasons: 1) water is a catalyst for corrosion, and 2) water is conductive enough that it can cause short circuits. Do not install any electronic equipment in areas of high humidity, without providing a means to remove the moisture.

Temperature

Heat is the enemy of electronics. The cooler you keep your system, the more reliably it will perform. If you cannot provide a properly cooled room for your system, at a minimum ensure that it is placed in a location that ensures a steady supply of clean, cool air. Also, keep the temperature steady. Changes in temperature can lead to condensation and other damaging changes.

Dust

There is an old adage in the computer industry that holds that dust bunnies inside of a computer are lucky. Let's consider some of the realities of dust bunnies:

- Significant buildup of dust can restrict airflow inside the system, leading to increased levels of heat.
- Dust can contain metal particles, which, in sufficient quantities, can contribute to signal degradation or shorts on circuit boards.

Put critical servers in a filtered environment, and clean out dust bunnies on a regular schedule.

Security

Server security naturally involves protecting against network-originated intrusions, but the environment also plays a part in the security of a system. Telephone equipment should always be locked away, and only persons who have a need to access the equipment should be allowed near it.

Telephony Hardware

If you are going to connect Asterisk to any legacy telecommunications equipment, you will need the correct hardware. The hardware you require will be determined by what it is you want to achieve.

Connecting to the PSTN

Asterisk allows you to seamlessly bridge circuit-switched telecommunications networks[*] with packet-switched data networks.[†] Because of Asterisk's open architecture (and open source code), it is ultimately possible to connect any standards-compliant interface hardware. The selection of open source telephony interface boards is currently limited, but as interest in Asterisk grows, that will rapidly change.[‡] At the moment, one of the most popular and cost-effective ways to connect to the PSTN is

---

[*] Often referred to as *TDM networks*, due to the Time Division Multiplexing used to carry traffic through the PSTN.

[†] Popularly called VoIP networks, although Voice over IP is not the only method of transmitting voice over packet networks (Voice over Frame Relay was very popular in the late 1990s).

[‡] The evolution of inexpensive, commodity-based telephony hardware is only slightly behind the telephony software revolution. New companies spring up on a weekly basis, each one bringing new and inexpensive standards-based devices into the market.

to use the interface cards that evolved from the work of the Zapata Telephony Project (*http://www.zapatatelephony.org*).

Analog interface cards

Unless you need a lot of channels (or a have lot of money to spend each month on telecommunications facilities), chances are that your PSTN interface will consist of one or more analog circuits, each of which will require a Foreign eXchange Office (FXO) port.

Digium, the company that sponsors Asterisk development, produces the most popular analog interface card for Asterisk, known as the TDM400P.* The TDM400P is a four-port base card that allows for the insertion of up to four daughter cards, which deliver either FXO or Foreign eXchange Station (FXS) ports. The TDM400P can be purchased with these cards preinstalled, and Digium has designated part numbers to describe these configurations. The naming convention is TDM *x* *y* B, where *x* and *y* are numbers representing the quantity of FXS and FXO† cards on the board, respectively. Check out Digium's web site (*http://www.digium.com*) for more information about this card.

An older card produced by Digium was known as the X100P. It is no longer available from Digium, but you may be able to find a clone of this card.

Another company that produces Asterisk-compatible analog cards is Voicetronix. They have three Asterisk cards in their analog lineup: OpenLine4, OpenSwitch6, and OpenSwitch12.

Digital interface cards

If you require more than 10 circuits, or require digital connectivity, chances are you're going to be in the market for a T1 or E1 card.‡ Bear in mind, though, that the monthly charges for a digital PSTN circuit vary widely. In some places, as few as five circuits can justify a digital circuit; in others, the technology may never be cost-justifiable. The more competition there is in your area, the better chance you have of finding a good deal. Be sure to shop around.

The Zapata Telephony Project originally produced a T1 card, the Tormenta, that is the ancestor of most Asterisk-compatible T1 cards. The original Tormenta cards are now considered obsolete, but they do still work with Asterisk. Currently, the only company known to be producing these cards is Varion.

---

* The TDM400P is not, in fact, a TDM card at all. It is analog.

† FXS and FXO refer to the opposing ends of an analog circuit. Which one you need will be determined by what you want to connect to. Chapter 7 discusses these in more detail.

‡ T1 and E1 are digital telephony circuits. We'll discuss them further in Chapter 7.

Digium makes several different digital circuit interface cards. The features on the cards are the same; the primary differences are whether they provide T1 or E1 interfaces, and how many interfaces each card provides. Although it's technically possible, the general consensus in the Asterisk community is that no more than one of these cards should be deployed in a single system.

Sangoma, who have been producing open source WAN cards for many years, have recently added Asterisk support for their T1/E1 cards.* Sangoma's cards contain powerful field-programmable gate arrays (FPGAs), which make them extremely flexible. In an Asterisk environment, for example, they have been programmed to interface with the Zapata channel driver.

Channel banks

A *channel bank* is loosely defined as a device that allows a digital circuit to be de-multiplexed into several analog circuits (and vice versa). More specifically, a channel bank lets you connect analog telephones and lines into a system across a T1 line. Figure 2-2 shows how a channel bank fits into a typical office phone system.

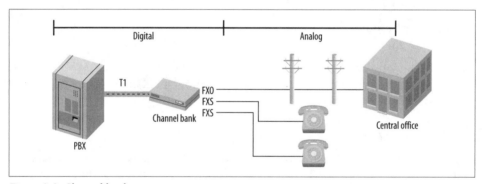

Figure 2-2. Channel bank

Although they can be expensive to purchase, many people feel very strongly that the only proper way to integrate analog circuits and devices into Asterisk is through a channel bank.

Other types of PSTN interfaces

Many VoIP gateways exist that can be configured to provide access to PSTN circuits. Generally speaking, these will be of most use in a smaller system (one or two lines). They can also be very complicated to configure, as the interaction between the various networks and devices requires a solid understanding of both telephony and VoIP

---

* It should be noted that a Sangoma Frame Relay card figured prominently in the original development of Asterisk (see *http://linuxdevices.com/articles/AT8678310302.html*)—Sangoma has a long history of supporting open source WAN interfaces with Linux.

fundamentals. For that reason, we will not discuss these devices in detail in this book. They are worth looking into, however—popular units are made by Sipura, Grandstream, Digium, and many other companies.

Another way to connect to the PSTN is through the use of Basic Rate Interface (BRI) ISDN circuits. BRI is a digital telecom standard that specifies a two-channel circuit that can carry up to 144 kbps of traffic. It is very rarely used in North America and most of the rest of the world, but it's quite popular in Europe. Due to the variety of different ways this technology has been implemented, we will not be discussing BRI in very much detail in this book.

Connecting Exclusively to a Packet-Based Telephone Network

If you do not need to connect to the PSTN, Asterisk requires no hardware other than a server with a Network Interface Card.

However, if you are going to be providing music on hold or conferencing and you have no physical timing source, you will need the *ztdummy* Linux kernel module. *ztdummy* is a clocking mechanism designed to provide a timing source to a system where no hardware timing source exists. In Version 2.4 of the Linux kernel, to use *ztdummy* you must have a UHCI-type USB controller on your motherboard. In Linux 2.6, that requirement is no more.

Types of Phone

Since the title of this book is *Asterisk: The Future of Telephony*, we would be remiss if we didn't discuss the devices that all of this technology ultimately has to interconnect: telephones!

We all know what a telephone is—but will it be the same five years from now? Part of the revolution that Asterisk is contributing to is the evolution of the telephone, from a simple audio communications device into a multimedia communications terminal providing all kinds of yet-to-be-imagined functions.

As an introduction to this exciting concept, we will briefly discuss the various kinds of devices we currently call "telephones" (any of which can easily be integrated with Asterisk). We will also discuss some ideas about what these devices may evolve into in the future (devices that will also easily integrate with Asterisk).

Physical Telephones

Any physical device whose primary purpose is terminating an on-demand audio communications circuit between two points can be classified as a physical telephone. At a minimum, such a device has a handset and a dial pad; it may also have feature keys, a display screen, and various audio interfaces.

This section takes a brief look at the various user (or endpoint) devices you might want to connect to your Asterisk system. We'll delve more deeply into the mechanics of analog and digital telephony in Chapter 7.

Analog telephones

Analog phones have been around since the invention of the telephone. Up until about 20 years ago, all telephones were analog. Although analog phones have some technical differences in different countries, they all operate on similar principles.

When a human being speaks, the vocal cords, tongue, teeth, and lips create a complex variety of sounds. The purpose of the telephone is to capture these sounds and convert them into a format suitable for transmission over wires. In an analog telephone, the transmitted signal is *analogous* to the sound waves produced by the person speaking. If you could see the sound waves passing from the mouth to the microphone, they would be proportional to the electrical signal you could measure on the wire.

 This contiguous connection is referred to as a *circuit*, which the telephone network used to use electromechanical switches to create; hence the term *circuit-switched network*.

Analog telephones are the only kind of phone that are commonly available in any retail electronics store. In the next few years, that can be expected to change dramatically.

Proprietary digital telephones

As digital switching systems developed in the 1980s and 1990s, telecommunications companies developed digital Private Branch eXchanges (PBXs) and Key Telephone Systems (KTSs). The proprietary telephones developed for these systems were completely dependent on the systems to which they were connected and could not be used on any other systems. Even phones produced by the same manufacturer were not cross-compatible (for example, a Nortel Norstar set will not work on a Nortel Meridian 1 PBX). The proprietary nature of digital telephones limits their future. In this emerging era of standards-based communications, they will quickly be relegated to the dustbin of history.

The handset in a digital telephone is generally identical in function to the handset in an analog telephone, and they are often compatible with each other. Where the digital phone is different is that inside the telephone, the analog signal is sampled and converted into a digital signal—that is, a numerical representation of the analog waveform. We'll leave a detailed discussion of digital signals until Chapter 7; for now, suffice it to say that the primary advantage of a digital signal is that it can be transmitted over limitless distances with no loss of signal quality.

The chances of anyone ever making a proprietary digital phone directly compatible with Asterisk are fairly small, but companies such as Citel (*http://www.citel.com*) have created gateways that convert the proprietary signals to SIP.*

ISDN telephones

Prior to VoIP, the closest thing to a standards-based digital telephone was an ISDN-BRI terminal. Developed in the early 1980s, ISDN was expected to revolutionize the telecommunications industry in exactly the same way that VoIP promises to finally achieve today.

 There are two types of ISDN: *Primary Rate Interface* (PRI) and *Basic Rate Interface* (BRI). PRI is commonly used to provide trunking facilities between PBXs and the PSTN, and is widely deployed. BRI is not at all common in North America, but has enjoyed some success in Europe.

While ISDN was widely deployed by the telephone companies, many consider the standard to have been a flop, as it generally failed to live up to its promises. The high costs of implementation, recurring charges, and lack of cooperation amongst the major players contributed to an environment that caused more problems than it solved.

BRI was intended to service terminal devices and smaller sites (a BRI loop provides two digital circuits). While a wealth of BRI devices have been developed, BRI has largely been deprecated in favor of faster, less expensive technologies such as ADSL, cable modems, and VoIP.

BRI is still very popular for use in video-conferencing equipment, as it provides a fixed bandwidth link. Also, BRI does not have the type of quality of service issues a VoIP connection might, as it is circuit-switched.

BRI is still sometimes used in place of analog circuits to provide trunking. Whether or not this is a good idea depends mostly on how your local phone company prices the service, and what features it is willing to provide.

IP telephones

IP telephones are heralds of the most exciting change in the telecommunications industry. In the very near future, standards-based IP telephones will be available in retail stores.† The wealth of possibilities inherent in these devices will cause an explosion of interesting applications, from video phones, to high-fidelity broadcasting devices, to wireless mobility solutions, to purpose-built sets for particular industries, to flexible all-in-one multimedia systems.

---

* The Session Initiation Protocol is currently the most well-known and popular protocol for VoIP. We will discuss it further in Chapter 8.

† As of this writing, Wal-Mart was offering a basic IP telephone on its web site (*http://www.walmart.com*).

The revolution that IP telephones will spawn has nothing to do with a new type of wire to connect your phone to, and everything to do with giving you the power to communicate the way you want.

The early-model IP phones that have been available for several years now do not represent the future of these exciting appliances. They are merely a stepping-stone; a familiar package in which to wrap a fantastic new way of thinking.

The future is far more promising.

Soft Phones

A *soft phone* is a software program that provides telephone functionality on a non-telephone device, such as a PC or PDA. So how do we recognize such a beast? What might at first glance seem a simple question actually raises many. A soft phone should probably have some sort of dial pad, and it should provide an interface that reminds users of a telephone. But will this always be the case?

The term "soft phone" can be expected to evolve rapidly, as our concept of what exactly a telephone is undergoes a revolutionary metamorphosis. As an example of this evolution, consider the following: would we correctly define popular communication programs such as Instant Messenger as soft phones? IM provides the ability to initiate and receive standards-based VoIP connections. Does this not qualify it as a soft phone? Answering that question requires knowledge of the future that we do not yet possess. Suffice it to say that while at this point in time, soft phones are expected to look and sound like traditional phones, that conception is likely to change in the very near future.

As standards evolve and we move away from the traditional telephone and toward a multimedia communications culture, the line between soft phones and physical telephones will become blurred indeed. For example, we might purchase a communications terminal to serve as a telephone, and install a soft phone program onto it to provide the functions we desire.

Having thus muddied the waters, the best we can do at this point is to define what the term "soft phone" will refer to in relation to this book, with the understanding that the meaning of the term can be expected to undergo a massive change over the next few years. For our purposes, we will define a soft phone: any device that runs on a personal computer, presents the look and feel of a telephone, and provides as its primary function the ability to make and receive full-duplex audio communications (formerly known as "phone calls")[*] through E.164 addressing.[†]

---

[*] OK, so you think you know what a phone call is? So did we. Let's just wait a few years, shall we?

[†] E.164 is the ITU standard that defines how phone numbers are assigned. If you've used a telephone, you've used E.164 addressing.

Telephony Adaptors

A *telephony adaptor* (usually referred to as an ATA, or Analog Terminal Adaptor) can loosely be described as an end-user device that converts communications circuits from one protocol to another. Most commonly, these devices are used to convert from some digital (IP or proprietary) signal to an analog connection that you can plug a standard telephone or fax machine into.

These adaptors could be described as gateways, for that is their function. However, popular usage of the term *telephony gateway* would probably best describe a multi-port telephony adaptor, generally with more complicated routing functions.

Telephony adaptors will be with us for as long as there is a need to connect incompatible standards and old devices to new networks. Eventually, our reliance on these devices will disappear, as did our reliance on the modem—obsolescence through irrelevance.

Communications Terminals

Communications terminal is an old term that disappeared for a decade or two and is being reintroduced here, very possibly for no other reason than that it needs to be discussed so that it can eventually disappear again—once it becomes ubiquitous.

First, a little history. When digital PBX systems were first released, manufacturers of these machines realized that they could not refer to their endpoints as telephones—their proprietary nature prevented them from connecting to the PSTN. They were therefore called *terminals*, or stations. Users, of course, weren't having any of it. It looked like a telephone and acted like a telephone, and therefore it *was* a telephone. You will still occasionally find PBX sets referred to as terminals, but for the most part they are called telephones.

The renewed relevance of the term "communications terminal" has nothing to do with anything proprietary—rather, it's the opposite. As we develop more creative ways of communicating with each other, we gain access to many different devices that will allow us to connect. Consider the following scenarios:

- If I use my PDA to connect to my voicemail and retrieve my voice messages (converted to text), does my PDA become a phone?
- If I attach a video camera to my PC, connect to a company's web site, and request a live chat with a customer service rep, is my PC now a telephone?
- If I use the IP phone in my kitchen to surf for recipes, is that a phone call?

The point is simply this: we'll probably always be "phoning" each other, but will we always be using "telephones" to do so?

Linux Considerations

If you ask anyone at the Free Software Foundation, they will tell you that what we know as Linux is in fact GNU/Linux. All etymological arguments aside, there is some valuable truth to this statement. While the kernel of the operating system is indeed Linux, the vast majority of the utilities installed on a Linux system and used regularly are in fact GNU utilities. "Linux" is probably only 5% Linux, possibly 75% GNU, and perhaps 20% everything else.

Why does this matter? Well, the flexibility of Linux is both a blessing and a curse. It is a blessing because with Linux you can truly craft your very own operating system from scratch. Since very few people ever do this, the curse is in large part due to the responsibility you must bear in determining which of the GNU utilities to install, and how to configure the system.

If this seems overwhelming, do not fear. In the next chapter, we will discuss the selection, installation, and configuration of the software environment for your Asterisk system.

Conclusion

In this chapter, we've discussed all manner of issues that can contribute to the stability and quality of an Asterisk installation. Before we scare you off, we should tell you that many people have installed Asterisk on top of a graphical Linux workstation, running a web server, a database, an X-windowing environment, and who knows what else, with no problems whatsoever. How much time and effort you should devote to following the best practices and engineering tips in this chapter all depends on how much work you expect the Asterisk server to perform, and how much quality and reliability your system must provide.

What we have attempted to do in this chapter is give you a feel for the kinds of best practices that will help to ensure that your Asterisk system will be built on a reliable, stable platform. Asterisk is quite willing to operate under far worse conditions, but the amount of effort and consideration you decide to give these matters will play a part in the stability of your PBX. Your decision should depend on how critical your Asterisk system will be.

## CHAPTER 3

Installing Asterisk

*I long to accomplish great and noble tasks, but it is my
chief duty to accomplish humble tasks as though they
were great and noble. The world is moved along, not
only by the mighty shoves of its heroes, but also by the
aggregate of the tiny pushes of each honest worker.*

—Helen Keller

In the previous chapter, we discussed preparing a system to install Asterisk. Now it's time to obtain, extract, compile, and install the software.

Although a large number of Linux distributions[*] and PC architectures are excellent candidates for Asterisk, we have chosen to focus on a single distribution in order to maintain brevity and clarity throughout the book. The instructions that follow have been made as generic as possible, but you may notice a leaning toward Red Hat structures and utilities. We have chosen to focus on Red Hat because its command set, directory structure, and so forth are likely to be familiar to the majority of users (we have found that most Linux administrators are familiar with Red Hat, even if they don't prefer it). However, this doesn't mean that Red Hat is the only choice, or even the best one for you. A question that often appears on the mailing lists is: "Which distribution of Linux is the best to use with Asterisk?" The multitude of answers generally boils down to "the one you like the best."

What Packages Do I Need?

Asterisk uses three main packages: the main Asterisk program (*asterisk*), the Zapata telephony drivers (*zaptel*), and the PRI libraries (*libpri*). If you plan on a pure VoIP network, the only real requirement is the *asterisk* package. The *zaptel* drivers are

---

[*] And some non-Linux operating systems as well, such as Solaris, *BSD, and OS X. However, while people have managed to successfully run Asterisk on these alternative systems, Asterisk was, and continues to be, actively developed for Linux.

required if you are using analog or digital hardware, or if you're using the *ztdummy* driver (discussed later in this chapter) as a timing interface. The *libpri* library is technically optional unless you're using ISDN PRI interfaces, and you may save a small amount of RAM if you don't load it, but we recommend that it be installed in conjunction with the *zaptel* package for completeness.

One other package you may want to install is *asterisk-sounds*. While Asterisk comes with many sound prompts in the main source distribution, the *asterisk-sounds* package will give you even more. If you would like to expand the number of professionally recorded prompts for use with your Asterisk system, this package is essential. Some of our examples in the following chapters will make use of files included in this package, so we will assume that you have it installed.

Package Requirements

To compile Asterisk, you must install the *GCC compiler* (Version 3.x or later) and its dependencies. While Version 2.96 of GCC may work for the time being, future versions will not support it. Asterisk also requires *bison*, a parser generator program that replaces *yacc*, and *ncurses* for CLI functionality. The cryptographic library in Asterisk requires *OpenSSL* and its development packages. If you want to use *ztdummy* for timing, or any of the hardware drivers provided by Zaptel, you'll need to install the *zaptel* package as well. If you are installing *libpri*, be sure to install it before *asterisk* (see "Compiling libpri").

Zaptel requires *libnewt* and its development packages for the *zttool* program (see "Using ztcfg and zttool," below) and the *usb-uhci* module for *ztdummy*. If you're using PRI interfaces, Zaptel also requires the *libpri* package (again, even if you aren't using PRI circuits, we recommend that you install *libpri* along with *zaptel*).

The following sections discuss how to obtain, extract, compile, and install the *asterisk*, *zaptel*, *libpri*, and *asterisk-sounds* packages.

Obtaining the Source Code

The Asterisk source code can be obtained either through FTP or CVS. We will show you how to acquire the source with both methods, although you only need to use one of them to retrieve the packages (FTP is the preferred method).

Obtaining Asterisk Source Code from FTP

The Asterisk source code can be obtained from the Digium FTP server, located at *ftp:// ftp.digium.com*. The easiest way to obtain the stable release is through the use of the program *wget*.

Stable and Head

Asterisk comes in two different flavors, generally referred to as *stable* and *head*. Stable, as the name implies, is the established branch of Asterisk for use in production systems. The head branch is what the developers use to test new features and bug fixes.

Bug fixes (not features) are merged over to the stable branch after a reasonable period of testing. It is entirely possible that the development branch may be broken at certain points during testing; thus, the stable branch is what you will want to run your production system on, and it is what we will be using throughout this book.

You can obtain stable releases via FTP. Both the stable and head branches of Asterisk can also be obtained from CVS, as explained later in this chapter. However, it is important to note the difference between *releases* and *CVS*. Releases are snapshots from the stable CVS tree, tagged with a version number and released via the FTP server when a new stable release is deemed ready. Note that the stable CVS branch is *not* a release— it's a work in a progress, and it may be buggy (i.e., not so stable after all). The FTP tarballs are the actual releases.

To summarize, use only stable releases obtained via the FTP server for production systems.

Note that we will be making use of the */usr/src/* directory to extract and compile the Asterisk source. Also be aware that you will need *root* access to write files to the */usr/src/* directory and to install Asterisk and its associated packages.

To obtain the latest stable source code via *wget*, enter the following commands on the command line:

```
# cd /usr/src/
# wget --passive-ftp ftp.digium.com/pub/asterisk/asterisk-1.*.tar.gz
# wget --passive-ftp ftp.digium.com/pub/asterisk/asterisk-sounds-*.tar.gz
# wget --passive-ftp ftp.digium.com/pub/zaptel/zaptel-*.tar.gz
# wget --passive-ftp ftp.digium.com/pub/libpri/libpri-*.tar.gz
```

 As long as Digium doesn't change the way they put things on the FTP site, the wget command will automagically get the latest version. You may also replace the wildcard mask (*) with the currently available software version.

Now that you've retrieved the files for Asterisk and the Digium hardware, you are ready to extract the code.

Extracting the Source Code

If you use wget to obtain the source code from the FTP server, you will need to extract it before compiling. If you didn't download the packages to */usr/src/*, either move them there now, or specify the full path to their location. We will be using the GNU *tar* application to extract the source code from the compressed archive. This is a simple process that can be achieved through the use of the following commands:

```
# cd /usr/src/
# tar zxvf zaptel-*.tar.gz
# tar zxvf libpri-*.tar.gz
# tar zxvf asterisk-*.tar.gz
# tar zxvf asterisk-sounds*.tar.gz
```

These commands will extract the packages and source code to their respective directories.

Obtaining Asterisk Source Code from CVS

The Concurrent Versioning System (CVS) is a tool that provides a central repository that large (and diverse) development teams can use to manage the multitude of files associated with a development project. When a change is made, it is committed to the CVS server, where it is immediately available for download and compilation. Another added benefit of using CVS is that the version for any particular file can be rolled back to a certain instance, so that if something was working at one point but a change causes it to break, you can easily revert to the working version. This is true for the entire tree as well. If you find that installing the latest version of Asterisk causes any part of the system to break, you can "roll back" to an earlier point in time and investigate the cause of the problem.

If you are a developer looking to obtain the latest updates to the source code, you will need to get them from the CVS servers. You can also download the stable branch via CVS:

- Export the *CVSROOT* path:
  ```
  # cd /usr/src/
  # export CVSROOT=:pserver:anoncvs:anoncvs@cvs.digium.com:/usr/cvsroot
  ```
- Download *HEAD* from CVS:
  ```
  # cvs checkout zaptel libpri asterisk
  ```
- Download *STABLE 1.0* from CVS:
  ```
  # cvs checkout -r v1-0 zaptel libpri asterisk
  ```
- Download *STABLE 1.2* from CVS:
  ```
  # cvs checkout -r v1-2 zaptel libpri asterisk
  ```
- Download optional modules from CVS:
  ```
  # cvs checkout asterisk-sounds asterisk-addons
  ```

Again, note that the stable branch available from CVS is not a release and should not be used for production systems.

Compiling Zaptel

Figure 3-1 shows the layers of interaction between Asterisk and the Linux kernel with respect to hardware control. On the Asterisk side is the Zapata channel module, *chan_zap*. Asterisk uses this interface to communicate with the Linux kernel, where the drivers for the hardware are loaded.

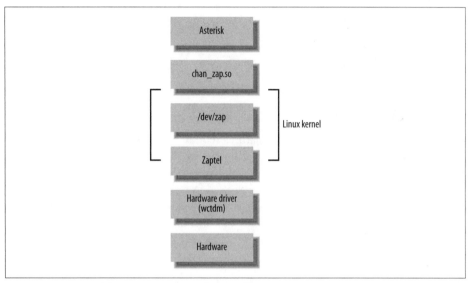

Figure 3-1. Layers of device interaction with Asterisk

The Zaptel interface is a kernel loadable module that presents an abstraction layer between the hardware drivers and the Zapata module in Asterisk. It is this concept that allows the device drivers to be modified without any changes being made to the Asterisk source itself. The device drivers are used to communicate with the hardware directly and to pass the information between Zaptel and the hardware.

 While Asterisk itself compiles on a variety of platforms, the Zaptel drivers are Linux-specific—they are written to interface directly with the Linux kernel. There are no official Zaptel drivers for other operating systems, although work has been going on to write drivers for FreeBSD.

We will discuss the Zaptel compile-time options momentarily, in "The zconfig.h File." First, let's take a look at compiling and installing the drivers. (The configuration of Zaptel drivers will be discussed in the next chapter.)

Before compiling the Zaptel drivers on a system running a Linux 2.4 kernel, you should verify that */usr/src/* contains a symbolic link named *linux-2.4* pointing to your kernel source. If the symbolic link doesn't exist, you can create it with the following command (assuming you've installed the source in */usr/src/*):

```
# ln -s /usr/src/`uname -r` /usr/src/linux-2.4
```

Computers running Linux 2.6 kernel–based distributions do not usually require the use of the symbolic link, as these distributions will search for the kernel build directory automatically. However, if you've placed the build directory in a nonstandard place (i.e., somewhere other than */lib/modules/<kernel version>/build/*), you will require the use of the symbolic link.

The ztdummy Driver

In Asterisk, certain applications and features require a timing device in order to operate (Asterisk won't even compile them if no timing device is found). All Digium PCI hardware provides a 1-kHz timing interface. If you lack the PCI hardware required to provide timing, the *ztdummy* driver can be used as a timing device. On Linux 2.4 kernel–based distributions, *ztdummy* must use the clocking provided by the UHCI USB controller. The driver looks to see that the *usb-uhci* module is loaded and that the kernel version is at least 2.4.5. Older kernel versions are incompatible with *ztdummy*.

On a 2.6 kernel–based distribution, *ztdummy* does not require the use of the USB controller. (As of v2.6.0, the kernel now provides 1-kHz timing with which the driver can interface; thus, the USB controller hardware requirement is no longer necessary.)

The default *Makefile* configuration does not create *ztdummy*. To compile *ztdummy*, you must remove a comment marker from the *Makefile*. Open it in your favorite text editor and look for the following line:

```
MODULES=zaptel tor2 torisa wcusb wcfxo wctdm \
        ztdynamic ztd-eth wct1xxp wct4xxp wcte11xp # ztdummy
```

Remove the hash[*] (#) symbol from in front of "ztdummy," save the file, and compile Zaptel as usual.

The Zapata Telephony Drivers

Compiling the Zapata telephony drivers for use with your Digium hardware is straightforward—simply run make for either the 2.4 or 2.6 Linux kernels (the *Make-*

---

[*] The # symbol is most widely known as "hash," so that is what we have chosen to call it. North Americans tend to call it a "pound sign," the ITU uses the term "square," and yet others call it a "crosshatch" or "number sign." Another term, made up by Don Macpherson to describe the # symbol during initial training on an early PBX system, is "octothorpe." This term eventually found its way into memos and letters at Bell Labs, then into other official documents, and from there leaked to the Internet.

file will determine the kernel version for you). Use these commands to compile Zaptel (replace *version* with your version of *zaptel*):

```
# cd /usr/src/zaptel-version
# make clean
# make
# make install
```

 While running make clean is not always necessary, it's a good idea to run it before recompiling any of the modules, as it will remove the compiled binary files from within the source code directory. You can also use it to clean up after installing, if you don't like to leave the compiled binaries floating around. Note that this removes the binaries only from the source directory, not from the system.

In addition to the executables, make clean also removes the intermediary files (i.e., the object files) after compilation. You don't need them occupying space on your hard drive.

If you're using a system that makes use of the */etc/rc.d/init.d/* or */etc/init.d/* directories, you may wish to run the make config command as well. This will install the startup scripts and configure the system, using the chkconfig command to load the *zaptel* module automatically at startup.

 The Debian equivalent of chkconfig is update-rc.d.

Using ztcfg and zttool

Two programs installed along with Zaptel are *ztcfg* and *zttool*. The *ztcfg* program is used to read the configuration in */etc/zaptel.conf* to configure the hardware. The *zttool* program can be used to check the status of your installed hardware. For instance, if you are using a T1 card and there is no communication between the endpoints, you will see a red alarm. If everything is configured correctly and communication is possible, you should see an "OK." The *zttool* application is also useful for analog cards, because it tells you their current state (configured, off-hook, etc.). The use of these programs will be explored further in the next chapter.

 The *libnewt* libraries and its development packages (*newt-devel* on Red Hat–based distributions) must be installed for *zttool* to be compiled.

The zconfig.h File

The *zconfig.h* file is where many of the Zaptel compile-time options lie. For the most part, you should not need to edit this file, but below are some of the options that may be of interest. To enable the options, remove the comment tags (/* */). If you decide to enable any of these options, be sure to do a make clean before recompiling and reinstalling Zaptel.

Boost ringer

By enabling the BOOST_RINGER option, you increase the amount of voltage supplied to a telephone during ringing from ~70V to ~89V. Some devices may not detect ringing below certain voltages, so this setting may be necessary. Note that upping the voltage requires more power, and that it will probably only be necessary on a telephone connected to a long loop. Basically, you should leave this alone unless the far end isn't detecting ringing properly. To enable this option, uncomment the following line:

```
/* #define BOOST_RINGER */
```

The BOOST_RINGER option can also be declared when loading the driver via modprobe, so it does not need to be compiled into the driver (recommended).

Disable μ-law/A-law precomputation

Defining CONFIG_CALC_XLAW tells Zaptel to not precompute μ-law/A-law into tables and to recalculate it for each sample. We haven't timed it, but the original coder felt that if you have a small number of channels and/or a small level-2 cache, it may be quicker to execute the calculation code than to actually do a lookup on the table loaded into memory.

To enable this option, uncomment the following line within *zconfig.h*:

```
/* #define CONFIG_CALC_XLAW */
```

Enable MMX optimization

You can enable MMX optimization (if your processor supports it) by removing the comment tags around the following line:

```
/* #define CONFIG_ZAPTEL_MMX */
```

Be aware that CONFIG_ZAPTEL_MMX is considered to be incompatible with AMD processors and can cause system instability.

Choose echo cancellation method

All the echo cancellers in Asterisk use a Finite Impulse Response (FIR) algorithm. The differences between them—mostly in code implementation and slight algorithm tweaks—are minimal. By default, the *MARK2* echo canceller is used, and it is generally considered the most robust. To change the default, add comment tags around the #define ECHO_CAN_MARK2 line and uncomment another line:

```
/* #define ECHO_CAN_STEVE */
/* #define ECHO_CAN_STEVE2 */
/* #define ECHO_CAN_MARK */
#define ECHO_CAN_MARK2
/* #define ECHO_CAN_MARK3 */
```

Enable aggressive suppression

Aggressive residual echo suppression with the *MARK2* echo canceller can be enabled by removing the comment tags around the following line:

```
/* #define AGGRESSIVE_SUPPRESSOR */
```

The aggressive suppressor makes the nonlinear processor (NLP) stronger. What the NLP essentially does is say, "If the sample is that quiet anyway, make the volume level about 0."

Disable echo cancellation

When echo cancellation is enabled in Asterisk, it is possible to disable it by sending a 2100-Hz tone at the beginning of a call. If you do not want Asterisk to disable echo cancellation even when it detects the echo cancel disable tone, uncomment the following line:

```
/* #define NO_ECHOCAN_DISABLE */
```

Fax machines and modems use the 2100-Hz tone during negotiation, and Asterisk monitors for this tone during call setup.

Enable HDLC

When using the Zaptel driver with T1 or E1 hardware, you can configure Zaptel to use TDM channels for data instead of voice. To enable HDLC functionality in the drivers, uncomment the following line:

```
/* #define CONFIG_ZAPATA_NET */
```

For this change to be meaningful, you must also use the *sethdlc* utility and perform some configuration in *zapata.conf*.

Enable ZapRAS

You can also make use of the *ZapRAS* program to turn Asterisk into a Remote Access Server (RAS) for use with your ISDN connections. To enable this functionality, you must uncomment the following line from within the *zconfig.h* file:

```
/* #define CONFIG_ZAPATA_PPP */
```

You must also patch Asterisk and configure a PPP daemon, so be aware that this task is nontrivial.

Enable Zaptel's watchdog

You can tell Zaptel to monitor the status of interfaces via its built-in "watchdog." It will check if the interfaces stop taking interrupts or otherwise misbehave. If this happens, the hardware will automatically be restarted. To enable the watchdog, uncomment this line:

```
/* #define CONFIG_ZAPTEL_WATCHDOG */
```

Set default tone zone

The tone zone info option is used to select which set of tones (e.g., dial tone, busy indication, ring tone, stutter, etc.), as defined in the *zonedata.c* file, should be used as the default. The *zonedata.c* file contains the frequencies and patterns that Asterisk uses to communicate on the PSTN networks in various countries and to signal connected telephones. The default tone zone (0) is used to indicate North American signaling frequencies. Other tone zones include Australia (1), France (2), Japan (7), Taiwan (14), and many others. You can change the default on the following line:

```
#define DEFAULT_TONE_ZONE 0
```

Enable CAC ground start signaling

Some devices, such as the FXO ports on a Carrier Access Corporation (CAC) channel bank, have nonstandard FXS ground start signaling start states (A=low, B=low). You can configure the drivers to use this state by removing the comment tags around the following line:

```
/* #define CONFIG_CAC_GROUNDSTART */
```

TDM400P Revision H PCI ID workaround

If you happen to be using an older TDM400P Revision H card, you may find that it sometimes forgets its PCI ID. To make the *wctdm* driver essentially match all subvendor IDs, uncomment the following line:

```
/* #define TDM_REVH_MATCHALL */
```

This may be required when using older revisions of TDM400P cards with newer versions of Asterisk, due to a change in the subvendor ID code. This has been known to cause the following type of error when loading the *wctdm* module:

```
# ZT_CHANCONFIG failed on channel 12: No such device or address (6)
```

Uncommenting the #define line above should resolve this problem.

Passing Module Parameters to Configure Zaptel

Some of the Zaptel options can also be enabled when loading the module, by passing module parameters to the *wctdm* driver. You can list these parameters at load time (as opposed to statically changing them in the *zconfig.h* file) with the modinfo command:

```
# modinfo -p wctdm
debug int
loopcurrent int
robust int
_opermode int
opermode string
timingonly int
lowpower int
boostringer int
fxshonormode int
```

You then pass the module parameters to the modprobe command. For example, you can use the following command to activate the boostringer parameter when the module is loaded, instead of statically defining its use with #define BOOST_RINGER in the *zconfig.h* file:

```
# modprobe wctdm boostringer=1
```

Another common parameter to pass to a module is opermode. By passing opermode to the *wctdm* driver, you can configure the TDM400P to better deal with line imped-ances for your country. opermode accepts a two-letter country code as its argument.

Compiling libpri

Compiling and installing *libpri* follows the same pattern as described above for *zap-tel*. *libpri* is used by various makers of Time Division Multiplexing (TDM) hardware, but even if you don't have the hardware installed it is safe to compile and install this library. You must compile and install *libpri* before Asterisk, as it will be detected and used when Asterisk is compiled. Here are the commands (replace *version* with your version of *libpri*):

```
# cd /usr/src/libpri-version
# make clean
# make
# make install
```

Compiling Asterisk

Once you've compiled and installed the *zaptel* and *libpri* packages (if you need them), you can move on to Asterisk. This section walks you through a standard installation and introduces some of the alternative make arguments that you may find useful. We'll also look at how you can edit the *Makefile* to optimize the compilation of Asterisk.

Standard Installation

Asterisk is compiled with *gcc* through the use of the GNU *make* program. Unlike many other programs, there is no need to run a configuration script for Asterisk. To get started compiling Asterisk, simply run the following commands (replace *version* with your version of Asterisk):

```
# cd /usr/src/asterisk-version
# make clean
# make
# make install
# make samples
```

Be aware that compile times will vary between systems. On a current-generation processor, you shouldn't need to wait more than five minutes. At Astricon, someone reported successfully compiling Asterisk on a 133-MHz Pentium, but it took approximately five hours. You do the math.

Run the make samples command to install the default configuration files. Installing these files (instead of configuring each file manually) will allow you to get your Asterisk system up and running much faster. Many of the default values are fine for Asterisk. Files that require editing will be explained in future chapters.

> If you already have configuration files installed in */etc/asterisk/* when you run the make samples command, *.old* will be appended to the end of each of your current configuration files—for example, *extensions. conf* will be renamed *extensions.conf.old*. Be careful, though, because if you run make samples more than once you will overwrite your original configuration files!
>
> The sample configuration files can also be found in the *configs/* subdirectory within your Asterisk sources directory.

If you're using a system that makes use of the */etc/rc.d/init.d/* or */etc/init.d/* directories, you may wish to run the make config command as well. This will install the startup scripts and configure the system (through the use of the chkconfig command) to execute Asterisk automatically at startup.

Alternative make Arguments

There are several other make arguments that you can pass at compile time. While some of these will be discussed here, the remainder are used internally within the file and really have no bearing or use for the end user. (Of course, new functions may have been added, so be sure to check the *Makefile* for other options.)

Let's take a look at some useful make arguments.

make clean

The make clean command is used to remove the compiled binaries from within the source directory. This command should be run before you attempt to recompile or, if space is an issue, if you would like to clean up the files.

make update

This command is used to update the existing code from the Digium CVS server. If you downloaded the source code from the FTP server, you will receive an error stating so.

 A common problem that you may find if you update with the cvs update command is that when you then do a show version at the Asterisk command-line interface (CLI), your version does not appear to have been updated. This problem can be resolved by removing the hidden *.version* file within the Asterisk source code directory before recompiling, or by using the make update command (which will remove the file for you).

make upgrade

If you run the make install command to install Asterisk after using the make update command to update from CVS, the *.version* file will not be updated. If you do not want to manually delete the *.version* file before running make and make install, you can use the make upgrade command instead.

make webvmail

The Asterisk Web Voicemail script is used to give a graphical interface to your voicemail account, allowing you to manage and interact with your voicemail remotely from a web browser.

When you run the make webvmail command, the Asterisk Web Voicemail script will be placed into the *cgi-bin/* directory of your HTTP daemon. If you have specific policies with respect to security, be aware that it uses a setuid root Perl script. This command will install only on a Red Hat or Fedora box, as other distributions may have different paths to their *cgi-bin/* directories. (This, of course, can be changed by editing the *Makefile*.)

make progdocs

This command will create documentation using the *doxygen* software from comments placed within the source code by the developers. You must have the appropriate *doxygen* software installed on your system in order for this to work. Note that *doxygen* assumes that the source code is well documented, which, sadly, is not always the case.

make mpg123

Asterisk uses the *mpg123* program to stream MP3s during the use of Music on Hold (MoH). Because Asterisk only works with *mpg123* v0.59r, this shortcut will determine if the correct version of *mpg123* is installed on your system and, if not, will attempt to download, extract, and compile it for you. Be aware that newer versions will not work, and some distributions even symbolically link *mpg321* and *mpg123*, which are entirely different programs. If you run the make install command after running this command, Asterisk will detect the directory and install it for you as well.

make config

The make config command will install Red Hat–style initialization scripts, if the */etc/rc.d/init.d* or */etc/init.d* directories are found to exist. If they do exist, the scripts are installed with file permissions equal to 755. If the script detects that */etc/rc.d/init.d/* exists, the chkconfig --add asterisk command will also be run to cause Asterisk to be started automatically at boot time. This is not the case, however, with distributions that only use the */etc/init.d/* directory. Running make config will not do anything to an already running Asterisk process, or start one if it's not running.

This script currently is only really useful on a Red Hat–based system, although initialization scripts are available for other distributions (such as Gentoo, Mandrake, and Slackware) in the *./contrib./init.d/* directory of your Asterisk source directory.

Editing the Makefile

At the top of the *Makefile* contained within the Asterisk source directory are several options for optimizing the compilation of Asterisk. You can enable GSM codec optimizations (with the use of MMX instructions), disable configuration file overwrites, add extra debugging information, change Asterisk's installation and staging directories, and modify which type of processor you are compiling for. While you may never edit or require any of these options, they are mentioned here for completeness.

Enabling GSM optimizations

Uncomment the following line in your Asterisk *Makefile* to enable GSM codec optimizations on x86 CPU architectures that support MMX instructions:

```
#K6OPT = -DK6OPT
```

This includes newer Pentium processors, Pentium Pros, and the AMD K6 and K7 processors; however, you may not want to enable MMX support unless you have a true Intel processor, as problems have been reported with the MMX instructions on non-Intel processors.

Disabling configuration file overwrites

By default, Asterisk will overwrite your configuration files if you run make samples more than once. To change this behavior, change the y in the line below to n:

```
OVERWRITE=y
```

Enabling debug profiling information

Debug symbols allow you to do symbolic debugging. The profiling information (-pg) flag will produce a file when you run Asterisk that can be processed in order to obtain information about how long (relatively) Asterisk spends in each function. Use of the –pg flag is not recommended for a normal build, but it may be useful during development. To enable profiling information, replace the –g in the following line with –pg:

```
DEBUG=-g
```

Specifying where to install Asterisk after compiling

You can change the directory where Asterisk is installed by specifying a path on the following line:

```
INSTALL_PREFIX=
```

Changing the staging directory

The staging directory is where Asterisk temporarily copies its files during the install process. You may want the files to be copied to a directory such as */tmp/asterisk/*. If no staging directory is specified (the default), Asterisk will use the source directory. To specify a staging directory, enter the desired directory on this line:

```
DESTDIR=
```

Compiling on VIA motherboards

On VIA-based motherboards, you need to set the processor to i586. If Asterisk detects the processor as i686, you may get random core dumps. To force Asterisk to compile using i586, remove the comment from the following PROC line in the *Makefile* (line 81, at the time of this writing):

```
# Pentium & VIA processors optimize
# PROC=i586
```

Using Precompiled Binaries

While the documented process of installing Asterisk expects you to compile the source code yourself, there are Linux distributions (such as Debian) that include precompiled Asterisk binaries. Failing that, you may be able to install Asterisk with the package managers that those distributions of Linux provide (such as *apt-get* for

Debian and *portage* for Gentoo). However, you may also find that many of these prebuilt binaries are quite out of date and do not follow the same furious development cycle as Asterisk.

Finally, there do exist basic, precompiled Asterisk binaries that can be downloaded and installed in whatever Linux distribution you have chosen. However, the use of precompiled binaries doesn't really save much time, and we have found that compiling Asterisk with each install is not a very cumbersome task. We believe that the best way to install Asterisk is to compile from the source code, so we won't discuss prebuilt binaries very much in this book. In the next chapter, we'll look at how to initially configure Asterisk and several kinds of channels.

Installing Additional Prompts

The *asterisk-sounds* package contains many useful professionally recorded prompts. It is highly recommended that you install it now, as we will be using some of the prompts from this package in later chapters. To do so, run the following commands:

```
# cd /usr/src/asterisk-sounds
# make install
```

Other Useful Add-ons

The *asterisk-addons* package contains code to allow the storage of Call Detail Records (CDRs) to a MySQL database and to natively play MP3s, as well as an interpreter for loading Perl code into memory for the life of an Asterisk process. Programs are placed into *asterisk-addons* when there are licensing issues preventing them from being implemented directly into the Asterisk source code, or when they are not yet ready for primetime.

The *g729/* directory contains the code and registration program for the proprietary G.729 codec. Even if your end devices have the G.729 codec installed, in order to allow the phones to communicate with Asterisk using G.729 (e.g., in voicemail or to allow attended transfers), you must purchase a license. Licenses for the codec can be purchased online from Digium and activated with the registration program contained in the *g729/* directory.

Updating Your Source Code

Instead of deleting the sources and downloading the entire tree every time you want to update, you can update just the files that have changed since the last revision. To do this, change into the directory containing the files you want to update and run the make update command:

```
# cd /usr/src/asterisk/
# make update
# make clean
# make upgrade
```

Note that this will work only with code obtained via the CVS method (see "make update," earlier in this chapter). The `make upgrade` command is used only in the Asterisk source directory. In other directories, use `make install`.

Common Compiling Issues

There are many common compiling issues that users often run into. Here are some of the more common problems, and how to resolve them.

Asterisk

First, let's take a look at some of the errors you may encounter when compiling Asterisk.

C compiler cannot create executables

If you receive the following error while attempting to compile Asterisk, you must install the *gcc* compiler and its dependencies:

```
checking whether the C compiler (gcc ) works... no
configure: error: installation or configuration problem: C compiler cannot create
executables.
make: *** [editline/libedit.a] Error 1
```

The following packages are required for *gcc*:

- *gcc*
- *glibc-kernheaders*
- *cpp*
- *binutils*
- *glibc-headers*
- *glibc-devel*

These can be installed manually, by copying the files off of your distribution disks, or through the *yum* package manager, with the command `yum install gcc`.

bison: command not found

The following error may be encountered if the *bison* parser, which is required for parsing expressions in the *extensions.conf* file, is not found:

```
bison ast_expr.y -name-prefix=ast_yy -o ast_expr.c
make: bison: Command not found
make: *** [ast_expr.c] Error 127
```

The following files are required in order to install Asterisk; they can be installed with the yum install bison command:

- *bison*
- *m4*

/usr/bin/ld: cannot find −lssl

The OpenSSL development packages are required by Asterisk within the *res_crypto.so* module for RSA key checks performed by the IAX2 protocol. If the OpenSSL development packages are not installed, the following error will occur:

```
/usr/bin/ld: cannot find -lssl
collect2: ld returned 1 exit status
make: *** [asterisk] Error 1
```

To install the OpenSSL development library, you'll require the following dependencies:

- *openssl-devel*
- *e2fsprogs-devel*
- *zlib-devel*
- *krb5-devel*
- *krb5-libs*

You can use the yum install openssl-devel command to install these files.

rpmbuild: command not found

To use the make rpm command, you must have the Red Hat Package Manager (RPM) development package installed. The following error will be encountered if it is absent:

```
make[1]: Leaving directory `/usr/src/asterisk-1.0.3'
/bin/sh: line 1: rpmbuild: command not found
make: *** [__rpm] Error 127
```

You can install the build environment with yum install rpmbuild.

Zaptel

You may also run into errors when compiling Zaptel. Here are some of the most commonly occurring problems, and what to do about them.

make: cc: Command not found

You will receive the following error if you attempt to build Zaptel without the *gcc* compiler installed:

```
make: cc: Command not found
make: *** [gendigits.o] Error 127
```

Be sure to install *gcc* and its dependencies. For more information, see "C compiler cannot create executables" in the previous section.

FATAL: Module wctdm/fxs/fxo not found

The TDM400P cards require the PCI bus to be Version 2.2. If you attempt to load the Zapata telephony drivers with an older version, you may get the following errors:

- When attempting to load the *wctdm* driver, you may see this error:

      FATAL: Module wctdm not found

- When attempting to load the *wctdm* or *wcfxo* driver, you may see an error such as this:

      ZT_CHANCONFIG failed on channel 1: No such device or address (6)
      FATAL: Module wctdm not found

The only way to resolve these errors is to use a newer motherboard that supports PCI Version 2.2.

> You may also encounter these errors if the power has not been attached to the Molex connector found on the TDM400P card.

Unresolved symbol link when loading ztdummy

The *ztdummy* driver requires that a UHCI USB controller be available on Linux 2.4 kernels (the USB controller is not a requirement on Linux 2.6 kernels, because they are capable of generating the 1-kHz timing reference). There exists a secondary kind of controller, known as OHCI, which is not compatible with the *ztdummy* driver. If the UHCI USB controller is not accessible on Linux 2.4 kernels, the following error will occur:

```
/lib/modules/2.4.22/misc/ztdummy.o: /lib/modules/2.4.22/misc/ztdummy.o: unresolved
symbol unlink_td
/lib/modules/2.4.22/misc/ztdummy.o: /lib/modules/2.4.22/misc/ztdummy.o: unresolved
symbol alloc_td
/lib/modules/2.4.22/misc/ztdummy.o: /lib/modules/2.4.22/misc/ztdummy.o: unresolved
symbol delete_desc
/lib/modules/2.4.22/misc/ztdummy.o: /lib/modules/2.4.22/misc/ztdummy.o: unresolved
symbol uhci_devices
/lib/modules/2.4.22/misc/ztdummy.o: /lib/modules/2.4.22/misc/ztdummy.o: unresolved
symbol uhci_interrupt
/lib/modules/2.4.22/misc/ztdummy.o: /lib/modules/2.4.22/misc/ztdummy.o: unresolved
symbol fill_td
/lib/modules/2.4.22/misc/ztdummy.o: /lib/modules/2.4.22/misc/ztdummy.o: unresolved
symbol insert_td_horizontal
/lib/modules/2.4.22/misc/ztdummy.o: insmod /lib/modules/2.4.22/misc/ztdummy.o failed
/lib/modules/2.4.22/misc/ztdummy.o: insmod ztdummy failed
```

You can verify that you have the correct style of USB controller and its associated drivers with the lsmod command:

```
# lsmod
Module                  Size  Used by
usb_uhci               26412  0
usbcore                79040  1 [hid usb-uhci]
```

As you can see in the example above, you are looking to make sure that the *usbcore* and *usb_uhci* modules are loaded. If these modules are not loaded, be sure that USB has been activated within your BIOS and that the modules exist and are being loaded.

If the USB drivers are not loaded, you can still check which type of USB controller you have with the dmesg command:

```
# dmesg | grep -i usb
```

To verify that you indeed have a UHCI USB controller, look for the following lines:

```
uhci_hcd 0000:00:04.2: new USB bus registered, assigned bus number 1
hub 1-0:1.0: USB hub found
uhci_hcd 0000:00:04.3: new USB bus registered, assigned bus number 2
hub 2-0:1.0: USB hub found
```

Depmod errors during compilation

If you experience depmod errors during compilation, you more than likely don't have a symbolic link to your Linux kernel sources. If you don't have your Linux kernel sources installed, retrieve the sources for your installed kernel, install them, and create a symbolic link against */usr/src/linux-2.4*. The following is an example of a depmod error:

```
depmod: *** Unresolved symbols in /lib/modules/2.4.22/kernel/drivers/block/loop.o
```

Loading Zaptel Modules

In this section, we'll take a quick look at how to load the *zaptel* and *ztdummy* modules. The *zaptel* module does not require any configuration if it's being used only for the *ztdummy* module. If you plan on loading the *ztdummy* module as your timing source (and thus, you will not be running any PCI hardware in your system), now is a good time to load both drivers.

Systems Running udevd

In the early days of Linux, the system's */dev/* directory was populated with a list of devices with which the system could potentially interact. At the time, nearly 18,000 devices were listed. That all changed when *devfs* was released, allowing dynamic creation of devices that are active within the system. Some of the recently released distributions have incorporated the *udev* daemon into their systems to dynamically populate */dev/* with device nodes.

To allow Zaptel and other device drivers to access the PCI hardware installed in your system, you must add some rules. Using your favorite text editor, open up your *udevd* rules file. On Fedora Core 3, for example, this file is located at */etc/udev/rules. d/50-udev.rules*. Add the following lines to the end of your rules file:

```
# Section for zaptel device
KERNEL="zapctl",     NAME="zap/ctl"
KERNEL="zaptimer",   NAME="zap/timer"
KERNEL="zapchannel", NAME="zap/channel"
KERNEL="zappseudo",  NAME="zap/pseudo"
KERNEL="zap[0-9]*",  NAME="zap/%n"
```

Save the file and reboot your system for the settings to take effect.

Loading Zaptel

The *zaptel* module must be loaded before any of the other modules are loaded and used. Note that if you will be using the *zaptel* module with PCI hardware, you must configure */etc/zaptel.conf* before you load it. (We will discuss how to configure *zaptel. conf* for use with hardware in Chapter 4.) If you are using *zaptel* only to access *ztdummy*, you can load it with the modprobe command, as follows:

```
# modprobe zaptel
```

If all goes well, you shouldn't see any output. To verify that the *zaptel* module loaded successfully, use the lsmod command. You should be returned a line showing the *zaptel* module and the amount of memory it is using:

```
# lsmod | grep zaptel
zaptel                 201988  0
```

Loading ztdummy

The *ztdummy* module is an interface to a device that provides timing, which in turn allows Asterisk to provide timing to various applications and functions that require it. Use the modprobe command to load the *ztdummy* module after *zaptel* has been loaded:

```
# modprobe ztdummy
```

If *ztdummy* loads successfully, no output will be displayed. To verify that *ztdummy* is loaded and is being used by *zaptel*, use the lsmod command. The following output is from a computer running the 2.6 kernel:

```
# lsmod | grep ztdummy
Module                 Size   Used by
ztdummy                3796   0
zaptel                 201988 1 ztdummy
```

If you happen to be running a 2.4 kernel-based computer, your output from lsmod will show that *ztdummy* is using the *usb-uhci* module:

```
# lsmod | grep ztdummy
Module              Size  Used by
ztdummy             3796  0
zaptel            201988  0 ztdummy
usb-uhci           24524  0 ztdummy
```

Loading libpri

The *libpri* libraries do not need to be loaded like modules. Asterisk looks for *libpri* at compile time and configures itself to use the libraries if they are found.

Loading Asterisk

Asterisk can be loaded in a variety of ways. The easiest way is to start Asterisk by running the binary file directly from the Linux command-line interface. If you are running a system that uses the *init.d* scripts, you can easily start and restart Asterisk that way as well. However, the preferred way of starting Asterisk is via the *safe_asterisk* script.

CLI Commands

The Asterisk binary is, by default, located at */usr/sbin/asterisk*. If you run */usr/sbin/asterisk*, it will be loaded as a daemon. There are also a few switches you should be aware of that allow you to (re)connect to the Asterisk CLI, set the verbosity of CLI output, and allow core dumps if Asterisk crashes (for debugging with *gdb*). To explore the full range of options, run Asterisk with the –h switch:

```
# /usr/sbin/asterisk -h
```

Here is a list of the most commonly used options:

-c

Console. This allows you to connect to the Asterisk CLI.

-v

Verbosity. This is used to set the amount of output for CLI debugging.

-g

Core dump. If Asterisk were to crash unexpectedly, this would cause a core file to be created for later tracing with *gdb*.

-r

Remote. This is used to reconnect remotely to an already running Asterisk process. (The process is remote from the standpoint of the console connecting to it but is actually a local process on the machine. This has nothing to do with connecting to a remote process over a network using a protocol such as IP, as this is not supported.)

-rx "restart now"

Execute. Using this command in combination with –r allows you to execute a CLI command without having to connect to the CLI and type it manually.

Let's look at some examples. To start Asterisk and connect to the CLI with a verbosity level of 3, use the following command:

```
# /usr/sbin/asterisk -cvvv
```

If the Asterisk process is already running (for example, if you started Asterisk with */usr/sbin/asterisk*), instead use the reconnect switch, like so:

```
# /usr/sbin/asterisk -vvvr
```

If you want Asterisk to dump a core file after a crash, you can use the –g switch when starting Asterisk:

```
# /usr/sbin/asterisk -g
```

To execute a command without connecting to the CLI and typing it (perhaps for use within a script), you can use the –x switch in combination with the –r switch:

```
# /usr/sbin/asterisk -rx "restart now"
```

If you are experiencing crashes and would like to output to a debug file, use the following command:

```
# /usr/sbin/asterisk -vvvvvvvvvc | tee /tmp/debug.log
```

Red Hat–Style Initialization Script

If you ran the make config command earlier (or manually copied the initialization scripts), you can start and restart Asterisk with the following commands:

```
# /etc/rc.d/init.d/asterisk start
# /etc/rc.d/init.d/asterisk stop
```

The safe_asterisk Script

The main purpose of the *safe_asterisk* script is to dump a core file if Asterisk fails and to automatically restart it. There is also a notify option within the script, which, if set, will send an email letting you know that Asterisk died unexpectedly. An added benefit of the script is that it will load the Asterisk CLI on terminal interface 9 (by default; this is configurable), so you can easily switch to that window to monitor your Asterisk system.

The default location of the *safe_asterisk* script is */usr/sbin/safe_asterisk*, and it can be executed as such. Let's review the various options contained in the *safe_asterisk* script:

```
CLIARGS="$*"              # Grab any args passed to safe_asterisk
TTY=9                     # TTY (if you want one) for Asterisk to run on
CONSOLE=yes               # Whether or not you want a console
#NOTIFY=ben@alkaloid.net  # Email address for crash notifications
```

The first line simply allows you to pass arguments to the *safe_asterisk* script from the Linux CLI; it should not be edited directly. TTY=9 specifies the Linux console on which to run the Asterisk CLI output. You can disable this feature by specifying CONSOLE=no. If you would like to be notified if Asterisk dies suddenly and requires a restart, uncomment the NOTIFY line and replace *ben@alkaloid.net* with your email address. Note that the crash notifications are sent with the mail command, so your system must be set up to process and send email.

Directories Used by Asterisk

Asterisk uses several directories on a Linux system to manage the various aspects of the system, such as voicemail recordings, voice prompts, and configuration files. This section discusses the necessary directories, all of which are created during installation and configured in the *asterisk.conf* file.

/etc/asterisk/

The */etc/asterisk/* directory contains the Asterisk configuration files. One file, how-ever—*zaptel.conf*—is located in the */etc/* directory. The Zaptel hardware was origi-nally designed by Jim Dixon of the Zapata Telephony Group as a way of bringing reasonable and affordable computer telephony equipment to the world. Asterisk makes use of this hardware, but any other software can also make use of the Zaptel hardware and drivers. Consequently, the *zaptel.conf* configuration file is not directly located in the */etc/asterisk/* directory.

/usr/lib/asterisk/modules/

The */usr/lib/asterisk/modules/* directory contains all the Asterisk loadable modules. Within this directory are the various applications, codecs, formats, and channels used by Asterisk. By default, Asterisk loads all of these modules at startup. You can disable any modules you are not using in the *modules.conf* file, but be aware that cer-tain modules are required by Asterisk or are dependencies of other modules. Attempting to load Asterisk without these modules will cause an error at startup.

/var/lib/asterisk

The */var/lib/asterisk/* directory contains the *astdb* file and a number of subdirectories. The *astdb* file contains the local Asterisk database information, which is somewhat like the Microsoft Windows Registry. The Asterisk database is a simple implementation based on v1 of the Berkeley database. The *db.c* file in the Asterisk source states that this version was chosen for the following reason: "DB3 implementation is released under an alternative license incompatible with the GPL. Thus in order to keep Asterisk licensing simplistic, it was decided to use version 1 as it is released under the BSD license."

The subdirectories within */var/lib/asterisk/* include:

agi-bin/
> The *agi-bin/* directory contains your custom scripts, which can interface with Asterisk via the various built-in AGI applications. For more information about AGI, see Chapter 8.

firmware/
> The *firmware/* directory contains firmware for various Asterisk-compatible devices. It currently contains only the *iax/* subdirectory, which holds the binary firmware image for Digium's IAXy.

images/
> Applications that communicate with channels supporting graphical images look in the *images/* directory. Most channels do not support the transmission of images, so this directory is rarely used. However, if more devices that support and make use of graphical images are released, this directory will become more relevant.

keys/
> Asterisk can use a public/private key system to authenticate peers connecting to your box via an RSA digital signature. If you place a peer's public key in your *keys/* directory, that peer can be authenticated by channels supporting this method (such as the IAX2 channels). The private key is never distributed to the public. The reverse is also true: you can distribute your public key to your peers, allowing you to be authenticated with the use of your private key. Both the public and private keys—ending in the *.pub* and *.key* file extensions, respectively—are stored in the *keys/* directory.

mohmp3/
> When you configure Asterisk for Music on Hold, applications utilizing this feature look for their MP3 files in the *mohmp3/* directory. Asterisk is a bit picky about how the MP3 files are formatted, so you should use constant bitrate (CBR) encoding and strip the ID3 tags from your files.

sounds/

> All of the available voice prompts for Asterisk reside in the *sounds/* directory. The contents of the basic prompts included with Asterisk are in the *sounds.txt* file located in your Asterisk source code directory. Contents of the additional prompts are located in the *sounds-extra.txt* file in the directory to which you extracted the *asterisk-sounds* package earlier in this chapter.

/var/spool/asterisk/

The Asterisk spool directory contains several subdirectories, including *outgoing/*, *qcall/*, *tmp/*, and *voicemail/* (see Figure 3-2). Asterisk monitors the *outgoing* and *qcall* directories for text files containing call request information. These files allow you to generate a call simply by copying or moving the correctly structured file into the *outgoing/* directory.

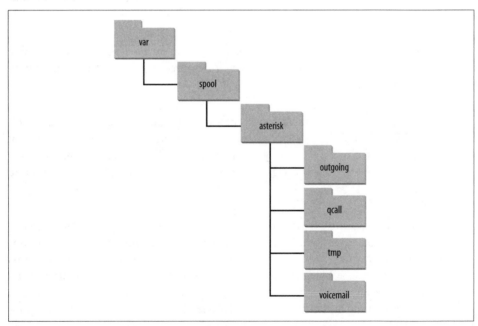

Figure 3-2. /var/spool/asterisk/ directory structure

The old (now deprecated) *qcall* method of generating calls utilized a single line of text within the call file. Call files for use within the *qcall* directory took the form of:

```
Dialstring Caller-ID Extension Maxsecs [Identifier] [Required-response]
```

This rather limited what you could do with the call file, and what kinds of information you could pass to Asterisk. Thus, a new spooling method was developed in Asterisk, using the *outgoing* directory. Call files being placed into this directory can contain much more valuable information, such as the Context, Extension, and Priority where the answered call should start, or simply the application and its arguments. You can also set variables and specify an account code for Call Detail Records. More information about the use of call files is presented in Chapter 9.

The *tmp/* directory is used, funny enough, to hold temporary information. Certain applications may require a place to write files to before copying the complete files to their final destinations. This prevents two processes from trying to write to and read from a file at the same time.

All voicemail and user greetings are contained within the *voicemail/* directory. Extensions configured in *voicemail.conf* that have been logged into at least once are created as subdirectories of *voicemail/*.

/var/run/

The */var/run/* directory contains the process ID (pid) information for all active processes on the system, including Asterisk (as specified in the *asterisk.conf* file). Note that */var/run/* is OS-dependent and may differ.

/var/log/asterisk/

The */var/log/asterisk/* directory is where Asterisk logs information. You can control the type of information being logged to the various files by editing the *logger.conf* file located in the */etc/asterisk/* directory. Basic configuration of the *logger.conf* file is covered in Appendix E.

/var/log/asterisk/cdr-csv

The */var/log/asterisk/cdr-csv* directory is used to store the CDRs in comma-separated value (CSV) format. By default information is stored in the *Master.csv* file, but individual accounts can store their own CDRs in separate files with the use of the accountcode option (see Appendix A for more information).

Conclusion

In this chapter, we have reviewed the procedures for obtaining, compiling, and installing Asterisk and the associated packages. In the following chapter, we will touch on the initial configuration of your system with regard to various communications channels, such as analog devices attached to FXS and FXO ports, SIP channels, and IAX2 endpoints.

## CHAPTER 4

Initial Configuration of Asterisk

*Perseverance is the hard work you do after you get
tired of doing the hard work you already did.*
—Newt Gingrich

The purpose of this chapter is to guide the user through the configuration of four channels: a Foreign eXchange Office (FXO) channel, a Foreign eXchange Station (FXS) channel, a Session Initiation Protocol (SIP) channel, and an Inter-Asterisk eXchange protocol (IAX)* channel. The purpose is not to give an exhaustive survey of all channel types or topologies, but rather to provide a base platform on which to build your telecommunications system. Further scenarios and channel configuration details can be found in Appendix D. We start by exploring the basic configuration of analog interfaces such as FXS and FXO ports with the use of a Digium Dev-Lite kit. We'll then configure two Voice over Internet Protocol (VoIP) interfaces: a local SIP channel connected to a soft phone, and a connection to Free World Dialup via IAX.

Once you've worked through this chapter, you will have a basic system consisting of many useful interfaces, and you will be ready to learn more about the *extensions.conf* file (discussed further in Chapter 5), which contains the instructions Asterisk needs to build the dialplan.

What Do I Really Need?

The asterisk character (*) is used as a wildcard in many different applications. It is the perfect name for this PBX for many reasons, one of which is the enormous number of interface types to which Asterisk can connect. These include:

- Analog interfaces, such as your telephone line and analog telephones
- Digital circuits, such as T-1 and E-1 lines
- VoIP protocols such as SIP and IAX

---

* Officially, the current version is IAX2, but all support for IAX1 has been dropped, so whether you say "IAX" or "IAX2," it is expected that you are talking about Version 2.

Asterisk doesn't need any specialized hardware—not even a sound card. Channel cards that connect Asterisk to analog phones or phone lines are available, but not essential. You can connect to Asterisk using the soft phones that are available for Windows, Linux, and other operating systems without using a special hardware interface. You can also use any IP phone that supports either SIP or IAX2. On the other side, if you don't connect directly to an analog phone line from your central office, you can route your calls over the Internet to a telephony service provider.

Working with Interface Configuration Files

In this chapter, we're finally going to "get our hands dirty" and start building an Asterisk configuration. For the first few sections on FXO and FXS channels, we'll assume that you have the Digium Dev-Lite kit with one FXO and one FXS interface, which allows you to connect to an analog phone line (FXO) and to an analog phone (FXS). Note that this hardware interface isn't necessary; if you want to build an IP-only configuration, you can skip to the section on configuring SIP.

The configuration we do in this chapter won't be particularly useful on its own, but it will be a kernel to build on. We're going to touch on the following files:

zaptel.conf
> Here, we'll do low-level configuration for the hardware interface. We'll set up one FXO channel and one FXS channel.

zapata.conf
> In this file, we'll configure Asterisk's interface to the hardware.

extensions.conf
> The dialplans we create will be extremely primitive, but they will prove that the system is working.

sip.conf
> This is where we'll configure the SIP protocol.

iax.conf
> This is where we'll configure incoming and outgoing IAX channels.

In the following sections, you will be editing several configuration files. You'll have to reload these files for your changes to take effect. After you edit the *zaptel.conf* file, you will need to reload the configuration for the hardware with /sbin/ztcfg -vv (you may omit the –vv if you don't need verbose output). Changes made in *zapata.conf* will require a reload from the Asterisk console; however, changing signaling methods requires a restart. You will need to perform a reload chan_iax2.so and a reload chan_sip.so after editing the *iax.conf* and *sip.conf* files, respectively.

FXO and FXS Channels

The difference between an FXO channel and an FXS channel is simply which end of the connection provides the dial tone. An FXO port does not generate a dial tone; it accepts one. A common example is the dial tone provided by your phone company. An FXS port provides both the dial tone and ringing voltage to alert the station user of an inbound call. Both interfaces provide bidirectional communication (i.e., communication that is transmitted and received in both directions simultaneously).

If your Asterisk server has a compatible FXO port, you can plug a telephone line from your telephone company (or "telco") into this port. Asterisk can then use the telco line to place and receive telephone calls. By that same token, if your Asterisk server has a compatible FXS port, you may plug an analog telephone into your Asterisk server, so that Asterisk may call the phone and you may place calls.

Ports are defined in the configuration by the signaling they use, as opposed to the physical type of port they are. For instance, a physical FXO port will be defined in configuration with FXS signaling, and an FXS port will be defined with FXO signaling. This can be confusing until you understand the reasons for it. FX_ cards are named not according to what they are, but rather according to what is connected to them. An FXS card, therefore, is a card that connects to a station. Since that is so, you can see that in order to do its job, an FXS card must *behave* like a central office and use FXO signaling. Similarly, an FXO card connects to a central *office* (CO), which means it will need to behave like a station and use FXS signaling. The modem in your computer is a classic example of an FXO device.

> The older X100P card used a Motorola chipset, and the X101P (which Digium sold before completely switching to the TDM400P) is based on the Ambient/Intel MD3200 chipset. These cards are modems with drivers adapted to utilize the card as a single FXO device (the telephone interface cannot be used as an FXS port). Support for the X101P card has been dropped in favor of the TDM series of cards. Use of these cards (or their clones) is not recommended in production environments.

Determining the FXO and FXS Ports on Your TDM400P

Figure 4-1 contains a picture of a TDM400P with an FXS module and an FXO module. You can't see the colors, but module 1 is a green FXS module and module 2 is an orange/red FXO module. In the bottom-right corner of the picture is the Molex connector, where power is supplied from computer's power supply.

> Plugging an FXS port (the green module) into the PSTN may destroy the module and the card!

Figure 4-1. A TDM400P with an FXS module (1 across) and an FXO module (2 across)

Be sure to connect your computer's power supply to the Molex connector on the TDM400P if you have FXS modules, as it is used to generate the voltage to produce ringing on the phone. The Molex connector is not required if you have only FXO modules.

Configuring an FXO Channel

We'll start by configuring an FXO channel. First we'll configure the Zaptel hardware, and then the Zapata hardware. We'll set up a very basic dialplan, and we'll show you how to test the channel.

Zaptel Hardware Configuration

The *zaptel.conf* file located in */etc/* is used to configure your hardware. The following minimal configuration defines an FXO port with FXS signaling:

```
fxsks=2
loadzone=us
defaultzone=us
```

In the first line, in addition to indicating whether we are using FXO or FXS signaling, we specify one of the following protocols for channel 2:

- Loop start (ls)
- Ground start (gs)
- Kewlstart (ks)

The difference between *loop start* and *ground start* has to do with how the equipment requests a dial tone: a ground start circuit signals the far end that it wants a dial tone by momentarily grounding one of the leads; a loop start circuit uses a short to request a dial tone. Though not common for new installations, analog ground start lines still exist in many areas of the country.* For example, ground start lines are predominately used to reduce a condition known as "glare"† that is associated with loop start lines and PBXs with high call volumes. All home lines (and analog telephones/modems/faxes) in North America use loop start signaling. *Kewlstart* is in fact the same as loop start, except that it has greater intelligence and is thus better able to detect far-end disconnects.‡ Kewlstart is the preferred signaling protocol for analog circuits in Asterisk.

To configure a signaling method other than kewlstart, replace the ks in fxsks with either ls or gs (for loop start or ground start, respectively).

loadzone configures the set of indications (as configured in *zonedata.c*) to use for the channel. The *zonedata.c* file contains information about all the various sounds that a phone system makes in a particular country: dial tone, ringing cycles, busy tone, and so on. When you apply a loaded tone zone to a Zap channel, that channel will mimic the indications for the specified country. Different indication sets can be configured for different channels. The defaultzone is used if no zone is specified for a channel.

After configuring *zaptel.conf*, you can load the drivers for the card. modprobe is used to load modules for use by the Linux kernel. For example, to load the *wctdm* driver, you would run:

```
# modprobe wctdm
```

---

* Yes, there is such a thing as ground start signaling on channelized T-1s, but that has nothing to do with an actual ground condition on the circuit (which is entirely digital).

† When a call is initiated from one end of a circuit at the same approximate time a call is initiated from the opposite end of the circuit.

‡ A *far-end disconnect* happens when the far end hangs up. In an unsupervised circuit, there is no method of telling the near end that the call has ended. If you are on the phone this is no problem, since you will know the call has ended and will manually hang up your end. If, however, your voicemail system is recording a message, it will have no way of knowing that the far end has terminated and will thus keep recording silence, or even the dial tone or reorder tone. Kewlstart can detect these conditions and disconnect the circuit.

If the drivers load without any output, they have loaded successfully.* You can verify that the hardware and ports were loaded and configured correctly with the use of the *ztcfg* program:

```
# /sbin/ztcfg -vv
```

The channels that are configured and the signaling method being used will be displayed. For example, a TDM400P with one FXO module has the following output:

```
Zaptel Configuration
======================


Channel map:

Channel 02: FXS Kewlstart (Default) (Slaves: 02)

1 channels configured.
```

If you receive the following error, you have configured the channel for the wrong signaling method:

```
ZT_CHANCONFIG failed on channel 2: Invalid argument (22)
Did you forget that FXS interfaces are configured with FXO signalling
and that FXO interfaces use FXS signalling?
```

To unload drivers from memory, use the rmmod (remove module) command, like so:

```
# rmmod wctdm
```

The *zttool* program is a diagnostic tool used to determine the state of your hardware. After running it, you will be presented with a menu of all installed hardware. You can then select the hardware and view the current state. A state of "OK" means the hardware is successfully loaded:

```
Alarms          Span
OK              Wildcard TDM400P REV E/F Board 1
```

Zapata Hardware Configuration

Asterisk uses the *zapata.conf* file to determine the settings and configuration for telephony hardware installed in the system. The *zapata.conf* file also controls the various features and functionality associated with the hardware channels, such as Caller ID, call waiting, echo cancellation, and a myriad of other options.

When you configure *zaptel.conf* and load the modules, Asterisk is not aware of anything you've configured. The hardware doesn't have to be used by Asterisk; it could very well be used by another piece of software that interfaces with the Zaptel

---

* It is generally safe to assume that the modules have loaded successfully, but to view the debugging output when loading the module, check the console output (by default this is located on TTY terminal 9, but this is configurable in the *safe_asterisk* script—see the previous chapter for details).

modules. You tell Asterisk about the hardware and control the associated features via *zapata.conf*:

```
[trunkgroups]
; define any trunk groups

[channels]
; hardware channels
; default
usecallerid=yes
hidecallerid=no
callwaiting=no
threewaycalling=yes
transfer=yes
echocancel=yes
echotraining=yes

; define channels
context=incoming          ; Incoming calls go to [incoming] in extensions.conf
signalling=fxs_ks         ; Use FXS signalling for an FXO channel
channel => 2              ; PSTN attached to port 2
```

The [trunkgroups] section is for NFAS and GR-303 connections, and it won't be discussed in this book. If you require this type of functionality, see the *zapata.conf.sample* file for more information.

The [channels] section determines the signaling method for hardware channels and their options. Once an option is defined, it is inherited down through the rest of the file. A channel is defined using channel =>, and each channel definition inherits all the options defined above that line. If you wish to configure different options for different channels, remember that the options should be configured *before* the channel => definition.

We've enabled Caller ID with usecallerid=yes and specified that it will not be hidden for outgoing calls with hidecallerid=no. Call waiting is deactivated on an FXO line with callwaiting=no. Enabling three-way calling with threewaycalling=yes allows an active call to be placed on hold with a hook switch flash (discussed in Chapter 7) to suspend the current call. You may then dial a third party and join them to the conversation with another hook switch. The default is to not enable three-way calling.

Allowing call transfer with a hook switch is accomplished by configuring transfer=yes; it requires that three-way calling be enabled. The Asterisk echo canceller is used to remove the echo that can be created on analog lines. You can enable the echo canceller with echocancel=yes. The echo canceller in Asterisk requires some time to learn the echo, but you can speed this up by enabling echo training (echotraining=yes). This tells Asterisk to send a tone down the line at the start of a call to measure the echo, and therefore learn it more quickly.

When a call comes in on an FXO interface, you will want to perform some action. The action to be performed is configured inside a block of instructions called a *context*. Incoming calls on the FXO interface are directed to the incoming context with context=incoming. The instructions to perform inside the context are defined within *extensions.conf*.

Finally, since an FXO channel uses FXS signaling, we define it as such with signalling=fxs_ks.

Dialplan Configuration

The following minimal dialplan makes use of the Echo( ) application to verify that bidirectional communications for the channel are working:

```
[incoming]
; incoming calls from the FXO port are directed to this context from zapata.conf
exten => s,1,Answer( )
exten => s,2,Echo( )
```

Whatever you say, the Echo( ) application will relay back to you.

Dialing in

Now that the FXO channel is configured, let's test it. Run the *zttool* application and connect your PSTN line to the FXO port on your TDM400P. Once you have a phone line connected to your FXO port, you can watch the card come out of a RED alarm.

Now dial the PSTN number from another external phone (such as a cell phone). Asterisk will answer the call and execute the Echo( ) application. If you can hear your voice being reflected back, you have successfully installed and configured your FXO channel.

Configuring an FXS Channel

The configuration of an FXS channel is similar to that of an FXO channel. Let's take a look.

Zaptel Hardware Configuration

The following is a minimal configuration for an FXS channel on a TDM400P. The configuration is identical to the FXO channel configuration above, with the addition of fxoks=1.

Recall from our earlier discussion that the opposite type of signaling is used for FXO and FXS channels, so we will be configuring FXO signaling for our FXS channel. In the example below we are configuring channel 1 to use FXO signaling, with the kewlstart signaling protocol:

```
fxoks=1
fxsks=2
loadzone=us
defaultzone=us
```

After loading the drivers for your hardware, you can verify their state with the use of /sbin/ztcfg –vv:

```
Zaptel Configuration
======================


Channel map:

Channel 01: FXO Kewlstart (Default) (Slaves: 01)
Channel 02: FXS Kewlstart (Default) (Slaves: 02)

2 channels configured.
```

Zapata Hardware Configuration

The following configuration is identical to that for the FXO channel, with the addition of a section for our FXS port and of the line immediate=no. The context for our FXS port is internal, the signaling is fxoks (kewlstart), and the channel number is set to 1.

FXS channels can be configured to perform one of two different actions when a phone is taken off the hook. The most common (and often expected) option is for Asterisk to produce a dial tone and wait for input from the user. This action is configured with immediate=no. The alternative action is for Asterisk to automatically perform a set of instructions configured in the dialplan instead of producing a dial tone, which you indicate by configuring immediate=yes.* The instructions to be performed are found in the context configured for the channel and will match the s extension (both of these topics will be discussed further in the following chapter).

Here's our new *zapata.conf*:

```
[trunkgroups]
; define any trunk groups

[channels]
; hardware channels
; default
usecallerid=yes
hidecallerid=no
callwaiting=no
threewaycalling=yes
transfer=yes
echocancel=yes
echotraining=yes
immediate=no
```

---

* Also referred to as the BatPhone method, and more formally known as an Automatic Ringdown or Private Line Automatic Ringdown (PLAR) circuit. This method is commonly used at rental car counters and airports.

```
; define channels
context=internal          ; Uses the [internal] context in extensions.conf
signalling=fxo_ks         ; Use FXO signalling for an FXS channel
channel => 1              ; Telephone attached to port 1

context=incoming          ; Incoming calls go to [incoming] in extensions.conf
signalling=fxs_ks         ; Use FXS signalling for an FXO channel
channel => 2              ; PSTN attached to port 2
```

Dialplan Configuration

To test our newly created Zap extension, we need to create a basic dialplan. The following dialplan contains a context called internal. This is the same context name that we configured in *zapata.conf* for channel 1. When we configure context=internal in *zapata.conf*, we are telling Asterisk where to look for instructions when a user presses digits on his telephone. In this case, the only extension number that will work is 611. When you dial 611 on your telephone, Asterisk will execute the Echo( ) application so that when you talk into the phone whatever you say will be played back to you, thereby verifying bidirectional voice.

The dialplan looks like this:

```
[internal]
exten => 611,1,Answer( )
exten => 611,2,Echo( )
```

Configuring SIP

The Session Initiation Protocol (SIP), often used in VoIP phones (either hard phones or soft phones), takes care of the setup and teardown of calls, along with any renegotiations during a call. Basically, it helps two endpoints talk to each other (if possible, *directly* to each other). SIP does not carry media; rather, it uses the Real-time Transport Protocol (RTP) to transfer the media* directly between phone A and phone B once the call has been set up.

SIP and RTP

SIP is an application-layer signaling protocol that uses the well-known port 5060 for communications. SIP can be transported with either the UDP or TCP transport-layer protocols. Asterisk does not currently have a TCP implementation for transporting SIP messages, but it is possible that future versions may support it (and patches to the code base are gladly accepted). SIP is used to "establish, modify, and terminate

---

* We use the term *media* to refer to the data transferred between endpoints and used to reconstruct your voice at the other end. It may also refer to music or prompts from the PBX.

multimedia sessions such as Internet telephony calls."* SIP does not transport media between endpoints.

RTP is used to transmit media (i.e., voice) between endpoints. RTP uses high-numbered, unprivileged ports in Asterisk (10,000 through 20,000, by default).

A common topology to illustrate SIP and RTP, commonly referred to as the "SIP trapezoid," is shown in Figure 4-2. When Alice wants to call Bob, Alice's phone contacts her proxy server, and the proxy tries to find Bob (often connecting through his proxy). Once the phones have started the call, they communicate directly with each other (if possible), so that the data doesn't have to tie up the resources of the proxy.

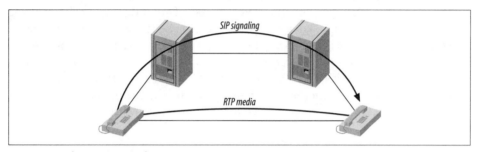

Figure 4-2. The SIP trapezoid

SIP was not the first, and is not the only, VoIP protocol in use today (others include H.323, MGCP, IAX, and so on), but currently it seems to have the most momentum with hardware vendors. The advantages of the SIP protocol lie in its wide acceptance and architectural flexibility (and, we used to say, simplicity!).

SIP Configuration

Here is a basic *sip.conf* file:

```
[general]
context=default
srvlookup=yes

[john]
type=friend
secret=welcome
qualify=yes          ; Qualify peer is no more than 2000 ms away
nat=no               ; This phone is not natted
host=dynamic         ; This device registers with us
canreinvite=no       ; Asterisk by default tries to redirect
context=internal     ; the internal context controls what we can do
```

---

* RFC 3261, SIP: Session Initiation Protocol, p. 9, Section 2.

The *sip.conf* file starts with a [general] section, which contains the channel settings and default options for all *users* and *peers* defined within *sip.conf*. You can override the default settings on a per-user/peer basis by configuring them within the user/peer definition.

Domain Name System Service records (DNS SRV records) are a way of setting up a logical, resolvable address where you can be reached. This allows calls to be forwarded to different locations without the need to change the logical address. By using SRV records, you gain many of the advantages of DNS, whereas disabling them breaks the SIP RFC and removes the ability to place SIP calls based on domain names. (Note that if multiple records are returned, Asterisk will use only the first.) DNS SRV record lookups are disabled by default in Asterisk, but it's highly recommended that you turn them on. To enable them, set srvlookup=yes in the [general] section of *sip.conf*.

Each connection is defined as a user, peer, or friend. A user type is used to authenticate incoming calls, a peer type is used for outgoing calls, and a friend type is used for both. The extension name is defined within square brackets ([]). In this case, we have defined the extension john as a friend.

A secret is a password used for authentication. Our secret is defined as welcome. We can monitor the latency between our Asterisk server and the phone with qualify=yes, thereby determining whether the remote device is reachable. qualify=yes can be used to monitor any end device, including other Asterisk servers. By default, Asterisk will consider an extension reachable if the latency is less than 2,000 ms (2 seconds). You can configure the time Asterisk should use when determining whether or not a peer is reachable by replacing yes with the number of milliseconds.

If an extension is behind a device performing Network Address Translation (NAT), such as a router or firewall, configure nat=yes to force Asterisk to ignore the contact information for the extension and use the address from which the packets are being received. Setting host=dynamic will require the extension to register so that Asterisk knows how to reach the phone. To limit an endpoint to a single IP address or fully qualified domain name (FQDN), replace dynamic with the IP address or domain name. Note that this limits only where you place calls *to*, as the user is allowed to place calls *from* anywhere (assuming she has authenticated successfully). If you set host=static, the end device is not required to register.

We've also set canreinvite=no. In SIP, *invites* are used to set up calls and to redirect media. Any invite issued after the initial invite in the same dialog is referred to as a *reinvite*. For example, suppose two parties are exchanging media traffic. If one client goes on hold and Asterisk is configured to play Music on Hold (MoH), Asterisk will issue a reinvite to the secondary client, telling it to redirect its media stream toward the PBX. Asterisk is then able to stream music or an announcement to the on-hold client.

The primary client then issues an off-hold command in a reinvite to the PBX, which in turn issues a reinvite to the secondary party requesting that it redirect its media stream toward the primary party, thereby ending the on-hold music and reconnecting the clients.

Normally, when two endpoints set up a call they pass their media directly from one to the other. Asterisk generally breaks this rule by staying within the media path, allowing it to listen for digits dialed on the phone's keypad. This is necessary because if Asterisk cannot determine the call length, inaccurate billing can occur. Configuring canreinvite=no forces Asterisk to stay in the media path, not allowing RTP messages to be exchanged directly between the endpoints.

Asterisk will not issue a reinvite in any of the following situations:

- If either of the clients is configured with canreinvite=no
- If the clients cannot agree on a common set of codecs and Asterisk needs to perform codec conversion
- If either of the clients is configured with nat=yes
- If Asterisk needs to listen to Dual Tone Multi-Frequency (DTMF) tones during the call (for transfers or any other features)

Lastly, context=internal specifies the location of the instructions used to control what the phone is allowed to do, and what to do with incoming calls for this extension. The context name configured in *sip.conf* matches the name of the context in *extensions.conf*, which contains the instructions. More information about contexts and dialplans will be presented in the following chapter.

If you are configuring a number of clients with similar configurations, you can place like commands under the [general] heading. Asterisk will use the defaults specified in the [general] section unless they are explicitly changed within a client's configuration block.

Client Configuration

While it would be impossible to show all the possible configurations for all the end devices that can communicate with Asterisk, we feel it beneficial to provide the configuration for at least one free soft phone, which you can use in determining if Asterisk is right for your organization. We've chosen to use X-ten's X-Lite client, which you can download from their web site (*http://www.xten.com*).

The configuration of the client is generally straightforward. The most important parts are the username and password for registration, plus the address of the Asterisk server with which you wish to register. Figure 4-3 shows a sample configuration for the X-Lite client. Be sure to modify the values of the fields to reflect your configuration.

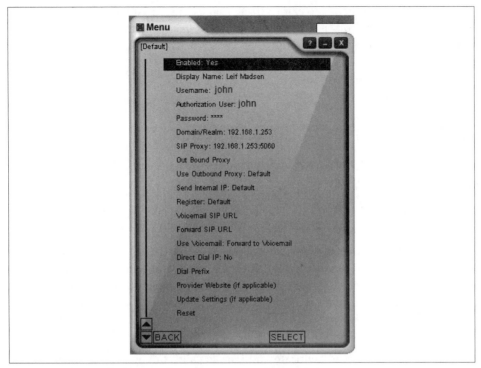

Figure 4-3. X-Lite soft phone configuration screen

The display name is the string that will be used for Caller ID. The username and authorization user are used for authentication, along with the password. The domain/realm should be the IP address or FQDN of your Asterisk server. The SIP proxy is the same as the one entered for the domain/realm, but with :5060 appended (this specifies the port number to use for SIP signaling—be sure it matches the port you have configured in *sip.conf*).

After entering all this information, verify that Enabled is set to Yes, and then close the configuration menu. X-Lite will then register to Asterisk. If X-Lite doesn't appear to register, simply restart the client. Because X-Lite is minimized to the task tray when you close the application with the X button, you will need to exit the program by right-clicking on the icon in the tray and then clicking "Exit" in the pop-up menu before restarting.

Dialplan Configuration

Many SIP phones, both soft and hard, are multi-line phones. This means they can accept multiple incoming calls at the same time. Thus, to test your X-Lite soft phone you can simply call yourself, and the call will loop back from the Asterisk server and onto line two of the client. To call yourself, dial extension 100. If your preferred client doesn't support multi-line functionality, you can use extension 611 to enter the Echo( ) test application.

```
[internal]
exten => 100,1,Dial(SIP/john)
exten => 611,1,Echo( )
```

Configuring Inbound IAX Connections

The Inter-Asterisk eXchange (IAX) protocol is usually used for server-to-server communication; more hard phones are available that talk SIP. However, there are several soft phones that support the IAX protocol, and work is progressing on several fronts for hard phone support in firmware. The primary difference between the IAX and SIP protocols is the way media (your voice) is passed between endpoints.

With SIP, the RTP (media) traffic is passed using different ports than those used by the signaling methods. For example, Asterisk receives the signaling of SIP on port 5060 and the RTP (media) traffic on ports 10,000 through 20,000, by default. The IAX protocol differs in that both the signaling and media traffic are passed via a single port: 4569. An advantage to this approach is that the IAX protocol tends to be better suited to topologies involving NAT.

An IAX *user* is used to authenticate and handle calls coming into the PBX system. For calls going out from the PBX, Asterisk uses an IAX peer entry in the *iax.conf* file to authenticate with the remote end. (IAX peers will be explored in the section "Configuring Outbound IAX Connections.")

This section explores the configuration of your system for a Free World Dialup (FWD) account via IAX. Free World Dialup is a free VoIP service provider that allows you to connect to any other member of the network, regardless of physical location, for free. FWD is also connected to over 100 other networks to which you can connect for free.

 Be sure to enable IAX2 support for your FWD account before you get started by visiting *http://www.fwdnet.net/index.php?section_id=112*.

This section sets up *iax.conf* and *extensions.conf* to allow you to accept calls from another FWD user. The section on outgoing IAX connections deals with placing calls.

iax.conf Configuration

In *iax.conf*, sections are defined with a name enclosed in square brackets ([ ]). Every *iax.conf* file needs at least one main section: [general]. Within the [general] section, you define the settings related to the use of the IAX protocol, such as default codecs and jitter buffering. You can override the default codecs you specify in the [general] section by specifying them within the user or peer definitions.

The following [general] section is the default from the *iax.conf.sample* configuration file (the same file that's installed when you perform a make samples). For more information about the options, see Appendix A.

```
[general]
bandwidth=low
disallow=lpc10
jitterbuffer=no
forcejitterbuffer=no
tos=lowdelay
autokill=yes

register => fwd_number:password@iax2.fwdnet.net

[iaxfwd]
type=user
context=incoming
auth=rsa
inkeys=freeworlddialup
```

Within the [general] section, you'll need to add a register statement. The purpose of the register statement is to tell the FWD IAX server where you are on the Internet (your IP address). When a call is placed to your FWD number, the FWD servers do a lookup in their database and forward the call to the IP address associated with the FWD number.

In the [iaxfwd] section, define the user for incoming calls with type=user. Then define where the incoming call will be handled within the dialplan, with context=incoming. To specify that the authentication for the incoming call will be done with an RSA public/private key pair, use auth=rsa. The public key is defined with inkeys=freeworlddialup. The freeworlddialup public key comes standard with Asterisk.

Dialplan Configuration

Handling an incoming call in the *extensions.conf* file is simple. First, create a context called incoming (the same context name configured for the iaxfwd user in *iax.conf*). The context is followed by a Dial( ) statement that will dial the SIP extension created earlier in this chapter. Replace the number 10001 with that of your FWD account:

```
[incoming]
exten => 10001,1,Dial(SIP/john)
```

Configuring Outbound IAX Connections

While an IAX *user* receives inbound calls; an IAX *peer* is used to place outbound calls. This section will set up *iax.conf* and *extensions.conf* so that you can place calls.

iax.conf Configuration

The following entry in *iax.conf* can be used to place a call on the FWD network:

```
[iaxfwd]
type=peer
host=iax2.fwdnet.net
username=<fwd-account-number>
secret=<fwd-account-password>
qualify=yes
disallow=all
allow=ulaw
allow=gsm
allow=ilbc
allow=g726
```

A peer is defined with type=peer. Use host to configure the server through which you will place calls (iax2.fwdnet.net). Your FWD account number and password will be used for authentication to the FWD network and are defined respectively with username and secret.

You can use the qualify=yes statement to occasionally check that the remote server is responding. The response time (latency) can be viewed from the Asterisk console with iax2 show peers. By default, a peer is considered unreachable after 2000 ms (2 seconds). You can customize the time period by replacing yes with the number of milliseconds.

The available codecs and the order of preference can be defined on a per-peer basis. disallow=all is used to reset any codec settings set previously. You can then allow the codecs you support and set their preference (from top to bottom), using the syntax allow=*codec*.

Use the iax2 show registry command from the Asterisk CLI to verify that you've registered successfully.

Dialplan Configuration

Let's define a section in *extensions.conf* so that we can place a call to the FWD echo test application. As in previous configurations, we will create a context, followed by the instructions to connect to the FWD echo test. Use either your telephone attached to the FXS port or your SIP phone to place the call by dialing 613.

```
[internal]
exten => 613,1,Dial(IAX2/iaxfwd/613)
```

Debugging

Several methods of debugging are available in Asterisk. Once you've connected to the console, you can enable different levels of verbosity and debugging output, as well as protocol packet tracing. We'll take a look at the various options in this section. (The Asterisk console is discussed in more detail in Appendix E.)

Connecting to the Console

To connect to the Asterisk console, you can either start the server in the console directly (in which case you will not be able to exit out of the console without killing the Asterisk process), or start Asterisk as a daemon and then connect to a remote console.

To start the Asterisk process directly in the console, use the console flag:

```
# /usr/sbin/asterisk -c
```

To connect to a remote Asterisk console, start the daemon first, then connect with the –r flag:

```
# /usr/sbin/asterisk
# /usr/sbin/asterisk -r
```

If you are having a problem with a specific module not loading, or a module causing Asterisk to not load, start the Asterisk process with the –c flag to monitor the status of modules loading. For example, if you attempt to load the OSS channel driver (which allows the use of the CONSOLE channel), and Asterisk is unable to open */dev/ dsp*, you will receive the following error on startup:

```
WARNING[32174]: chan_oss.c:470 soundcard_init: Unable to open /dev/dsp: No such file
or directory
  == No sound card detected -- console channel will be unavailable
  == Turn off OSS support by adding 'noload=chan_oss.so' in /etc/asterisk/modules.
conf
```

Enabling Verbosity and Debugging

Asterisk can output debugging information in the form of WARNING, NOTICE, and ERROR messages. These messages will give you information about your system, such as registrations, status and progression of calls, and various other useful bits of information. Note that WARNING and NOTICE messages are not errors; however, ERROR messages should be investigated. To enable various levels of verbosity, use set verbose followed by a numerical value. Useful values range from 3 to 10. For example, to set the highest level of verbosity, use:

```
# set verbose 10
```

You can also enable core debugging messages with set debug followed by a numerical value. To enable DEBUG output on the console, you may need to enable it in the *logger.conf* file by adding debug to the console => statement, as follows:

```
console => warning,notice,error,event,debug
```

Useful values for set debug range from 3 to 10. For example:

```
# set debug 10
```

Conclusion

If you've worked through all of the sections in this chapter, you will have configured a pair of analog interfaces, a local SIP channel connected to a soft phone, and a connection to Free World Dialup via IAX2. These configurations are quite basic, but they give us functional channels to work with. We will make use of them in the following chapters, while we learn to build more useful dialplans.

## CHAPTER 5

Dialplan Basics

Everything should be made as simple as possible,
but not simpler.
—Albert Einstein (1879–1955)

The dialplan is truly the heart of any Asterisk system, as it defines how Asterisk handles inbound and outbound calls. In a nutshell, it consists of a list of instructions or steps that Asterisk will follow. Unlike traditional phone systems, Asterisk's dialplan is fully customizable. To successfully set up your own Asterisk system, you will need to understand the dialplan.

If writing a dialplan sounds overwhelming, don't worry. This chapter explains how dialplans work in a step-by-step manner and teaches the skills necessary to create your own. The examples have been designed to build upon one another, so feel free to go back and re-read a section if something doesn't quite make sense. Please also note that this chapter is by no means an exhaustive survey of all the possible things dialplans can do; our aim is to cover just the fundamentals. We'll cover more advanced dialplan topics in later chapters.

Dialplan Syntax

The Asterisk dialplan is specified in the configuration file named *extensions.conf*.

> The *extensions.conf* file usually resides in the */etc/asterisk/* directory, but its location may vary depending on how you installed Asterisk. Other common locations for this file include */usr/local/asterisk/etc/* and */opt/asterisk/etc/*.

The dialplan is made up of four main parts: contexts, extensions, priorities, and applications. In the next few sections, we'll cover each of these parts and explain how they work together to create a dialplan. After explaining the role each of these elements plays in the dialplan, we will step you though the process of creating a basic, functioning dialplan.

Sample Configuration Files

If you installed the sample configuration files when you installed Asterisk, you will most likely have an existing *extensions.conf* file. Instead of starting with the sample file, we suggest that you build your *extensions.conf* file from scratch. This will be very beneficial, as it will give you a better understanding of dialplan concepts and fundamentals.

That being said, the sample *extensions.conf* file remains a fantastic resource, full of examples and ideas that you can use after you've learned the basic concepts. We suggest you rename the sample file to something like *extensions.conf.sample*. That way, you can refer to it in the future. You can also find the sample configuration files in the */configs/* directory of the Asterisk source.

Contexts

Dialplans are broken into sections called *contexts*. Contexts are named groups of extensions. Simply put, they keep different parts of the dialplan from interacting with one another. An extension that is defined in one context is completely isolated from extensions in any another context, unless interaction is specifically allowed. (We'll cover how to allow interaction between contexts near the end of the chapter.)

As a simple example, let's imagine we have two companies sharing an Asterisk server. If we place each company's voice menu in its own context, they are effectively separated from each other. This allows us to independently define what happens when, say, extension 0 is dialed: people pressing 0 at Company A's voice menu will get Company A's receptionist, and callers pressing 0 at Company B's voice menu will get Company B's receptionist. (This example assumes, of course, that we've told Asterisk to transfer the calls to the receptionists when callers press 0.)

Contexts are denoted by placing the name of the context inside square brackets ([ ]). The name can be made up of the letters A through Z (upper- and lowercase), the numbers 0 through 9, and the hyphen and underscore.* For example, a context for incoming calls looks like this:

    [incoming]

All of the instructions placed after a context definition are part of that context, until the next context is defined. At the beginning of the dialplan, there are two special

---

* Please note that the space is conspicuously absent from the list of allowed characters. Don't use spaces in your context names—you won't like the result!

contexts named [general] and [globals]. We will discuss the [globals] context later in this chapter; for now it's just important to know that these two contexts are special.

One of the most important uses of contexts is to enforce security. By using contexts correctly, you can give certain callers access to features (such as long-distance calling) that aren't made available to others. If you don't design your dialplan carefully, you may inadvertently allow others to fraudulently use your system. Please keep this in mind as you build your Asterisk system.

> The Asterisk source contains a very important file named *SECURITY*, which outlines several steps you should take to keep your Asterisk system secure. It is vitally important that you read and understand this file. If you ignore the security precautions outlined there, you may end up allowing anyone and everyone to make long-distance or toll calls at your expense!
>
> If you don't take the security of your Asterisk system seriously, you may end up paying—literally! *Please* take the time and effort to secure your system from toll fraud.

Extensions

Within each context, we define one or more extensions. An *extension* is an instruction that Asterisk will follow, triggered by an incoming call or by digits being dialed on a channel. Extensions specify what happens to calls as they make their way through the dialplan. Although extensions can be used to specify phone extensions in the traditional sense (i.e., please call John at extension 153), they can be used for much more in Asterisk.

The syntax for an extension is the word exten, followed by an arrow formed by the equals sign and the greater-than sign, like this:

```
exten =>
```

This is followed by the name of the extension. When dealing with telephone systems, we tend to think of extensions as the numbers you would dial to make another phone ring. In Asterisk, you get a whole lot more—for example, extension names can be any combination of numbers and letters. Over the course of this chapter and the next, we'll use both numeric and alphanumeric extensions.

> Assigning names to extensions may seem like a revolutionary concept, but when you realize that many Voice-over-IP transports support (or even actively encourage) dialing by name or email address instead of by number, it makes perfect sense. This is one of the features that make Asterisk so flexible and powerful.

A complete extension is composed of three components:

- The name (or number) of the extension
- The priority (each extension can include multiple steps; the step number is called the "priority")
- The application (or command) that performs some action on the call

These three components are separated by commas, like this:

```
exten => name,priority,application()
```

Here's a simple example of what a real extension might look like:

```
exten => 123,1,Answer()
```

In this example, the extension name is 123, the priority is 1, and the application is Answer(). Now, let's move ahead and explain priorities and applications.

Priorities

Each extension can have multiple steps, called *priorities*. Each priority is numbered sequentially, starting with 1. (Actually, there is one exception to this rule, as discussed in the sidebar "Unnumbered Priorities.") Each priority executes one specific application. As an example, the following extension would answer the phone (in priority number 1), and then hang it up (in priority number 2):

```
exten => 123,1,Answer()
exten => 123,2,Hangup()
```

 You must make sure that your priorities start at 1 and are numbered consecutively. If you skip a priority, Asterisk will not continue past it. If you find that Asterisk is not following all the priorities in a given extension, you may want to make sure you haven't skipped or misnumbered a priority.

Don't worry if you don't understand what Answer() and Hangup() are—we'll cover them shortly. The key point to remember here is that for a particular extension, Asterisk follows the priorities in numerical order.

Applications

Applications are the workhorses of the dialplan. Each application performs a specific action on the current channel, such as playing a sound, accepting touch-tone input, or hanging up the call. In the previous example, you were introduced to two simple applications: Answer() and Hangup(). You'll learn more about how these work momentarily.

Unnumbered Priorities

There's nothing like telling you that priorities have to be numbered sequentially, and then contradicting ourselves. Oh well, it needs to be done.

Version 1.2 of Asterisk adds a new twist to priority numbering. It introduces the use of the n priority, which stands for "next." Each time Asterisk encounters a priority named n, it takes the number of the previous priority and adds 1. This makes it easier to make changes to your dialplan, as you don't have to keep renumbering all your steps. For example, your dialplan might look something like this:

```
exten => 123,1,Answer( )
exten => 123,n,do something
exten => 123,n,do something else
exten => 123,n,do one last thing
exten => 123,n,Hangup( )
```

Version 1.2 also allows you to assign text labels to priorities. To assign a text label to a priority, simply add the label inside parentheses after the priority, like this:

```
exten => 123,n(label),do something
```

In the next chapter, we'll cover how to jump between different priorities based on dialplan logic.

Some applications, such as Answer( ) and Hangup( ), need no other instructions to do their jobs. Other applications require additional information. These pieces of information, called *arguments*, can be passed on to the applications to affect how they perform their actions. To pass arguments to an application, place them between the parentheses that follow the application name, separated by commas.

> Occasionally, you may also see the pipe character (|) being used as a separator between arguments, instead of a comma. Feel free to use whichever you prefer. For the examples in this book, however, we'll be using the comma to separate arguments to an application.

As we build our first dialplan in the next section, you'll learn to use applications (and their associated arguments) to your advantage.

A Simple Dialplan

Now we're ready to create our first dialplan. We'll start with a very simple example. We will design this dialplan so that as a call comes in, Asterisk will answer the call, play a sound file, and then hang up the call. We'll use this simple example to point out the most important dialplan fundamentals.

For the examples in this chapter to work correctly, we're assuming that at least one Zap channel has been created and configured (as described in the previous chapter), and that all incoming calls are sent to the [incoming] context. If you're using other types of channels, you may need to adjust these examples to fit your particular circumstances.

The s Extension

Before we get started with our dialplan, we ought to explain a special extension called s. When calls enter a context without a specific destination extension (for example, a ringing FXO line), they are handled automatically by the s extension. (The s stands for "start," as most calls start in the s extension.) Since this is exactly what we need for our dialplan, let's begin to fill in the pieces. We will be performing three actions on the call (answer it, play a sound file, and hang it up), so we need to create an extension called s with three priorities. We'll place the three priorities inside [incoming], as all incoming calls should start in this context:

```
[incoming]
exten => s,1,application( )
exten => s,2,application( )
exten => s,3,application( )
```

Now all we need to do is fill in the applications, and we've created our first dialplan.

The Answer( ), Playback( ), and Hangup( ) Applications

If we're going to answer the call, play a sound file, and then hang up, we'd better learn how to do just that. The Answer( ) application is used to answer a channel that is ringing. This does the initial setup for the channel that receives the incoming call. (A few applications don't require that you answer the channel first, but properly answering the channel before performing any other actions is a very good habit.) As we mentioned earlier, Answer( ) takes no arguments.

The Playback( ) application is used for playing a previously recorded sound file over a channel. When using the Playback( ) application, input from the user is simply ignored.

> Asterisk comes with many professionally recorded sound files, which should be found in the default sounds directory (usually /var/lib/asterisk/sounds/). They have been recorded in the GSM format, so they have a .gsm file extension. We'll be using these files in many of our examples. Several of the files in our examples come from the asterisk-sounds module, so please take the time to install it (see Chapter 3).

To use Playback( ), specify a filename (without a file extension) as the argument. For example, Playback(filename) would play the sound file called *filename.gsm*, assuming it was located in the default sounds directory. Note that you can include the full path to the file if you want, like this:

```
Playback(/home/john/sounds/filename)
```

This example would play *filename.gsm* from the */home/john/sounds/* directory. You can also use relative paths from the Asterisk sounds directory:

```
Playback(custom/filename)
```

This example would play *filename.gsm* from the *custom/* subdirectory of the default sounds directory. Note that if the specified directory contains more than one file with that filename but with different file extensions, Asterisk automatically plays the best file.*

The Hangup( ) application does exactly as its name implies: it hangs up the active channel. The caller will receive an indication that the call has been hung up. You will use this application at the end of a context when you want to end the current call, to ensure that callers don't continue on in the dialplan. This application takes no arguments.

Our First Dialplan

Now that we have created our extension, given it three different priorities, and learned about the applications we are going to use, let's put together all the pieces to create our first dialplan. As is typical in many technology books (especially computer programming books), our first example will be called "Hello World!"

In the first priority of our extension, we'll answer the call. In the second, we'll play a sound file named *hello-world.gsm*, and in the third we'll hang up the call. Here's what the dialplan looks like:

```
[incoming]
exten => s,1,Answer( )
exten => s,2,Playback(hello-world)
exten => s,3,Hangup( )
```

If you have a channel or two configured, go ahead and try it out! Simply make a new *extensions.conf* file with this short dialplan. (If it doesn't work, check the Asterisk console for error messages, and make sure your channels are configured to send inbound calls to the [incoming] context.)

Even though this example is very short and simple, it emphasizes the core concepts of contexts, extensions, priorities, and applications. Now that we've covered these

---

* Asterisk selects the best file based on translation cost; that is, it selects the file that is the least CPU-intensive to convert to its native audio format. When you start Asterisk, it calculates the translation costs between the different audio formats (they often vary from system to system). You can see these translation costs by typing **show translation** at the Asterisk command-line interface. We'll cover more about the different audio formats (known as *codecs*) in Chapter 8.

basic concepts, let's build upon our example. After all, a phone system that simply plays a sound file and then hangs up the channel isn't that useful!

Adding Logic to the Dialplan

The dialplan we just built was static—it always performs the same actions on every call. Now we'll start adding some logic to our dialplan so that it will perform different actions based on input from the user. We'll start by introducing a few more applications.

The Background( ) and Goto( ) Applications

One important key to building interactive Asterisk systems is the Background( ) application. Like Playback( ), it plays a recorded sound file. Unlike Playback( ), however, when the caller presses a key (or series of keys) on her telephone keypad, it interrupts the playback and goes to the extension that corresponds with the pressed digit(s). If a caller presses 5, for example, Asterisk will stop playing the sound file and send control of the call to the first priority of extension 5.

The most common use of the Background( ) application is to create voice menus (often called *auto-attendants* or *phone trees*). Many companies use voice menus to direct callers to the proper extensions, thus relieving their receptionists from having to answer every single call.

Background( ) has the same syntax as Playback( ):

```
exten => 123,1,Background(hello-world)
```

Another useful application is Goto( ). As its name implies, it is used to send the call to another context, extension, and priority. The Goto( ) application makes it easy to programmatically move a call between different parts of the dialplan. The syntax for the Goto( ) application calls for us to pass the destination context, extension, and priority as arguments to the application, like this:

```
exten => 123,1,Goto(context,extension,priority)
```

In our next example, we'll use the Background( ) and Goto( ) applications to create a slightly more complex dialplan, allowing the caller to interact with the system by pressing digits on the keypad. Let's begin by using Background( ) to accept input from the caller:

```
[incoming]
exten => s,1,Answer( )
exten => s,2,Background(enter-ext-of-person)
```

In this example, we'll play the sample sound file named *enter-ext-of-person.gsm*. While it's not the perfect fit for an auto-attendant greeting, it will certainly work for this example. Now let's add two extensions that will be triggered by the caller entering either 1 or 2 at the prompt:

```
[incoming]
exten => s,1,Answer( )
exten => s,2,Background(enter-ext-of-person)
exten => 1,1,Playback(digits/1)
exten => 2,1,Playback(digits/2)
```

Before going on, let's review what we've done so far. When users call into our dialplan, they will hear a greeting saying, "Please enter the number you wish to call." If they press 1, they will hear the number one, and if they press 2, they will hear the number two. While that's a good start, let's embellish it a little. We'll use the Goto( ) application to make the dialplan repeat the greeting after playing back the number:

```
[incoming]
exten => s,1,Answer( )
exten => s,2,Background(enter-ext-of-person)
exten => 1,1,Playback(digits/1)
exten => 1,2,Goto(incoming,s,1)
exten => 2,1,Playback(digits/2)
exten => 2,2,Goto(incoming,s,1)
```

These two new lines (highlighted in bold) will send the call control back to the s extension after playing back the selected number.

> If you look up the details of the Goto( ) application, you'll find that you can actually pass either one, two, or three arguments to the application. If you pass a single argument, it'll assume it's the destination priority in the current extension. If you pass two, it'll treat them as the extension and priority to go to in the current context.
>
> In this example, we've passed all three arguments for the sake of clarity, but passing just the extension and priority would have had the same effect.

Handling Invalid Entries and Timeouts

Now that our first voice menu is fairly complete, let's add some additional special extensions. First, we need an extension for invalid entries, so that when a caller presses an invalid entry (e.g., pressing 3 in the above example), the call is sent to the i extension. Second, we need an extension to handle situations when the caller doesn't give input in time (the default timeout is 10 seconds). Calls will be sent to the t extension if the caller takes too long to press a digit after Background( ) has finished playing the sound file. Here is what our dialplan will look like after we've added these two extensions:

```
[incoming]
exten => s,1,Answer( )
exten => s,2,Background(enter-ext-of-person)
exten => 1,1,Playback(digits/1)
```

```
exten => 1,2,Goto(incoming,s,1)
exten => 2,1,Playback(digits/2)
exten => 2,2,Goto(incoming,s,1)
exten => i,1,Playback(pbx-invalid)
exten => i,2,Goto(incoming,s,1)
exten => t,1,Playback(vm-goodbye)
exten => t,2,Hangup( )
```

Using the i and t extensions makes our dialplan a little more robust and user-friendly. That being said, it is still quite limited, because outside callers have no way of connecting to a live person. To do that, we'll need to learn about another application, called Dial( ).

Using the Dial( ) Application

One of Asterisk's most valuable features is its ability to connect different callers to each other. This is especially useful when callers are using different methods of communication. For example, caller A might be communicating over the standard analog telephone network, while user B might be sitting in a café halfway around the world and speaking on an IP telephone. Luckily, Asterisk takes most of the hard work out of connecting and translating between disparate networks. All you have to do is learn how to use the Dial( ) application.

The syntax of the Dial( ) application is a little more complex than that of the other applications we've used so far, but don't let that scare you off. Dial( ) takes up to four arguments. The first is the destination you're attempting to call, which is made up of a technology (or transport) across which to make the call, a forward slash, and the remote resource (usually a channel name or number). For example, let's assume that we want to call a Zap channel named Zap/1, which is an FXS channel with an analog phone plugged into it. The technology is "Zap," and the resource is "1." Similarly, a call to a SIP device might have a destination of SIP/1234, and a call to an IAX device might have a destination of IAX/fred. If we want Asterisk to ring the Zap/1 channel when extension 123 is reached in the dialplan, we'd add the following extension:

```
exten => 123,1,Dial(Zap/1)
```

When this extension is executed, Asterisk will ring the phone connected to channel Zap/1. If that phone is answered, Asterisk will bridge the inbound call with the Zap/1 channel. We can also dial multiple channels at the same time, by concatenating the destinations together with an ampersand (&), like this:

```
exten => 123,1,Dial(Zap/1&Zap/2&Zap/3)
```

The Dial( ) application will bridge the inbound call with whichever destination channel is answered first.

The second argument to the Dial( ) application is a timeout, specified in seconds. If a timeout is given, Dial( ) will attempt to call the destination(s) for that number of seconds before giving up and moving on to the next priority in the extension. If no timeout is specified, Dial( ) will continue to dial the called channel(s) until someone answers or the caller hangs up. Let's add a timeout of 10 seconds to our extension:

```
exten => 123,1,Dial(Zap/1,10)
```

If the call is answered before the timeout, the channels are bridged and the dialplan is done. If the destination simply does not answer, Dial( ) goes on to the next priority in the extension. If, however, the destination channel is busy, Dial( ) will go to priority n+101, if it exists (where n is the priority where the Dial( ) application was called). This allows us to handle unanswered calls differently from calls whose destinations were busy.

Let's put what we've learned so far into another example:

```
exten => 123,1,Dial(Zap/1,10)
exten => 123,2,Playback(vm-nobodyavail)
exten => 123,3,Hangup( )
exten => 123,102,Playback(tt-allbusy)
exten => 123,103,Hangup( )
```

As you can see, this example will play the *vm-nobodyavail.gsm* sound file if the call goes unanswered, or the *tt-allbusy.gsm* sound file if the Zap/1 channel is currently busy.

The third argument to Dial( ) is an option string. It may contain one or more characters that modify the behavior of the Dial( ) application. While the list of possible options is too long to cover here, the most popular option is the letter r. If you place the letter r as the third argument, the calling party will hear a ringing tone while the destination channel is being notified of an incoming call.

It should be noted that the r option isn't always required to indicate ringing, as Asterisk will automatically generate a ringing tone when it is attempting to establish a channel. However, you can use the r option to force Asterisk to indicate ringing even when no connection is being attempted. To add the r option to our last example, we simply change the first line:

```
exten => 123,1,Dial(Zap/1,10,r)
exten => 123,2,Playback(vm-nobodyavail)
exten => 123,3,Hangup( )
exten => 123,102,Playback(tt-allbusy)
exten => 123,103,Hangup( )
```

Since the extensions numbered 1 and 2 in our dialplan are somewhat useless now that we know how to use the Dial( ) application, let's replace them with extensions 101 and 102, which will allow outside callers to reach John and Jane:

```
[incoming]
exten => s,1,Answer( )
exten => s,2,Background(enter-ext-of-person)
exten => 101,1,Dial(Zap/1,10)
exten => 101,2,Playback(vm-nobodyavail)
```

```
exten => 101,3,Hangup( )
exten => 101,102,Playback(tt-allbusy)
exten => 101,103,Hangup( )
exten => 102,1,Dial(SIP/Jane,10)
exten => 102,2,Playback(vm-nobodyavail)
exten => 102,3,Hangup( )
exten => 102,102,Playback(tt-allbusy)
exten => 102,103,Hangup( )
exten => i,1,Playback(pbx-invalid)
exten => i,2,Goto(incoming,s,1)
exten => t,1,Playback(vm-goodbye)
exten => t,2,Hangup( )
```

The fourth and final argument to the Dial( ) application is a URL. If the destination channel supports receiving a URL at the time of the call, the specified URL will be sent (for example, if you have an IP telephone that supports receiving a URL, it will appear on the phone's display; likewise, if you're using a soft phone, the URL might pop up on your computer screen). This argument is very rarely used.

If you are making outbound calls on an FXO Zap channel, you can use the following syntax to dial a number on that channel:

```
exten => 123,1,Dial(Zap/4/5551212)
```

This example would dial the number 555-1212 on the Zap/4 channel. For other channel types, such as SIP and IAX, simply put the destination as the resource, as shown in these two lines:

```
exten => 123,1,Dial(SIP/1234)
exten => 124,1,Dial(IAX2/john@asteriskdocs.org)
```

Note that any of these arguments may be left blank. For example, if you want to specify an option but not a timeout, simply leave the timeout argument blank, like this:

```
exten => 123,1,Dial(Zap/1,,r)
```

Adding a Context for Internal Calls

In our examples thus far we have limited ourselves to a single context, but it is probably fair to assume that almost all Asterisk installations will have more than one context in their dialplans. As we mentioned at the beginning of this chapter, one important function of contexts is to separate privileges (such as making long-distance calls or calling certain extensions) for different classes of callers. In our next example, we'll add to our dialplan by creating two internal phone extensions, and we'll set up the ability for these two extensions to call each other. To accomplish this, we'll create a new context called [internal].

As in previous examples, we've assumed that an FXS Zap channel (Zap/1, in this case) has already been configured, and that your *zapata. conf* file is configured so that any calls originated by Zap/1 begin in the [internal] context. For a few examples at the end of the chapter, we'll also assume that an FXO Zap channel has been configured as Zap/4, with calls coming in on this channel being sent to the [incoming] context. This channel will be used for outbound calling.

We've also assumed you have at least one SIP channel (named SIP/ jane) that is configured to originate in the [internal] context. We've done this to introduce you to using other types of channels.

If you don't have hardware for the channels listed above (such as Zap/ 4), or if you're using hardware with different channel names (e.g., not SIP/jane), don't worry—you can change the examples to match your particular system configuration.

Our dialplan now looks like this:

```
[incoming]
exten => s,1,Answer( )
exten => s,2,Background(enter-ext-of-person)
exten => 101,1,Dial(Zap/1,10)
exten => 101,2,Playback(vm-nobodyavail)
exten => 101,3,Hangup( )
exten => 101,102,Playback(tt-allbusy)
exten => 101,103,Hangup( )
exten => 102,1,Dial(SIP/Jane,10)
exten => 102,2,Playback(vm-nobodyavail)
exten => 102,3,Hangup( )
exten => 102,102,Playback(tt-allbusy)
exten => 102,103,Hangup( )
exten => i,1,Playback(pbx-invalid)
exten => i,2,Goto(incoming,s,1)
exten => t,1,Playback(vm-goodbye)
exten => t,2,Hangup( )

[internal]
exten => 101,1,Dial(Zap/1,,r)
exten => 102,1,Dial(SIP/jane,,r)
```

In this example, we have added two new extensions to the [internal] context. This way, the person using channel Zap/1 can pick up the phone and dial the person at channel SIP/jane by dialing 102. By that same token, the phone registered as SIP/ jane can dial Zap/1 by dialing 101.

We've arbitrarily decided to use extensions 101 and 102 for our examples, but feel free to use whatever numbering convention you wish for your extensions. You should also be aware that you're not limited to three-digit extensions—you can use as few or as many digits as you like. (Well, almost. Extensions must be shorter than

80 characters long, and you shouldn't use single-character extensions for your own use, as they're reserved.) Don't forget that you can use names as well, like so:

```
[incoming]
exten => s,1,Answer( )
exten => s,2,Background(enter-ext-of-person)
exten => 101,1,Dial(Zap/1,10)
exten => 101,2,Playback(vm-nobodyavail)
exten => 101,3,Hangup( )
exten => 101,102,Playback(tt-allbusy)
exten => 101,103,Hangup( )
exten => 102,1,Dial(SIP/Jane,10)
exten => 102,2,Playback(vm-nobodyavail)
exten => 102,3,Hangup( )
exten => 102,102,Playback(tt-allbusy)
exten => 102,103,Hangup( )
exten => t,1,Playback(vm-goodbye)
exten => t,2,Hangup( )

[internal]
exten => 101,1,Dial(Zap/1,,r)
exten => john,1,Dial(Zap/1,,r)
exten => 102,1,Dial(SIP/jane,,r)
exten => jane,1,Dial(SIP/jane,,r)
```

It certainly wouldn't hurt to add named extensions if you think your users might be dialed via a VoIP transport that supports names.

Now that our internal callers can call each other, we're well on our way toward having a complete dialplan. Next, we'll see how we can make our dialplan more scalable and easier to modify in the future.

Using Variables

Variables can be used in an Asterisk dialplan to help reduce typing, add clarity, or add additional logic to a dialplan. If you have some computer programming experience, you probably already understand what a variable is. If not, don't worry; we'll explain what variables are and how they are used.

You can think of a variable as a container that can hold one value at a time. So, for example, we might create a variable called JOHN and assign it the value of Zap/1. This way, when we're writing our dialplan, we can refer to John's channel by name, instead of remembering that John is using Zap/1. To assign a value to a variable, simply type the name of the variable, an equals sign, and the value, like this:

```
JOHN=Zap/1
```

There are two ways to reference a variable. To reference the variable's name, simply type the name of the variable, such as JOHN. If, on the other hand, you want to reference its value, you must type a dollar sign, an opening curly brace, the name of the variable, and a closing curly brace. Here's how we'd reference the variable inside the Dial( ) application:

```
exten => 555,1,Dial(${JOHN},,r)
```

In our dialplan, whenever we write ${JOHN}, Asterisk will automatically replace it with whatever value has been assigned to the variable named JOHN.

 Note that variable names don't have to be capitalized, but we're doing so in this book for readability's sake.

There are three types of variables we can use in our dialplan: global variables, channel variables, and environment variables. Let's take a moment to look at each type.

Global variables

As their name implies, *global* variables apply to all extensions in all contexts. Global variables are useful in that they can be used anywhere within a dialplan to increase readability and manageability. Suppose for a moment that you had a large dialplan and several hundred references to the Zap/1 channel. Now imagine you had to go through your dialplan and change all those references to Zap/2. It would be a long and error-prone process, to say the least.

On the other hand, if you had defined a global variable with the value Zap/1 at the beginning of your dialplan and then referenced that instead, you would only have to change one line.

Global variables should be declared in the [globals] context at the beginning of the *extensions.conf* file. They can also be defined programmatically, using the SetGlobalVar( ) application. Here is how both methods look inside of a dialplan:

```
[globals]
JOHN=Zap/1

[internal]
exten => 123,1,SetGlobalVar(JOHN=Zap/1)
```

Channel variables

A *channel* variable is a variable (such as the Caller*ID number) that is associated only with a particular call. Unlike global variables, channel variables are defined only for the duration of the current call and are available only to the channel participating in that call.

There are many predefined channel variables available for use within the dialplan, which are explained in the *README.variables* file in the *doc* subdirectory of the Asterisk source. Channel variables are set via the Set( ) application:

```
exten => 123,1,Set(MAGICNUMBER=42)
```

We'll use several of these channel variables in the next chapter.

Environment variables

Environment variables are a way of accessing Unix environment variables from within Asterisk. These are referenced in the form of ${ENV(*var*)}, where *var* is the Unix environment variable you wish to reference.

Adding variables to our dialplan

Now that we've learned about variables, let's put them to work in our dialplan. We'll add variables for two people, John and Jane:

```
[globals]
JOHN=Zap/1
JANE=SIP/jane

[incoming]
exten => s,1,Answer( )
exten => s,2,Background(enter-ext-of-person)
exten => 101,1,Dial(${JOHN},10)
exten => 101,2,Playback(vm-nobodyavail)
exten => 101,3,Hangup( )
exten => 101,102,Playback(tt-allbusy)
exten => 101,103,Hangup( )
exten => 102,1,Dial(${JANE},10)
exten => 102,2,Playback(vm-nobodyavail)
exten => 102,3,Hangup( )
exten => 102,102,Playback(tt-allbusy)
exten => 102,103,Hangup( )
exten => i,1,Playback(pbx-invalid)
exten => i,2,Goto(incoming,s,1)
exten => t,1,Playback(vm-goodbye)
exten => t,2,Hangup( )

[internal]
exten => 101,1,Dial(${JOHN},,r)
exten => 102,1,Dial(${JANE},,r)
```

Pattern Matching

Often, it would be tedious to add every possible extension to a dialplan. This is especially the case for outbound calls. Can you imagine a dialplan with an extension for every possible number you could dial? Luckily, Asterisk has just the thing for situations like this: *pattern matching* to allow you to use one section of code for many different extensions.

Pattern-matching syntax

When using pattern matching, we use different letters and symbols to represent the possible digits we want to match. Patterns always start with an underscore (_). This tells Asterisk that we're matching on a pattern, and not on an extension name. (This means, of course, that you should never start your extension names with an underscore.)

> If you forget the underscore on the front of your pattern, Asterisk will think it's just a named extension and won't do any pattern matching.

After the underscore, you can use one or more of the following characters:

X

Matches any digit from 0 to 9.

Z

Matches any digit from 1 to 9.

N

Matches any digit from 2 to 9.

[15-7]

Matches any digit or range of digits specified. In this case, matches a 1, 5, 6, or 7.

. *(period)*

Wildcard match; matches one or more characters.

> If you're not careful, wildcard matches can make your dialplans do things you're not expecting. You should only use the wildcard match in a pattern after you've matched as many other digits as possible. For example, the following pattern match should probably never be used:
>
> _.
>
> In fact, Asterisk will warn you if you try to use it. Instead, use this one, if possible:
>
> _X.

To use pattern matching in your dialplan, simply put the pattern in the place of the extension name (or number):

```
exten => _NXX,1,Playback(auth-thankyou)
```

In this example, the pattern would match any 3-digit extension from 200 through 999 (the N matches any digit between 2 and 9, and each X matches a digit between 0 and 9). That is to say, if a caller dialed any 3-digit extension between 200 and 999 in this context, he would hear the sound file *auth-thankyou.gsm*.

One other important thing to know about pattern matching is that if Asterisk finds more than one pattern that matches the dialed extension, it will use the *most specific* one. Say you had defined the following two patterns, and a caller dialed 888-555-1212:

```
exten => _555XXXX,1,Playback(digits/1)
exten => _55512XX,1,Playback(digits/2)
```

In this case the second extension would be selected, because it is more specific.

Pattern-matching examples

Before we go on, let's look at a few more pattern-matching examples. In each one, see if you can tell what the pattern would match before reading the explanation. We'll start with an easy one:

```
_NXXXXXX
```

Got it? This pattern would match any seven-digit number, as long as the first digit was two or higher. According to the North American Numbering Plan, this pattern would match any local number.

The NANP and Toll Fraud

The North American Number Plan (NANP) is a shared telephone numbering scheme used by 19 countries in North America and the Caribbean.

In the United States and Canada, telecom regulations are similar (and sensible) enough that you can place a long-distance call to most numbers in country code 1 and expect to pay a reasonable toll. What many people don't realize, however, is that 19 countries, many of which have very different telecom regulations, share the NANP. (More information can be found at *http://www.nanpa.com*.)

Many toll-fraud schemes trick naive North Americans into calling shockingly expensive per-minute toll numbers in a Caribbean country—the callers believe that since they dialed 1-NPA-NXX-XXXX to reach the number, they'll be paying their standard national long-distance rate for the call. Since the country in question may have regulations that allow for this form of extortion, the caller is ultimately held responsible for the call charges.

The only way to prevent this sort of activity is to block calls to certain area codes (809, for example) and remove the restrictions only on an as-needed basis. Please take extra caution to make sure users can't abuse your phone system!

Let's try another:

```
_1NXXNXXXXXX
```

This one is slightly more difficult. This would match the number 1, followed by an area code between 200 and 999, then any 7-digit number. In the NANP, you would use this pattern to match any long-distance number.

Now for an even trickier example:

```
_011.
```

If that one left you scratching your head, look at it again. Did you notice the period on the end? This pattern matches any number that starts with 011 and has at least one more digit. In the NANP, this indicates an international phone number.[*]

Using the ${EXTEN} channel variable

We know what you're thinking... You're sitting there asking yourself, "So what happens if I want to use pattern matching, but I need to know which digits were actually dialed?" Luckily, Asterisk has just the answer. Whenever you dial an extension, Asterisk sets the ${EXTEN} channel variable to the digits that were dialed. We can use an application called SayDigits( ) to test it out:

```
exten => _XXX,1,SayDigits(${EXTEN})
```

In this example, the SayDigits( ) application will read back to you the three-digit extension you dialed.

Often, it's useful to manipulate the ${EXTEN} by stripping a certain number of digits off the front of the extension. This is accomplished by using the syntax ${EXTEN:*x*}, where *x* is the number of digits you'd like to remove. For example, if the value of EXTEN is 95551212, ${EXTEN:1} equals 5551212. Let's take a look at another example:

```
exten => _XXX,1,SayDigits(${EXTEN:1})
```

In this example, the SayDigits( ) application would read back only the last two digits of the dialed extension.

If *x* is negative, SayDigits( ) gives you the last *x* digits of the dialed extension. In this next example, SayDigits( ) will read back only the last digit of the dialed extension:

```
exten => _XXX,1,SayDigits(${EXTEN:-1})
```

Enabling Outbound Dialing

Now that we've introduced pattern matching, we can go about the process of allowing users to make outbound calls. The first thing we'll do is add a variable to the [globals] context to define which channel will be used for outbound calls:

```
[globals]
JOHN=Zap/1
JANE=SIP/jane
OUTBOUNDTRUNK=Zap/4
```

---

[*] If you find it peculiar that we've chosen patterns that are used to dial outbound numbers in the NANP, you're on to something! We'll be using these patterns in the next section to add outbound dialing capabilities to our dialplan.

Next, we will add contexts to our dialplan for outbound dialing.

You may be asking yourself at this point, "Why do we need separate contexts for outbound calls?" This is so that we can regulate and control who has permission to make outbound calls, and which types of outbound calls they are allowed to make.

First, let's make a context for local calls. To be consistent with most traditional phone switches, we'll put a 9 on the front of our patterns, so that users have to dial 9 before calling an outside number:

```
[outbound-local]
exten => _9NXXXXXX,1,Dial(${OUTBOUNDTRUNK}/${EXTEN:1})
exten => _9NXXXXXX,2,Congestion( )
exten => _9NXXXXXX,102,Congestion( )
```

 Note that dialing 9 doesn't actually give you an outside line, unlike with many traditional PBX systems. Once you dial 9 on an FXS line, the dial tone will stop. If you'd like the dial tone to continue even after dialing 9, add the following line (right after your context definition):

> ```
> ignorepat => 9
> ```
>
> This directive tells Asterisk to continue to provide a dial tone, even after the caller has dialed the indicated pattern.

Let's review what we've just done. We've added a global variable called OUTBOUNDTRUNK, which will control which channel to use for outbound calls. We've also added a context for local outbound calls. In priority 1, we take the dialed extension, strip off the 9 with the ${EXTEN:1} syntax, and then attempt to dial that number on the channel signified by the variable OUTBOUNDTRUNK. If the call is successful, the caller is bridged with the outbound channel. If the call is unsuccessful (because either the channel is busy or the number can't be dialed for some reason), the Congestion( ) application is called, which plays a "fast busy signal" (congestion tone) to let the caller know that the call was unsuccessful.

Before we go any farther, let's make sure our dialplan allows outbound emergency numbers:

```
[outbound-local]
exten => _9NXXXXXX,1,Dial(${OUTBOUNDTRUNK}/${EXTEN:1})
exten => _9NXXXXXX,2,Congestion( )
exten => _9NXXXXXX,102,Congestion( )

exten => 911,1,Dial(${OUTBOUNDTRUNK}/911)
exten => 9911,1,Dial(${OUTBOUNDTRUNK}/911)
```

Again, we're assuming for the sake of these examples that we're inside the United States or Canada. If you're outside of this area, please replace 911 with the emergency services number in your particular location. This is something you never want to forget to put in your dialplan!

Next, let's add a context for long-distance calls:

```
[outbound-long-distance]
exten => _91NXXNXXXXXX,1,Dial(${OUTBOUNDTRUNK}/${EXTEN:1})
exten => _91NXXNXXXXXX,2,Congestion( )
exten => _91NXXNXXXXXX,102,Congestion( )
```

Now that we have these two new contexts, how do we allow internal users to take advantage of them? We need a way for contexts to be able to use other contexts.

Includes

Asterisk enables us to use a context within another context via the `include` directive. This is used to grant access to different sections of the dialplan. We'll use the include functionality to allow users in our [internal] context the ability to make outbound phone calls. But first, let's cover the syntax.

The include statement takes the following form, where *context* is the name of the remote context we want to include in the current context:

```
include => context
```

When we include other contexts within our current context, we have to be mindful of the order in which we including them. Asterisk will first try to match the extension in the current context. If unsuccessful, it will then try the first included context, and then continue to the other included contexts in the order in which they were included.

As it sits, our current dialplan has two contexts for outbound calls, but there's no way for people in the [internal] context to use them. Let's remedy that by including the two outbound contexts in the [internal] context, like this:

```
[globals]
JOHN=Zap/1
JANE=SIP/jane
OUTBOUNDTRUNK=Zap/4

[incoming]
exten => s,1,Answer( )
exten => s,2,Background(enter-ext-of-person)
exten => 101,1,Dial(${JOHN},10)
exten => 101,2,Playback(vm-nobodyavail)
exten => 101,3,Hangup( )
exten => 101,102,Playback(tt-allbusy)
exten => 101,103,Hangup( )
exten => 102,1,Dial(${JANE},10)
exten => 102,2,Playback(vm-nobodyavail)
exten => 102,3,Hangup( )
exten => 102,102,Playback(tt-allbusy)
exten => 102,103,Hangup( )
exten => i,1,Playback(pbx-invalid)
exten => i,2,Goto(incoming,s,1)
```

```
exten => t,1,Playback(vm-goodbye)
exten => t,2,Hangup( )

[internal]
include => outbound-local
include => outbound-long-distance

exten => 101,1,Dial(${JOHN},,r)
exten => 102,1,Dial(${JANE},,r)

[outbound-local]
exten => _9NXXXXXX,1,Dial(${OUTBOUNDTRUNK}/${EXTEN:1})
exten => _9NXXXXXX,2,Congestion( )
exten => _9NXXXXXX,102,Congestion( )

exten => 911,1,Dial(${OUTBOUNDTRUNK}/911)
exten => 9911,1,Dial(${OUTBOUNDTRUNK}/911)

[outbound-long-distance]
exten => _91NXXNXXXXXX,1,Dial(${OUTBOUNDTRUNK}/${EXTEN:1})
exten => _91NXXNXXXXXX,2,Congestion( )
exten => _91NXXNXXXXXX,102,Congestion( )
```

These two include statements make it possible for callers in the [internal] context to make outbound calls. We should also note that for security's sake you should always make sure that your [incoming] context never allows outbound dialing. (If by chance it did, people could dial into your system, and then make outbound toll calls that would be charged to you!)

Conclusion

And there we have it—a basic but functional dialplan. It's not exactly fully featured, but we've covered all of the fundamentals. In the following chapters, we'll continue to add features to this foundation.

If parts of this dialplan don't make sense, you may want to go back and re-read a section or two before continuing on to the next chapter. It's imperative that you understand these principles and how to apply them, or the following chapters will only confuse you more. And we don't want you to be confused!

## CHAPTER 6

More Dialplan Concepts

For a list of all the ways technology has failed to
improve the quality of life, please press three.
—Alice Kahn

Alrighty. You've got the basics of dialplans down, and you're hoping there's more to come. Fear not; there is more—much more. If you don't have the last chapter sorted out yet, please go back and give it another read. We're building on what we've covered so far, and we need you to be comfortable with the material, as we're about to get into more advanced topics.

Expressions and Variable Manipulation

Before we dive further into dialplans, we need to introduce you to a few tricks that will greatly add to the power you can exercise with your dialplan. These constructs add incredible intelligence to your dialplan, by enabling it to make decisions based on all sorts of different criteria. Put on your thinking cap, and let's get started.

Basic Expressions

Expressions are combinations of variables, operators, and values that you put together to get a result. An expression can test values, alter strings, or perform mathematical calculations. Let's say we have a variable called COUNT. In plain English, two expressions using that variable might be "COUNT plus 1" and "COUNT divided by 2." Each of these expressions has a particular result or value, depending on the value of the given variable.

In Asterisk, expressions always begin with a dollar sign and an opening square bracket and end with a closing square bracket, as shown below:

```
$[expression]
```

Thus, we would write the above two examples like this:

```
$[${COUNT} + 1]
$[${COUNT} / 2]
```

When Asterisk encounters an expression in a dialplan, it replaces the entire expression with the resulting value. It is important to note that this takes place *after* variable substitution. To demonstrate, let's look at the following code:[*]

```
exten => 321,1,Set(COUNT=3)
exten => 321,2,Set(NEWCOUNT=$[${COUNT} + 1])
exten => 321,3,SayNumber(${NEWCOUNT})
```

In the first priority, we assign the value of 3 to the variable named COUNT.

In the second priority, only one application—Set( )—is involved, but three things actually happen:

1. Asterisk substitutes ${COUNT} with the number 3 in the expression. The expression effectively becomes this:

   ```
   exten => 321,2,Set(NEWCOUNT=$[3 + 1])
   ```

2. Next, Asterisk evaluates the expression, adding 1 to 3, and replaces it with its computed value of 4:

   ```
   exten => 321,2,Set(NEWCOUNT=4)
   ```

3. Finally, the value 4 is assigned to the NEWCOUNT variable by the Set( ) application.

The third priority simply invokes the SayNumber( ) application, which speaks the current value of the variable ${NEWCOUNT} (set to the value 4 in priority two).

Try it out in your own dialplan.

Operators

When you create an Asterisk dialplan, you're really writing code in a specialized scripting language. This means that the Asterisk dialplan—like any programming language—recognizes symbols called *operators* that allow you to manipulate variables. Let's look at the types of operators that are available in Asterisk:

Boolean operators

These operators evaluate the "truth" of a statement. In computing terms, that essentially refers to whether the statement is something or nothing (non-zero or zero, true or false, on or off, and so on). The Boolean operators are:

---

[*] Remember that when you *reference* a variable, you can call it by its name, but when you refer to a variable's *value*, you have to use the dollar sign and brackets around the variable name.

expr1 | expr2

> This operator (called the "or" operator, or "pipe") returns the evaluation of *expr1* if it is true (neither an empty string nor zero). Otherwise, it returns the evaluation of *expr2*.

expr1 & expr2

> This operator (called "and") returns the evaluation of *expr1* if both expressions are true (i.e., neither expression evaluates to an empty string or zero). Otherwise, it returns zero.

expr1 {=, >, >=, <, <=, !=} expr2

> These operators return the results of an integer comparison if both arguments are integers; otherwise, they return the results of a string comparison. The result of each comparison is 1 if the specified relation is true, or 0 if the relation is false. (If you are doing string comparisons, they will be done in a manner that's consistent with the current locale settings of your operating system.)

Mathematical operators

> Want to perform a calculation? You'll want one of these:

expr1 {+, -} expr2

> These operators return the results of the addition or subtraction of integer-valued arguments.

expr1 {, /, %} expr2*

> These return, respectively, the results of the multiplication, integer division, or remainder of integer-valued arguments.

Regular expression operator

> You can also use the regular expression operator in Asterisk:

expr1 : expr2

> This operator matches *expr1* against *expr2*, where *expr2* must be a regular expression.[*] The regular expression is anchored to the beginning of the string with an implicit ^.[†]

> If the match succeeds and the pattern contains at least one regular expression subexpression—\(...\)—the string corresponding to \1 is returned; otherwise, the matching operator returns the number of characters matched. If the match fails and the pattern contains a regular expression subexpression, the null string is returned; otherwise, 0 is returned.

---

[*] For more on regular expressions, grab a copy of the ultimate reference, Jeffrey Friedl's *Mastering Regular Expressions* (O'Reilly; *http://www.oreilly.com/catalog/regex2/*) or visit *http://www.regular-expressions.info*.

[†] If you don't know what a ^ has to do with regular expressions, you simply must obtain a copy of *Mastering Regular Expressions*. It will change your life!

The Asterisk parser is quite simple, so it requires that you put at least one space between the operator and any other values. Consequently, the following may not work as expected:

```
exten => 123,1,Set(TEST=$[2+1])
```

This would assign the variable TEST the string "2+1", instead of the value 3. Instead, put spaces around the operator, like this:

```
exten => 234,1,Set(TEST=$[2 + 1])
```

To concatenate text onto the beginning or end of a variable, simply place them together in an expression, like this:

```
exten => 234,1,Set(NEWTEST=$[blah${TEST}])
```

Dialplan Functions

Dialplan functions are not a new concept. In Asterisk 1.2, they should be used where possible. Many applications that perform the same operation as a corresponding function will eventually be removed in favor of the function. Functions allow you to add more power to your expressions—you can think of them as being similar to operators, but more advanced. For example, dialplan functions allow you to calculate string lengths, dates and times, MD5 checksums, and so on, all from within a dialplan expression.

Syntax

Dialplan functions have the following basic syntax:

```
FUNCTION_NAME(argument)
```

Much like with variables, you reference a function's *name* as above, but you reference a function's *value* with the addition of a dollar sign, an opening curly brace, and a closing curly brace:

```
${FUNCTION_NAME(argument)}
```

Functions can also encapsulate other functions, like so:

```
${FUNCTION_NAME(${FUNCTION_NAME(argument)})}
 ^              ^ ^              ^      ^^^^
 1              2 3              4      4321
```

As you've probably already figured out, you must be very careful about making sure you have matching parentheses and braces. In the above example, we have labeled the opening parentheses and curly braces with numbers and their corresponding closing counterparts with the same numbers.

Examples of Dialplan Functions

Functions are often used in conjunction with the Set( ) application to either get or set the value of a variable. As a simple example, let's look at the LEN( ) function. This function calculates the string length of its argument. Let's calculate the string length of a variable and read back the length to the caller:

```
exten => 123,1,Set(TEST=example)
exten => 123,2,SayNumber(${LEN(${TEST})})
```

The above example would evaluate the string example as having seven characters, assign the number of characters to the variable length, and then speak the number to the user with the SayNumber( ) application.

Let's look at another simple example. If we wanted to set one of the various channel timeouts, we could use the TIMEOUT( ) function. The TIMEOUT( ) function accepts one of three arguments: absolute, digit, and response. Their corresponding applications are AbsoluteTimeout( ), DigitTimeout( ), and ResponseTimeout( ). To set the digit timeout with the TIMEOUT( ) function, we could use the Set( ) application, like so:

```
exten => s,1,Set(TIMEOUT(digit)=30)
```

Notice the lack of ${ } surrounding the function. Just as if we were assigning a value to a variable, we assign a value to a function without the use of the ${ } encapsulation.

A complete list of available functions can be found by typing **show functions** at the Asterisk command-line interface.

Conditional Branching

Now that you've learned a bit about expressions and functions, it's time to put them to use. By using expressions and functions, you can add even more advanced logic to your dialplan. To allow your dialplan to make decisions, you'll use *conditional branching*. Let's take a closer look.

The GotoIf( ) Application

The key to conditional branching is the GotoIf( ) application. GotoIf( ) evaluates an expression and sends the caller to a specific destination based on whether the expression evaluates to true or false.

GotoIf( ) uses a special syntax, often called the *conditional syntax*:

```
GotoIf(expression?destination1:destination2)
```

If the expression evaluates to true, the caller is sent to the first destination. If the expression evaluates to false, the caller is sent to the second destination. So, what is true and what is false? An empty string and the number 0 evaluate as false. Anything else evaluates as true.

The destinations can each be one of the following:

- A priority within the same extension, such as 10
- An extension and a priority within the same context, such as 123,10
- A context, extension, and priority, such as incoming,123,10
- A named priority within the same extension, such as passed

Either of the destinations may be omitted, but not both. If the omitted destination is to be followed, Asterisk simply goes on to the next priority in the current extension.

Let's use GotoIf( ) in an example:

```
exten => 345,1,Set(TEST=1)
exten => 345,2,GotoIf($[${TEST} = 1]?10:20)
exten => 345,10,Playback(weasels-eaten-phonesys)
exten => 345,20,Playback(office-iguanas)
```

By changing the value assigned to TEST in the first line, you should be able to have your Asterisk server play a different greeting.

Let's look at another example of conditional branching. This time, we'll use both Goto( ) and GotoIf( ) to count down from 10 and then hang up:

```
exten => 123,1,Set(COUNT=10)
exten => 123,2,GotoIf($[${COUNT} > 0]?:10)
exten => 123,3,SayNumber(${COUNT})
exten => 123,4,Set(COUNT=$[${COUNT} - 1])
exten => 123,5,Goto(2)
exten => 123,10,Hangup( )
```

Let's analyze this example. In the first priority, we set the variable COUNT to 10. Next, we check to see if COUNT is greater than 0. If it is, we move on to the next priority. (Don't forget that if we omit a destination in the GotoIf( ) application, control goes to the next priority.) From there we speak the number, subtract 1 from COUNT, and go back to priority two. If COUNT is less than or equal to 0, control goes to priority 10, and the call is hung up.

The classic example of conditional branching is affectionately known as the anti-girl-friend logic. If the Caller ID number of the incoming call matches the phone number of the recipient's ex-girlfriend, Asterisk gives a different message than it ordinarily would to any other caller. While somewhat simple and primitive, it's a good example for learning about conditional branching within the Asterisk dialplan.

This example uses a channel variable called CALLERIDNUM, which is automatically set by Asterisk to the Caller ID number of the inbound call. Let's assume for the sake of this example that the victim's phone number is 885-555-1212:

```
exten => 123,1,GotoIf($[${CALLERIDNUM} = 8885551212]?20:10)
exten => 123,10,Dial(Zap/4)
exten => 123,20,Playback(abandon-all-hope)
exten => 123,21,Hangup( )
```

In priority one, we call the GotoIf( ) application. It tells Asterisk to go to priority 20 if the Caller ID number matches 8885551212, and otherwise to go to priority 10. If the Caller ID number matches, control of the call goes to priority 20, which plays back an uninspiring message to the undesired caller. Otherwise, the call attempts to dial the recipient on channel Zap/4.

Time-Based Conditional Branching with GotoIfTime( )

Another way to use conditional branches in your dialplan is with the GotoIfTime( ) application. Whereas GotoIf( ) evaluates an expression to decide what to do, GotoIfTime( ) looks at the current system time and uses that to decide whether or not to follow a different branch in the dialplan.

The most obvious use of this application is to give your callers a different greeting before and after normal business hours.

The syntax for the GotoIfTime( ) application looks like this:

```
GotoIfTime(times,days_of_week,days_of_month,months?label)
```

In short, GotoIfTime( ) sends the call to the specified *label* if the current date and time match the criteria specified by *times*, *days_of_week*, *days_of_month*, and *months*. Let's look at each argument in more detail:

times
> This is a list of one or more time ranges, in 24-hour format. As an example, 9:00 a.m. through 5:00 p.m. would be specified as 09:00-17:00. The day starts at 0:00 and ends at 23:59.

days_of_week
> This is a list of one or more days of the week. The days should be specified as mon, tue, wed, thu, fri, sat, and/or sun. Monday through Friday would be expressed as mon-fri. Tuesday and Thursday would be expressed as tue,thu.

days_of_month
> This is a list of the numerical days of the month. Days are specified by the numbers 1 through 31. The 7th through the 12th would be expressed as 7-12, and the 15th and 30th of the month would be written as 15,30.

months
> This is a list of one or more months of the year. The months should be written as jan, feb, mar, apr, and so on.

If you wish to match on all possible values for any of these arguments, simply put an * in for that argument.

The *label* argument can be any of the following:

- A priority within the same extension, such as 10
- An extension and a priority within the same context, such as 123,10

- A context, extension, and priority, such as `incoming,123,10`
- A named priority within the same extension, such as `passed`

Now that we've covered the syntax, let's look at a couple of examples. The following example would match from *9:00 a.m. to 5:59 p.m.*, on *Monday through Friday*, on *any day of the month*, in *any month of the year*:

```
exten => s,1,GotoIfTime(09:00-17:59,mon-fri,*,*?open,s,1)
```

If the caller calls during these hours, the call will be sent to the first priority of the s extension in the context named open. If the call is made outside of the specified times, it will be sent to the next priority of the current extension. This allows you to easily branch on multiple times, as shown in the next example (note that you should always put your most specific time matches before the least specific ones):

```
; If it's any hour of the day, on any day of the week,
; during the fourth day of the month, in the month of of July,
; we're closed
exten => s,1,GotoIfTime(*,*,4,jul?closed,s,1)

; During business hours, send calls to the open context
exten => s,2,GotoIfTime(09:00-17:59|mon-fri|*|*?open,s,1)
exten => s,3,GotoIfTime(09:00-11:59|sat|*|*?open,s,1)

; Otherwise, we're closed
exten => s,4,Goto(closed,s,1)
```

Voicemail

One of the most popular (or, actually, unpopular) features of any modern telephone system is voicemail. Naturally, Asterisk has a very flexible voicemail system. Some of the features of Asterisk's voicemail system include:

- Unlimited password-protected voicemail boxes, each containing mailbox folders for organizing voicemail
- Different greetings for busy and unavailable states
- Default and custom greetings
- The ability to associate phones with more than one mailbox and mailboxes with more than one phone
- Email notification of voicemail, with the voicemail optionally attached as a sound file[*]
- Voicemail forwarding and broadcasts

---

[*] No, you really don't have to pay for this; and yes, it really does work.

- Message-waiting indicator (flashing light or stuttered dial tone) on many types of phones
- Company directory of employees, based on voicemail boxes

And that's just the tip of the iceberg! In this section, we'll introduce you to the fundamentals of a typical voicemail setup.

The voicemail configuration is defined in the configuration file called *voicemail.conf*. This file contains an assortment of settings that you can use to customize the voicemail system to your needs. Covering all the available options in *voicemail.conf* would be beyond the scope of this chapter, but the sample configuration file is well documented and quite easy to follow. For now, look near the bottom of the file, where voicemail contexts and voicemail boxes are defined.

Just as dialplan contexts keep different parts of your dialplan separate, voicemail contexts allow you to define different sets of mailboxes that are separate from one another. This allows you to host voicemail for several different companies or offices on the same server. Voicemail contexts are defined in the same way as dialplan contexts, with square brackets surrounding the name of the context. For our examples, we'll be using the [default] voicemail context.

Creating Mailboxes

Inside each voicemail context, we define different mailboxes. The syntax for defining a mailbox is:

```
mailbox => password,name[,email[,pager_email[,options]]]
```

Let's explain what each part of the mailbox definition does:

mailbox
> This is the mailbox number. It usually corresponds with the extension number of the associated set.

password
> This is the numeric password the mailbox owner will use to access her voicemail. If the user changes her password, the system will update this field in the *voicemail.conf* file.

name
> This is the name of the mailbox owner. The company directory uses the text in this field to allow callers to spell usernames.

email
> This is the email address of the mailbox owner. Asterisk can send voicemail notifications (including the voicemail message itself) to the specified email box.

pager_email

> This is the email address of the mailbox owner's pager or cell phone. Asterisk can send a short voicemail notification message to the specified email address.

options

> This field is a list of options that sets the mailbox owner's time zone and overrides the global voicemail settings. There are nine valid options: attach, serveremail, tz, saycid, review, operator, callback, dialout, and exitcontext. These options should be in *option=value* pairs, separated by the pipe character (|). The tz option sets the user's time zone to a time zone previously defined in the [zonemessages] section of *voicemail.conf*, and the other eight options override the global voicemail settings with the same names.

A typical mailbox definition might look something like this:

```
101 => 1234,Joe Public,jpublic@somedomain.com,jpublic@pagergateway.net,tz=central|attach=yes
```

Continuing with our dialplan from the last chapter, let's set up voicemail boxes for John and Jane. We'll give John a password of 1234 and Jane a password of 4444 (remember, these go in *voicemail.conf*, not *extensions.conf*):

```
[default]
101 => 1234,John Doe,john@asteriskdocs.org,jdoe@pagergateway.tld
102 => 4444,Jane Doe,jane@asteriskdocs.org,jane@pagergateway.tld
```

Adding Voicemail to the Dialplan

Now that we've created mailboxes for Jane and John, let's allow callers to leave messages for them if they don't answer the phone. To do this, we'll use the VoiceMail() application.

The VoiceMail() application sends the caller to the specified mailbox, so that he can leave a message. The mailbox should be specified as *mailbox@context*, where *context* is the name of the voicemail context. The mailbox number can also be prefixed by the letter b or the letter u. If the letter b is used, the caller will hear the mailbox owner's *busy* message. If the letter u is used, the caller will hear the mailbox owner's *unavailable* message (if one exists).

Let's use this in our sample dialplan. Previously, we had a line like this in our [internal] context, which allowed us to call John:

```
exten => 101,1,Dial(${JOHN},,r)
```

Now, let's change it so that if John is busy (on another call), it'll send us to his voicemail, where we'll hear his busy message (don't forget that the Dial() application sends the caller to priority n+101 if the dialed line is busy):

```
exten => 101,1,Dial(${JOHN},,r)
exten => 101,102,VoiceMail(b101@default)
```

Next, let's add an unavailable message that the caller will be played if John doesn't answer the phone within 10 seconds. Remember, the second argument to the Dial( ) application is a timeout. If the call is not answered before the timeout expires, the call is sent to the next priority. Let's add a 10-second timeout, and a priority to send the caller to voicemail if John doesn't answer in time:

```
exten => 101,1,Dial(${JOHN},10,r)
exten => 101,2,VoiceMail(u101@default)
exten => 101,102,VoiceMail(b101@default)
```

If we add these two new priorities and a timeout argument to the Dial( ) application, callers will get John's voicemail (with the appropriate greeting) if John is either busy or unavailable. A slight problem remains, however, in that John has no way of retrieving his messages. Let's remedy that.

Accessing Voicemail

Users can retrieve their voicemail messages, change their voicemail options, and record their voicemail greetings by using the VoiceMailMain( ) application. In its typical form, VoiceMailMain( ) is called without any arguments. Let's add extension 500 to the [internal] context of our dialplan so that internal users can dial it to access their voicemail messages:

```
exten => 500,1,VoiceMailMain( )
```

Creating a Dial-by-Name Directory

One last feature of the Asterisk voicemail system we should cover is the dial-by-name directory. This is created with the Directory( ) application. This application uses the names defined in the mailboxes in *voicemail.conf* to present the caller with a dial-by-name directory of the users.

Directory( ) takes up to three arguments: the voicemail context from which to read the names, the optional dialplan context in which to dial the user, and an option string (which is also optional). By default, Directory( ) searches for the user by last name, but passing the f option forces it to search by first name instead. Let's add two dial-by-name directories to the [incoming] context of our sample dialplan, so that callers can search by either first or last name:

```
exten => 8,1,Directory(default,incoming,f)
exten => 9,1,Directory(default,incoming)
```

If callers press 8, they'll get a directory by first name. If they dial 9, they'll get the directory by last name.

Macros

Macros are a very useful construct designed to avoid repetition in the dialplan. They also help in making changes to the dialplan. To illustrate this point, let's look at our sample dialplan again. If you remember the changes we made for voicemail, we ended up with the following for John's extension:

```
exten => 101,1,Dial(${JOHN},10,r)
exten => 101,2,VoiceMail(u101@default)
exten => 101,102,VoiceMail(b101@default)
```

Now imagine you have a hundred users on your Asterisk system—setting up the extensions would involve a lot of copying and pasting. Then imagine that you need to make a change to the way your extensions work. That would involve a lot of editing, and you'd be almost certain to have errors.

Instead, you can define a macro that contains a list of steps to take, and then have all of the phone extensions refer to that macro. All you need to change is the macro, and everything in the dialplan that references that macro will change as well.

 If you're familiar with computer programming, you'll recognize that macros are similar to subroutines in many modern programming languages. If you're not familiar with computer programming, don't worry—we'll walk you through creating a macro.

The best way to appreciate macros is to see one in action, so let's move right along.

Defining Macros

For our first macro, let's take the dialplan logic we used above to set up voicemail for John and turn it into a macro. Then we'll use the macro to give John and Jane (and the rest of their coworkers) the same functionality.

Macro definitions look a lot like contexts. (In fact, you could argue that they really are small, limited contexts.) You define a macro by placing macro- and the name of your macro in square brackets, like this:

```
[macro-voicemail]
```

Macro names must start with macro-. This distinguishes them from regular contexts. The commands within the macro are built pretty nearly identically to anything else in the dialplan—the only limiting factor is that macros use only the s extension. Let's add our voicemail logic to the macro, changing the extension to s as we go:

```
[macro-voicemail]
exten => s,1,Dial(${JOHN},10,r)
exten => s,2,VoiceMail(u101@default)
exten => s,102,VoiceMail(b101@default)
```

That's a start, but it's not perfect, as it's still specific to John and his mailbox number. To make the macro generic so that it will work not only for John but also for all his coworkers, we'll take advantage of another property of macros: arguments. But first, let's see how we call macros in our dialplan.

Calling Macros from the Dialplan

To use a macro in our dialplan, we use the Macro( ) application. This application calls the specified macro and passes it any arguments. For example, to call our voicemail macro from our dialplan, we can do the following:

```
exten => 101,1,Macro(voicemail)
```

The Macro( ) application also defines several special variables for our use. They include:

${MACRO_CONTEXT}

The original context in which the macro was called.

${MACRO_EXTEN}

The original extension in which the macro was called.

${MACRO_PRIORITY}

The original priority in which the macro was called.

${ARG*n*}

The *n*th argument passed to the macro. For example, the first argument would be ${ARG1}, the second ${ARG2}, and so on.

As we explained earlier, the way we initially defined our macro was hard-coded for John, instead of being generic. Let's change our macro to use ${MACRO_EXTEN} instead of 101 for the mailbox number. That way, if we call the macro from extension 101 the voicemail messages will go to mailbox 101, if we call the macro from extension 102 messages will go to mailbox 102, and so on:

```
[macro-voicemail]
exten => s,1,Dial(${JOHN},10,r)
exten => s,2,VoiceMail(u${MACRO_EXTEN}@default)
exten => s,102,VoiceMail(b${MACRO_EXTEN}@default)
```

Using Arguments in Macros

Okay, now we're getting closer to having the macro the way we want it, but we still have one thing left to change—we need to pass in the channel to dial, as it's currently still hard-coded for ${JOHN} (remember that we defined the variable JOHN as the channel to call when we want to reach John). Let's pass in the channel as an argument, and then our first macro will be complete:

```
[macro-voicemail]
exten => s,1,Dial(${ARG1},10,r)
exten => s,2,VoiceMail(u${MACRO_EXTEN}@default)
exten => s,102,VoiceMail(b${MACRO_EXTEN}@default)
```

Now that our macro is done, we can use it in our dialplan. Here's how we can call our macro to provide voicemail to John, Jane, and Jack:

```
exten => 101,1,Macro(voicemail,${JOHN})
exten => 102,1,Macro(voicemail,${JANE})
exten => 103,1,Macro(voicemail,${JACK})
```

With 50 or more users, this dialplan will still look neat and organized—we'll simply have one line per user, referencing a macro that can be as complicated as required. We could even have a few different macros for various user types, such as executives, courtesy_phones, call_center_agents, analog_sets, sales_department, and so on.

A more advanced version of the macro might look something like this:

```
[macro-voicemail]
exten => s,1,Dial(${ARG1},20)
exten => s,2,Goto(s-${DIALSTATUS},1)
exten => s-NOANSWER,1,Voicemail(u${MACRO_EXTEN})
exten => s-NOANSWER,2,Goto(incoming,s,1)
exten => s-BUSY,1,Voicemail(b${MACRO_EXTEN})
exten => s-BUSY,2,Goto(incoming,s,1)
exten => _s-.,1,Goto(s-NOANSWER,1)
```

This macro depends on a nice side effect of the Dial( ) application: when you use the Dial( ) application, it sets the DIALSTATUS variable to indicate whether the call was successful or not. In this case, we're handling the NOANSWER and BUSY cases, and treating all other result codes as a NOANSWER.

Using the Asterisk Database (AstDB)

Having fun yet? It gets even better!

Asterisk provides a powerful mechanism for storing values, called the Asterisk database (AstDB). The AstDB provides a simple way to store data for use within your dialplan.

 For those of you with experience using relational databases such as PostgreSQL or MySQL, the Asterisk database is not a traditional relational database. It is a Berkeley DB Version 1 database. There are several ways to store data from Asterisk in a relational database, but this book will not delve into them.

The Asterisk database stores its data in groupings called *families*, with values identified by *keys*. Within a family, a key may be used only once. For example, if we had a family called test, we could store only one value with a key called count. Each stored value must be associated with a family.

Storing Data in the AstDB

To store a new value in the Asterisk database, we use the Set( ) application,* but instead of using it to set a channel variable, we use it to set an AstDB variable. For example, to assign the count key in the test family the value of 1:

```
exten => 456,1,Set(DB(test/count)=1)
```

If a key named count already exists in the test family, its value will be overwritten with the new value. You can also store values from the Asterisk command line, by running the command **database put *family key value***. For our example, you would type **database put test count 1**.

Retrieving Data from the AstDB

To retrieve a value from the Asterisk database and assign it to a variable, we use the Set( ) application again. Let's retrieve the value of count (again, from the test family), assign it to a variable called COUNT, and then speak the value to the caller:

```
exten => 456,1,Set(DB(test/count)=1)
exten => 456,2,Set(COUNT=${DB(test/count)})
exten => 456,3,SayNumber(${COUNT})
```

You may also check the value of a given key from the Asterisk command line by running the command **database get *family key***. To view the entire contents of the AstDB, use the **database show** command.

Deleting Data from the AstDB

There are two ways to delete data from the Asterisk database. To delete a key, use the DBdel( ) application. It takes the family and key as arguments, like this:

```
exten => 457,1,DBdel(test/count)
```

You can also delete an entire key family by using the DBdeltree( ) application. The DBdeltree( ) application takes a single argument: the name of the key family to delete. To delete the entire test family, do the following:

```
exten => 457,1,DBdeltree(test)
```

---

* Previous versions of Asterisk had applications called DBput( ) and DBget( ) that were used to set values in and retrieve values from the AstDB. If you're using an old version of Asterisk, you'll want to use them instead.

To delete keys and key families from the AstDB via the command-line interface, use the **database del** *key* and **database deltree** *family* commands, respectively.

Using the AstDB in the Dialplan

There are an infinite number of ways to use the Asterisk database in a dialplan. To introduce the AstDB, we'll show two simple examples. The first is a simple counting example to show that the Asterisk database is persistent (meaning that it survives system reboots). In the second example, we'll use the LookupBlacklist( ) application to evaluate whether or not a number is on the blacklist and should be blocked.

To begin the counting example, let's first retrieve a number (the value of the count key) from the database and assign it to a variable named COUNT. If the key doesn't exist, DBget( ) will send us to priority n+101, where we will set the value to 1. The next priority will send us back to priority 1. This will happen the very first time we dial this extension:

```
exten => 678,1,Set(COUNT=${DB(test/count)})
exten => 678,102,Set(DB(test/count)=1)
exten => 678,103,Goto(1)
```

Next, we'll say the current value of COUNT, and then increment COUNT:

```
exten => 678,1,Set(COUNT=${DB(test/count)})
exten => 678,2,SayNumber(${COUNT})
exten => 678,3,Set(COUNT=$[${COUNT} + 1])
exten => 678,102,Set(DB(test/count)=1)
exten => 678,103,Goto(1)
```

Now that we've incremented COUNT, let's put the new value back into the database. Remember that storing a value for an existing key overwrites the previous value:

```
exten => 678,1,Set(COUNT=${DB(test/count)})
exten => 678,2,SayNumber(${COUNT})
exten => 678,3,Set(COUNT=$[${COUNT} + 1])
exten => 678,4,Set(DB(test/count)=${COUNT})
exten => 678,102,Set(DB(test/count)=1)
exten => 678,103,Goto(1)
```

Finally, we'll loop back to the first priority. This way, the application will continue counting:

```
exten => 678,1,Set(COUNT=${DB(test/count)})
exten => 678,2,SayNumber(${COUNT})
exten => 678,3,SetVar(COUNT=$[${COUNT} + 1])
exten => 678,4,Set(DB(test/count)=${COUNT})
exten => 678,5,Goto(1)
exten => 678,102,Set(DB(test/count)=1)
exten => 678,103,Goto(1)
```

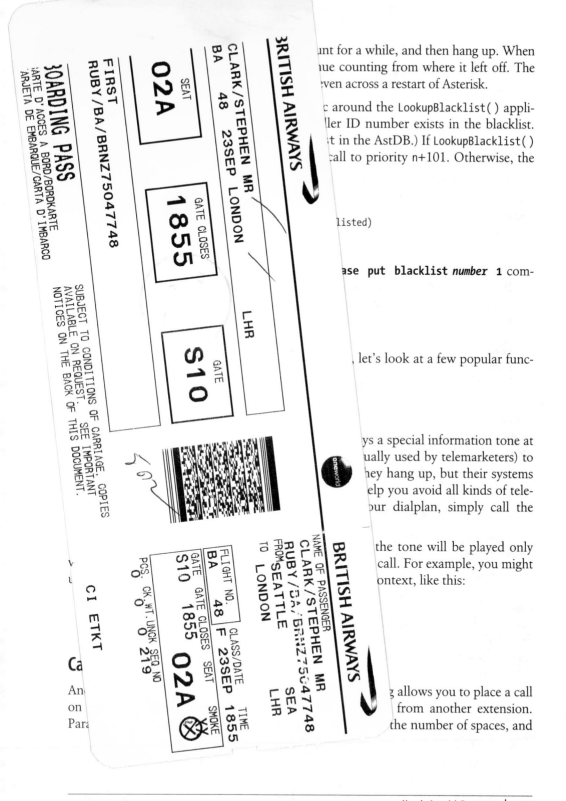

...nt for a while, and then hang up. When ...ue counting from where it left off. The ...ven across a restart of Asterisk.

...c around the `LookupBlacklist( )` appli-...ler ID number exists in the blacklist. ...t in the AstDB.) If `LookupBlacklist( )` ...all to priority n+101. Otherwise, the

...listed)

...ase put blacklist *number* **1** com-

..., let's look at a few popular func-

...ys a special information tone at ...ually used by telemarketers) to ...hey hang up, but their systems ...elp you avoid all kinds of tele-...ur dialplan, simply call the

...the tone will be played only ...call. For example, you might ...ontext, like this:

v...
u...

Ca...

An...g allows you to place a call
on...from another extension.
Para...the number of spaces, and

Handy Asterisk Features | **115**

so on) are all controlled within the *features.conf* configuration file. The [general] section of the *features.conf* file contains four settings related to call parking:

parkext

This is the parking lot extension. Transfer a call to this extension, and the system will tell you which parking position the call is in. By default, the parking extension is 700.

parkpos

This option defines the number of parking slots. For example, setting it to 701-720 creates 20 parking positions, numbered 701 through 720.

context

This is the name of the parking context. To be able to park calls, you must include this context.

parkingtime

If set, this option controls how long (in seconds) a call can stay in the parking lot. If the call isn't picked up within the specified time, the extension that parked the call will be called back.

 You must restart Asterisk after editing *features.conf*, as the file is read only on startup. Running the reload command will not cause the *features.conf* file to be read.

Also note that because the user needs to be able to transfer the calls to the parking lot extension, you should make sure you're using the t and/or T options to the Dial( ) application.

So, let's create a simple dialplan to show off call parking:

```
[internal]
include => parkedcalls

exten=103,1,Dial(SIP/Bob,,tT)
exten=104,1,Dial(SIP/Charlie,,tT)
```

To illustrate how call parking works, say that Alice calls into the system and dials extension 103, to reach Bob. After a while, Bob transfers the call to extension 700, which tells him that the call from Alice has been parked in position 701. Bob then dials Charlie at extension 104, and tells him that Alice is at extension 701. Charlie then dials extension 701, and begins to talk to Alice. This is a simple and effective way of allowing callers to be transferred between users.

Conferencing with MeetMe( )

Last but not least, let's cover setting up an audio conference bridge with the MeetMe( ) application.* This application allows multiple callers to converse together, as if they were all in the same physical location. Some of the main features include:

- The ability to create password-protected conferences
- Conference administration (mute conference, lock conference, kick participants)
- The option of muting all but one participant (useful for company announcements, broadcasts, etc.)
- Static or dynamic conference creation

Let's walk through setting up a basic conference room. The configuration options for the MeetMe conferencing system are found in *meetme.conf*. Inside the configuration file, you define conference rooms and optional numeric passwords. (If a password is defined here, it will be required to enter all conferences using that room.) For our example, let's set up a conference room at extension 600. First, we'll set up the conference room in *meetme.conf*. We'll call it 600, and we won't assign a password at this time:

```
[rooms]
conf => 600
```

Now that the configuration file is complete, we'll need to restart Asterisk so that it can re-read the *meetme.conf* file. Next, we'll add support for the conference room to our dialplan with the MeetMe( ) application. MeetMe( ) takes three arguments: the name of the conference room (as defined in *meetme.conf*), a set of options, and the password the user must enter to join this conference. Let's set up a simple conference using room 600, the i option (which announces when people enter and exit the conference), and a password of 54321:

```
exten => 600,1,MeetMe(600,i,54321)
```

That's all there is to it! When callers enter extension 600, they will be prompted for the password. If they correctly enter 54321, they will be added to the conference. See Appendix B for a list of all the options supported by the MeetMe( ) application.

Another useful application is MeetMeCount( ). As its name suggests, this application counts the number of users in a particular conference room. It takes up to two arguments: the conference room in which to count the number of participants, and optionally a variable name to assign the count to. If the variable name is not passed as the second argument, the count is read to the caller:

```
exten => 601,1,Playback(conf-thereare)
exten => 601,2,MeetMeCount(600)
exten => 601,3,Playback(conf-peopleinconf)
```

---

* In the world of legacy PBXs, this type of functionality is very expensive. Either you have to pay big bucks for a dial-in service, or you have to add an expensive conferencing bridge to your proprietary PBX.

If you pass a variable as the second argument to MeetMeCount( ), the count is assigned to the variable and playback of the count is skipped. You might use this to limit the number of participants, like this:

```
; limit the conference room to 10 participants
exten => 600,1,MeetMeCount(600,CONFCOUNT)
exten => 600,2,GotoIf($[${CONFCOUNT} <= 10]?3:100)
exten => 600,3,MeetMe(600,i,54321)
exten => 600,100,Playback(conf-full)
```

Isn't Asterisk fun?

Conclusion

In this chapter, we've covered a few more of the many applications in the Asterisk dialplan, and hopefully we've given you the seeds from which you can explore the creation of your own dialplans. As with the previous chapter, we invite you to go back and re-read any sections that require clarification.

The following chapters take us away from Asterisk for a bit, in order to talk about some of the technologies that all telephone systems use. We'll be referring to Asterisk a lot, but much of what we want to discuss are things that are common to many telecom systems.

## CHAPTER 7

Understanding Telephony

*Utility is when you have one telephone, luxury is when
you have two, opulence is when you have three—and
paradise is when you have none.*

—Doug Larson

We're now going to take a break from Asterisk for a chapter or two, because we want to spend some time discussing the technologies with which your Asterisk system will need to interface. In this chapter, we are going to talk about some of the technologies of the traditional telephone network—especially those that people most commonly want to connect to Asterisk.* While tomes could be written about the technologies in use in telecom networks, the material in this chapter was chosen based on our experiences in the community, which helped us define the specific items that might be most useful. Although this knowledge may not be strictly required in order to configure your Asterisk system, it will be of great benefit when interconnecting to systems (and talking with people) from the world of traditional telecommunications.

Analog Telephony

The purpose of the Public Switched Telephone Network (PSTN) is to establish and maintain audio connections between two endpoints.

Although humans can perceive sound vibrations in the range of 20–20,000 Hz,† most of the sounds we make when speaking tend to be in the range of 250–3,000 Hz. Since the purpose of the telephone network is to transmit the sounds of people speaking, it

---

* We'll discuss Voice over IP in the next chapter.

† If you want to play around with what different frequencies look like on an oscilloscope, grab a copy of Sound Frequency Analyzer, from Reliable Software. It's a really simple and fun way to visualize what sounds "look" like. The spectrograph gives a good picture of the complex harmonics our voices can generate, as well as an appreciation for the background sounds that always surround us. You should also try the delightfully annoying NCH Tone Generator, from NCH Swift Sound.

was designed with a bandwidth of somewhere in the range of 300–3,500 Hz. This limited bandwidth means that some sound quality will be lost (as anyone who's had to listen to music on hold can attest to), especially in the higher frequencies.

Parts of an Analog Telephone

An analog phone is composed of five parts: the ringer, the dial pad, the hybrid (or network), and the hook switch and handset (both of which are considered parts of the hybrid). The ringer, the dial pad, and the hybrid can operate completely independently from one another.

Ringer

When the central office (CO) wants to signal an incoming call, it will connect an alternating current (AC) signal of roughly 90 volts to your circuit. This will cause the bell in your telephone to produce a ringing sound. (In electronic telephones, this ringer may be a small electronic warbler rather than a bell. Ultimately, a ringer can be anything that is capable of reacting to the ringing voltage—for example, strobe lights are often employed in noisy environments such as factories.)

 Ringing voltage can be hazardous. Be very careful to take precautions when working with an in-service telephone line.

Many people confuse the AC voltage that triggers the ringer with the direct current (DC) voltage that powers the phone. Remember that the ringer will not respond to DC voltage, and you've got it.

In North America, the number of ringers you can connect to your line is dependent on the Ringer Equivalence Number (REN) of your various devices. (The REN must be listed on each device.) The total REN for all devices connected to your line cannot exceed 5.0. An REN of 1.0 is equivalent to an old-fashioned analog set with an electromechanical ringer. Some electronic phones have an REN of 0.3 or even less.

Dial pad

When you place a telephone call, you need some way of letting the network know the address of the party you wish to reach. The dial pad is the portion of the phone that provides this functionality. In the early days of the PSTN, dial pads were rotary devices that used pulses to indicate digits. This was a rather slow process, so the telephone companies eventually introduced touch-tone dialing. With touch-tone—also known as Dual-Tone Multi Frequency (DTMF)—dialing, the dial pad consists of 12 buttons. Each button has two frequencies assigned to it (see Table 7-1).

Table 7-1. DTMF digits

	1209 Hz	1336 Hz	1477 Hz	1633 Hz[a]
697 Hz	1	2	3	A
770 Hz	4	5	6	B
852 Hz	7	8	9	C
941 Hz	*	0	#	D

[a] Notice that this column contains letters that are not typically present as keys on a telephone dial pad. They are part of the DTMF standard nonetheless, and any proper telephone contains the electronics required to create them, even if it doesn't contain the buttons themselves. (These buttons actually do exist on some telephones, which are mostly used in military and government applications.)

When you press a button on your dial pad, the two corresponding frequencies are transmitted down the line.

We assume that you've used a telephone, so we won't spend any more time on DTMF.

Hybrid (or network)

The hybrid is a type of transformer that handles the need to combine the signals transmitted and received across a single pair of wires in the PSTN and two pairs of wires in the handset. One of the functions the hybrid performs is regulating *sidetone*, which is the amount of your transmitted signal that is returned to your earpiece—its purpose is to provide a more natural-sounding conversation. Too much sidetone, and your voice will sound too loud; too little, and you'll think the line has gone dead.

Hook switch (or switch hook). This device signals the state of the telephone circuit to the CO. When you pick up your telephone, the hook switch closes the loop between you and the CO, which is seen as a request for a dial tone. When you hang up, the hook switch opens the circuit, which indicates that the call has ended.*

The hook switch can also be used for signaling purposes. Some electronic analog phones have a button labeled "Link" that causes an event called a *flash*. You can perform a flash manually by depressing the hook switch for a duration of between 200 and 1,200 milliseconds. If you leave it down for longer than that, the carrier will assume you've hung up. The purpose of the Link button is to handle this timing for you. If you've ever used call waiting or three-way calling on an analog line, you have performed a hook switch flash for the purpose of signaling the network.

---

* When referring to the state of an analog circuit, people often speak in terms of "off-hook" and "on-hook." When your line is "off-hook," your telephone is "on" a call. If your phone is "on-hook," the telephone is essentially "off."

Handset. The handset is composed of the transmitter and receiver. It performs the conversion between the sound energy humans use and the electrical energy the telephone network uses.

Tip and Ring

In an analog telephone circuit, there are two wires. In North America, these wires are referred to as Tip and Ring.* This terminology comes from the days when telephone calls were connected by live operators sitting at cord boards. The plugs they used had two contacts, one located at the tip of the plug and the other connected to the ring around the middle (Figure 7-1).

Figure 7-1. Tip and Ring

The Tip lead is the positive polarity wire. In North America, this wire is typically green and provides the return path. The Ring wire is the negative polarity wire. In North America, this wire is normally red. When your telephone is on-hook, this wire will have a potential of –48V DC with respect to Tip. Off-hook, this voltage drops to roughly –7V DC.

Digital Telephony

Analog telephony is almost dead.

In the PSTN, the famous Last Mile is the final remaining piece of the telephone network still using technology pioneered well over a hundred years ago.†

One of the primary challenges when transmitting analog signals is that all sorts of things can interfere with those signals, causing low volume, static, and all manner of other undesired effects. Instead of trying to preserve an analog waveform over distances that may span thousands of miles, why not simply measure the characteristics

---

* They may have other names elsewhere in the world (such as "A" and "B").

† "The Last Mile" is a term that was originally used to describe the only portion of the PSTN that had not been converted to fiber optics: the connection between the central office and the customer. The Last Mile is more than that, however, as it also has significance as a valuable asset of the traditional phone companies—they own a connection into your home. The Last Mile is becoming more and more difficult to describe in technical terms, as there are now so many ways to connect the network to the customer. As a thing of strategic value to telecom, cable, and other utilities, its importance is obvious.

of the original sound and send that information to the far end? The original wave-form wouldn't get there, but all the information needed to reconstruct it would.

This is the principle of all digital audio (including telephony): sample the characteristics of the source waveform, store the measured information, and send that data to the far end. Then, at the far end, use the transmitted information to generate a completely new audio signal that has the same characteristics as the original. The reproduction is so good that the human ear can't tell the difference.

The principle advantage of digital audio is that the sampled data can be mathematically checked for errors all along the route to its destination, ensuring that a perfect duplicate of the original arrives at the far end. Distance no longer affects quality, and interference can be detected and eliminated.

Pulse-Code Modulation

There are several ways to digitally encode audio, but the most common method (and the one used in telephony systems) is known as *Pulse-Code Modulation* (PCM). To illustrate how this works, let's go through a few examples.

Digitally encoding an analog waveform

The principle of PCM is that the amplitude of the analog waveform is sampled at specific intervals so that it can later be recreated. The amount of detail that is captured is dependent both on the bit-resolution of each sample and on how frequently the samples are taken. A higher bit-resolution and a higher sampling rate will provide greater accuracy, but more bandwidth will be required to transmit this more detailed information.

To get a better idea of how PCM works, consider the waveform displayed in Figure 7-2.

To digitally encode the wave, it must be sampled on a regular basis, and the amplitude of the wave at each moment in time must be measured. The process of slicing up a waveform into moments in time and measuring the energy at each moment is called *quantization*, or *sampling*.

The samples will need to be taken frequently enough and will need to capture enough information to ensure that the far end can recreate a sufficiently similar waveform. To achieve a more accurate sample, more bits will be required. To explain this concept, we will start with a very low resolution, using four bits to represent our amplitude. This will make it easier to visualize both the quantization process itself and the effect that resolution has on quality.

Figure 7-3 shows the information that will be captured when we sample our sine wave at four-bit resolution.

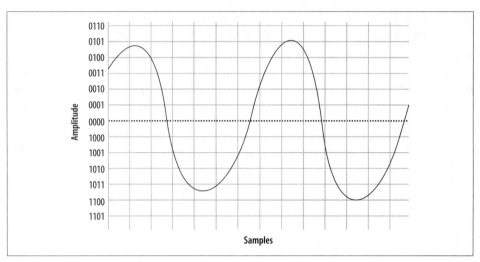

Figure 7-2. A simple sinusoidal (sine) wave

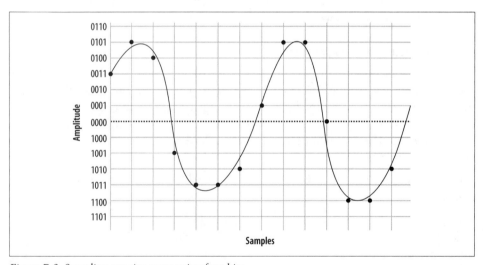

Figure 7-3. Sampling our sine wave using four bits

At each time interval, we measure the amplitude of the wave and record the corresponding intensity—in other words, we sample it. You will notice that the four-bit resolution limits our accuracy. The first sample has to be rounded to 0011, and the next quantization yields a sample of 0101. Then comes 0100, followed by 1001, 1011, and so forth. In total, we have 14 samples (in reality, several thousand samples must be taken per second). If we string together all the values, we can send them to the other side as:

    0011 0101 0100 1001 1011 1011 1010 0001 0101 0101 0000 1100 1100 1010

On the wire, this code might look something like Figure 7-4.

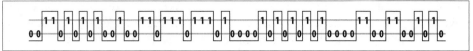

Figure 7-4. PCM encoded waveform

When the far end's digital-to-analog (D/A) converter receives this signal, it can use the information to plot the samples, as shown in Figure 7-5.

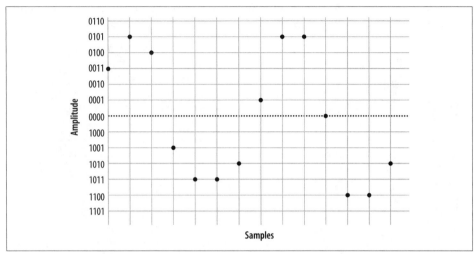

Figure 7-5. Plotted PCM signal

From this information, the waveform can be reconstructed (see Figure 7-6).

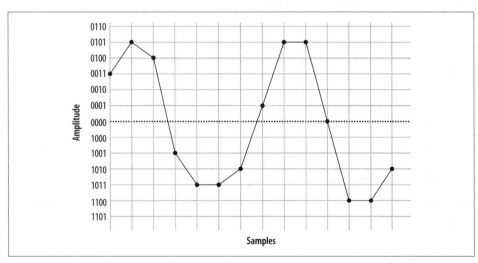

Figure 7-6. Delineated signal

As you can see if you compare Figure 7-2 with Figure 7-6, this reconstruction of the waveform is not very accurate. This was done intentionally, to demonstrate an important point: the quality of the digitally encoded waveform is affected by the resolution and rate at which it is sampled. At too low a sampling rate, and with too low a sample resolution, the audio quality will not be acceptable.

Increasing the sampling resolution and rate

Let's take another look at our original waveform, this time using five bits to define our quantization intervals (Figure 7-7).

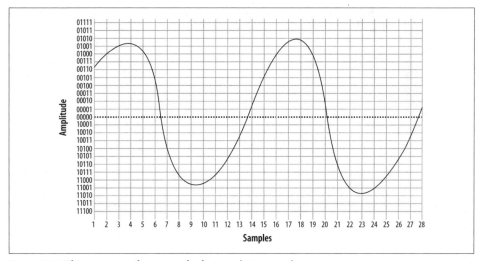

Figure 7-7. The same waveform, on a higher-resolution overlay

In reality, there is no such thing as five-bit PCM. In the telephone network, PCM samples are encoded using eight bits.[*]

We'll also double our sampling frequency. The points plotted this time are shown in Figure 7-8.

We now have twice the number of samples, at twice the resolution. Here they are:

```
00111 01000 01001 01001 01000 00101 10110 11000 11001 11001 11000 10111 10100 10001
00010 00111 01001 01010 01001 00111 00000 11000 11010 11010 11001 11000 10110 10001
```

When received at the other end, that information can now be plotted as shown in Figure 7-9.

---

[*] Other digital audio methods may employ 16 bits or more.

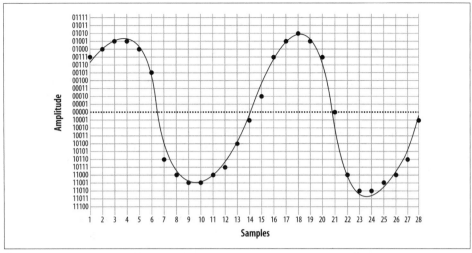

Figure 7-8. The same waveform at double the resolution

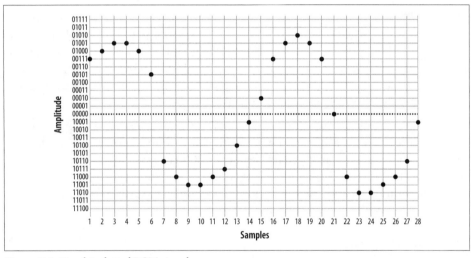

Figure 7-9. Five-bit plotted PCM signal

From this information, the waveform shown in Figure 7-10 can then be generated.

As you can see, the resultant waveform is a far more accurate representation of the original. However, you can also see that there is still room for improvement.

> Note that 40 bits were required to encode the waveform at 4-bit resolution, while 156 bits were needed to send the same waveform using 5-bit resolution (and also doubling the sampling rate). The point is, there is a tradeoff: the higher the quality of audio you wish to encode, the more bits will be required to do it, and the more bits you wish to send (in real time, naturally), the more bandwidth you will need to consume.

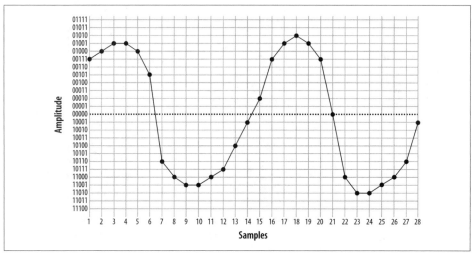

Figure 7-10. Waveform delineated from five-bit PCM

Nyquist's Theorem

So how much sampling is enough? That very same question was considered in the 1920s by an electrical engineer (and AT&T/Bell employee) named Harry Nyquist. Nyquist's Theorem[*] states: "When sampling a signal, the *sampling frequency* must be greater than twice the bandwidth of the input signal in order to be able to reconstruct the original perfectly from the sampled version."

In essence, what this means is that to accurately encode an analog signal you have to sample it twice as often as the total bandwidth you wish to reproduce. Since the telephone network will not carry frequencies below 300 Hz and above 4,000 Hz, a sampling frequency of 8,000 samples per second will be sufficient to reproduce any frequency within the bandwidth of an analog telephone. Keep that 8,000 samples per second in mind; we're going to talk about it more later.

Logarithmic companding

So, we've gone over the basics of quantization, and we've discussed the fact that more quantization intervals (i.e., a higher sampling rate) give better quality but also require more bandwidth. Lastly, we've discussed the minimum sample rate needed to accurately measure the range of frequencies we wish to be able to transmit (in the case of the telephone, it's 8,000 Hz). This is all starting to add up to a fair bit of data being sent on the wire, so we're going to want to talk about companding.

---

[*] Nyquist published two papers, "Certain Factors Affecting Telegraph Speed" (1924) and "Certain Topics in Telegraph Transmission Theory" (1928), in which he postulated what became known as Nyquist's Theorem. Proven in 1949 by Claude Shannon ("Communication in the Presence of Noise"), it is also referred to as the Nyquist-Shannon sampling theorem.

Companding is a method of improving the dynamic range of a sampling method without losing important accuracy. It works by quantizing higher amplitudes in a much coarser fashion than lower amplitudes. In other words, if you yell into your phone, you will not be sampled as cleanly as you will be when speaking normally. Yelling is also not good for your blood pressure, so it's best to avoid it.

Two companding methods are commonly employed: μ-law* in North America, and A-law in the rest of the world. They operate on the same principles but are otherwise not compatible with each other.

Companding divides the waveform into *cords*, each of which has several *steps*. Quantization involves matching the measured amplitude to an appropriate step within a cord. The value of the band and cord numbers (as well as the sign—positive or negative) becomes the signal. The following diagrams will give you a visual idea of what companding does. They are not based on any standard, but rather were made up for the purpose of illustration (again, in the telephone network companding will be done at an eight-bit, not five-bit, resolution).

Figure 7-11 illustrates five-bit companding. As you can see, amplitudes near the zero-crossing point will be sampled far more accurately than higher amplitudes (either positive or negative). However, since the human ear, the transmitter, and the receiver will also tend to distort loud signals, this isn't really a problem.

A quantized sample might look like Figure 7-12. It yields the following bit stream:

```
00000 10011 10100 10101 01101 00001 00011 11010 00010 00001 01000 10011 10100 10100
00101 00100 00101 10101 10011 10001 00011 00001 00000 10100 10010 10101 01101 10100
00101 11010 00100 00000 01000
```

Aliasing

If you've ever watched the wheels on a wagon turn backward in an old Western movie, you've seen the effects of aliasing. The frame rate of the movie cannot keep up with the rotational frequency of the spokes, and a false rotation is perceived.

In a digital audio system (which the modern PSTN arguably is), aliasing always occurs if frequencies that are greater than one-half the sampling rate are presented to the analog-to-digital (A/D) converter. In PSTN, that is any audio frequencies above 4,000 Hz (half the sampling rate of 8,000 Hz). This problem is easily corrected by passing the audio through a low-pass filter† before presenting it to the A/D converter.

---

* μ-law is often referred to as "u-law" because, let's face it, how many of us have μ keys on our keyboards? μ is in fact the Greek letter Mu; thus, you will also see μ-law written (more correctly) as "Mu-law." When spoken, it is correct to confidently say "Mew-law," but if folks look at you strange, and you're feeling generous, you can help them out and tell them it's "u-law." Many people just don't appreciate trivia.

† A low-pass filter, as its name implies, allows through only frequencies that are lower than its cut-off frequency. Other types of filters are high-pass filters (which remove low frequencies) and band-pass filters (which filter out both high and low frequencies).

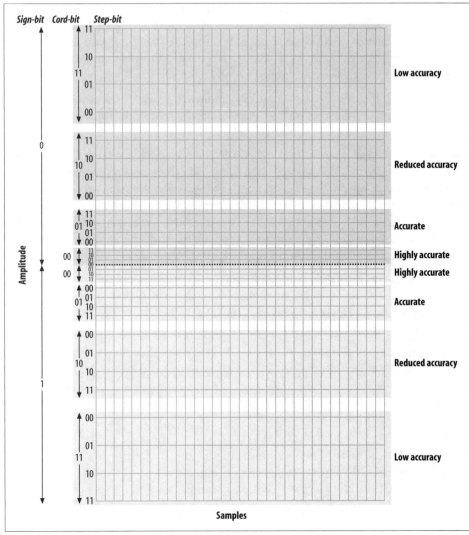

Figure 7-11. Five-bit companding

The Digital Circuit-Switched Telephone Network

For over a hundred years, telephone networks were exclusively circuit-switched. What this meant was that for every telephone call made, a dedicated connection was established between the two endpoints, with a fixed amount of bandwidth allocated to that circuit. Creating such a network was costly, and where distance was concerned, using that network was costly as well. Although we are all predicting the end of the circuit-switched network, many people still use it every day, and it really does work rather well.

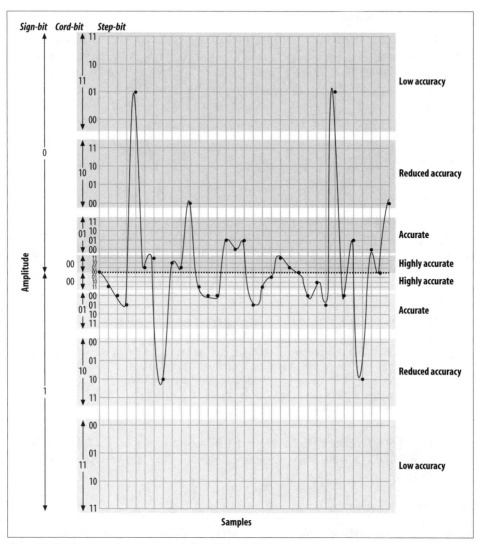

Figure 7-12. Quantized and companded at 5-bit resolution

Circuit Types

In the PSTN, there are many different sizes of circuits serving the various needs of the network. Between the central office and a subscriber, one or more analog circuits, or a few dozen channels delivered over a digital circuit, generally suffice. Between PSTN offices (and with larger customers), fiber-optic circuits are generally used.

The humble DS-0, the foundation of it all

Since the standard method of digitizing a telephone call is to record an 8-bit sample 8,000 times per second, we can see that a PCM-encoded telephone circuit will need a bandwidth of 64,000 bps. This 64-kbps channel is referred to as a DS-0 (that's Dee-Ess-Zero). The DS-0 is the fundamental building block of all digital telecommunications circuits.

Even the ubiquitous analog circuit is sampled into a DS-0 as soon as possible. Sometimes this happens where your circuit terminates at the central office, and sometimes well before.

T-carrier circuits

The venerable T-1 is one of the more recognized digital telephony terms. A T-1 is a digital circuit consisting of 24 DS-0s multiplexed together into a 1.544-Mbps bit stream.* This bit stream is properly defined as a *DS-1*. Voice is encoded on a T-1 using the μ-law companding algorithm.

> The European version of the T-1 was developed by the European Conference of Postal and Telecommunications Administrations† (CEPT), and was first referred to as a *CEPT-1*. It is now called an *E-1*.
>
> The E-1 is comprised of 32 DS-0s, but the method of PCM encoding is different—E-1s use A-law companding. This means that connecting between an E-1-based network and a T-1-based network will always require a transcoding step. Note that an E-1, although it has 32 channels, is also considered a *DS-1*.

The various other T-carriers (T-2, T-3, and T-4) are multiples of the T-1, each based on the humble DS-0. Table 7-2 illustrates the relationships between the different T-carrier circuits.

Table 7-2. T-carrier circuits

Carrier	Equivalent data bitrate	Number of DS-0s	Data bitrate
T-1	24 DS-0s	24	1.544 Mbps
T-2	4 T-1s	96	6.312 Mbps
T-3	7 T-2s	672	44.736 Mbps
T-4	6 T-3s	4032	274.176 Mbps

At densities above T-3, it is very uncommon to see a T-carrier circuit. For these speeds, optical carrier (OC) circuits may be used.

---

* The 24 DS-0s use 1.536 Mbps, and the remaining .008 Mbps is used by framing bits.

† Conférence Européenne des Administrations des Postes et des Télécommunications.

SONET and OC circuits

The Synchronous Optical Network (SONET) was developed out of a desire to take the T-carrier system to the next technological level: fiber optics. SONET is based on the bandwidth of a T-3 (44.736Mbps), with a slight overhead making it 51.84 Mbps. This is referred to as an *OC-1* or *STS-1*. As Table 7-3 shows, all higher-speed OC circuits are multiples of this base rate.

Table 7-3. OC circuits

Carrier	Equivalent data bitrate	Number of DS-0s	Data bitrate
OC-1	1 DS-3 (plus overhead)	672	51.840 Mbps
OC-3	3 DS-3s	2016	155.520 Mbps
OC-12	12 DS-3s	8064	622.080 Mbps
OC-48	48 DS-3s	32256	2488.320 Mbps
OC-192	192 DS-3s	129024	9953.280 Mbps

SONET was created in an effort to standardize optical circuits, but due to its high cost, coupled with the value offered by many newer schemes, such as Dense Wave Division Multiplexing (DWDM), there is some controversy surrounding its future.

Digital Signaling Protocols

As with any circuit, it is not enough for the circuits used in the PSTN to just carry (voice) data between endpoints. Mechanisms must also be provided to pass information about the state of the channel between each endpoint. (Disconnect and answer supervision are two examples of basic signaling that might need to take place; Caller ID is an example of a more complex form of signaling.)

Channel Associated Signaling (CAS)

Also known as robbed-bit signaling, CAS is what you will use to transmit voice on a T-1 when ISDN is not available. Rather than taking advantage of the power of the digital circuit, CAS simulates analog channels. CAS signaling works by stealing bits from the audio stream for signaling purposes. Although the effect on audio quality is not really noticeable, the lack of a powerful signaling channel limits your flexibility.

When configuring a CAS T-1, the signaling options at each end must match. E&M (Ear & Mouth or recEive & transMit) signaling is generally preferred, as it offers the best supervision.

CAS is very rarely used on PSTN circuits anymore, due to the superiority of ISDN-PRI. One of the limitations of CAS is that it does not allow the dynamic assignment of channels to different functions. Also, Caller ID information (which may not even be supported) has to be sent as part of the audio stream. CAS is commonly used on the T-1 link in channel banks, although PRI is sometimes available (and preferable).

ISDN

The Integrated Services Digital Network (ISDN) has been around for over 20 years. Because it separates the channels that carry the traffic (the bearer channels, or B-channels) from the channel that carries the signaling information (the D-channel), ISDN allows for the delivery of a much richer set of features than CAS. In the beginning, ISDN promised to deliver much the same sort of functionality that the Internet has given us, including advanced capabilities for voice, video, and data transfer.

Unfortunately, rather than ratifying a standard and sticking to it, the respective telecommunications manufacturers all decided to add their own tweaks to the protocol, in the belief that their versions were superior and would eventually dominate the market. As a result, getting two ISDN-compliant systems to connect to each other was often a painful and expensive task. The carriers who had to implement and support this expensive technology in turn priced it so that it was not rapidly adopted. Currently, ISDN is rarely used for much more than basic trunking—in fact, the acronym ISDN has become a joke in the industry: "It Still Does Nothing."

Having said that, ISDN has become quite popular for trunking, and it is now (mostly) standards-compliant. If you have a PBX with more than a dozen lines connected to the PSTN, there's a very good chance that you'll be running an ISDN-PRI circuit. Also, in places where DSL and cable access to the Internet are not available (or too expensive), an ISDN-BRI circuit might provide you with an affordable 128-kbps connection. In much of North America, the use of ISDN-BRI for Internet connectivity has been deprecated in favor of DSL and cable modems, but it's still very popular in other parts of the world.

ISDN-BRI/BRA. Basic Rate Interface (or Basic Rate Access) is the flavor of ISDN designed to service small endpoints such as workstations.

The BRI flavor of the ISDN specification is often referred to simply as "ISDN," but this can be a source of confusion, as ISDN is a protocol, not a type of circuit (not to mention that PRI circuits are also correctly referred to as ISDN!).

A Basic Rate ISDN circuit consists of two 64-kbps B-channels controlled by a 16-kbps D-channel, for a total of 144 kbps.

Basic Rate ISDN has been a source of much confusion during its life, due to problems with standards compliance, technical complexity, and poor documentation. Still, in European countries ISDN-BRI circuits remain quite a popular way of connecting to the PSTN.

ISDN-PRI/PRA. The Primary Rate Interface (or Primary Rate Access) flavor of ISDN is used to provide ISDN service over larger network connections. A Primary Rate ISDN circuit uses a single DS-0 channel as a signaling link (the D-channel); the remaining channels serve as B-channels.

In North America, Primary Rate ISDN is commonly carried on one or more T-1 circuits. Since a T-1 has 24 channels, a North American PRI circuit typically consists of 23 B-channels and 1 D-channel. For this reason, PRI is often referred to as 23B+D.*

 In Europe, a 32-channel E-1 circuit is used, so a Primary Rate ISDN circuit is referred to as 30B+D (the final channel is used for synchronization).

Primary Rate ISDN is very popular, due to its technical benefits and generally competitive pricing. If you believe you will require more than a dozen or so PSTN lines, you should look into Primary Rate ISDN pricing.

From a technical perspective, ISDN-PRI is always preferable to CAS.

Signaling System 7

SS7 is the signaling system used by carriers. It is conceptually similar to ISDN, and it is instrumental in providing a mechanism for the carriers to transmit the additional information ISDN endpoints typically need to pass. However, the technology of SS7 is different from that of ISDN—one big difference is that SS7 runs on a completely separate network from the actual trunks that carry the calls.

SS7 support in Asterisk is on the horizon, as there is much interest in making Asterisk compatible with the carrier networks. An open source version of SS7 (*http://www.openss7.org*) exists, but work is still needed for full SS7 compliance, and as of this writing it is not known whether this will be integrated with Asterisk. Another promising source of SS7 support comes from Sangoma Technologies, who offer SS7 functionality in many of their products.

It should be noted that adding support for SS7 in Asterisk is not going to be as simple as writing a proper driver. Connecting equipment to an SS7 network will not be possible without that equipment having passed an extremely rigorous certification processes. Even then, it seems doubtful that any traditional carrier is going to be in a hurry to allow such a thing to happen.

Packet-Switched Networks

In the mid-1990s, network performance improved to the point where it became possible to send a stream of media information in real time across a network connection. Because the media stream is chopped up into segments, which are then

---

* PRI is actually quite a bit more flexible than that, as it is possible to span a single PRI circuit across multiple T-1 spans. This can give rise, for example, to a 47B+D circuit (where a single D-channel serves two T-1s) or a 46B+2D circuit (where primary and backup D-channels serve a pair of T-1s). You will sometimes see PRI described as nB+nD, because the number of B- and D-channels is, in fact, quite variable.

wrapped in an addressing envelope, such connections are referred to as *packet-based*. The challenge, of course, is to send thousands of these packets between two end-points, ensuring that the packets arrive in the same order in which they were sent, in less than 300 milliseconds, with none lost. This is the essence of Voice over IP.

Conclusion

This chapter has explored the technologies currently in use in the PSTN. In the next chapter, we will discuss protocols for VoIP: the carrying of telephone connections across IP-based networks. These protocols define different mechanisms for carrying telephone conversations, but their significance is far greater than just that. Bringing the telephone network into the data network will finally erase the line between telephones and computers, which holds the promise of a revolutionary evolution in the way we communicate.

## CHAPTER 8

Protocols for VoIP

The Internet is a telephone system that's gotten uppity.
—Clifford Stoll

The telecommunications industry spans over 100 years, and Asterisk integrates most—if not all—of the major technologies that it has made use of over the last century. To make the most out of Asterisk, you need not be a professional in all areas, but understanding the differences between the various codecs and protocols will give you a greater appreciation and understanding of the system as a whole.

This chapter explains Voice over IP and what makes VoIP networks different from the traditional circuit-switched voice networks that were the topic of the last chapter. We will explore the need for VoIP protocols, outlining the history and potential future of each. We'll also look at security considerations and these protocols' abilities to work within topologies such as Network Address Translation (NAT). The following VoIP protocols will be discussed:

- IAX
- SIP
- H.323
- MGCP
- Skinny/SCCP
- UNISTIM

Codecs are the means by which analog voice can be converted to a digital signal and carried across the Internet. Bandwidth at any location is finite, and the number of simultaneous conversations any particular connection can carry is directly related to the type of codec implemented. In this chapter, we'll also explore the differences between the following codecs in regards to bandwidth requirements (compression level) and quality:

- G.711
- G.726

- G.723.1
- G.729A
- GSM
- iLBC
- Speex
- MP3

We will then conclude the chapter with a discussion of how voice traffic can be routed reliably, what causes echo and how to minimize it, and how Asterisk controls the authentication of inbound and outbound calls.

The Need for VoIP Protocols

The basic premise of VoIP is the packetization* of audio streams for transport over Internet Protocol–based networks. The challenges to accomplishing this relate to the manner in which humans communicate. Not only must the signal arrive in essentially the same form that it was transmitted in, but it needs to do so in less than 300 milliseconds. If packets are lost or delayed, there will be degradation in the quality of the communications experience.

The transport protocols that collectively are called "the Internet" were not originally designed with real-time streaming of media in mind. Endpoints were expected to resolve missing packets by waiting longer for them to arrive, requesting retransmission, or, in some cases, considering the information to be gone for good and simply carrying on without it. In a typical voice conversation, these mechanisms will not serve. Our conversations do not adapt well to the loss of letters or words, nor to any appreciable delay between transmittal and receipt.

The traditional PSTN was designed specifically for the purpose of voice transmission, and it is perfectly suited to the task from a technical standpoint. From a flexibility standpoint, however, its flaws are obvious to even people with a very limited understanding of the technology. VoIP holds the promise of incorporating voice communications into all the other protocols we carry on our networks, but due to the special demands of a voice conversation, special skills are needed to design, build, and maintain these networks.

The problem with packet-based voice transmission stems from the fact that the way in which we speak is totally incompatible with the way in which IP transports data. Speaking and listening consist of the relaying of a stream of audio, whereas the

---

* This word hasn't quite made it into the dictionary, but it is a term that is becoming more and more common. It refers to the process of chopping a steady stream of information into discreet chunks (or *packets*), suitable for delivery independently of one another.

Internet protocols are designed to chop everything up, encapsulate the bits of information into thousands of packages, and then deliver each package in whatever way possible to the far end. Clearly, some sort of bridge was required.

VoIP Protocols

The mechanism for carrying a VoIP connection generally involves a series of signaling transactions between the endpoints (and gateways in between), culminating in two persistent media streams (one for each direction) that carry the actual conversation. There are several protocols in existence to handle this. In this section, we will discuss some of those that are important to VoIP in general and to Asterisk specifically.

IAX (The "Inter-Asterisk eXchange" Protocol)

The test of your Asterisk-ness comes when you have to pronounce the name of this protocol. Newbies say "eye-ay-ex"; those in the know say "eeks." IAX* is an open protocol, meaning that anyone can download and develop for it, but it is not yet a standard of any kind.

In Asterisk, IAX is supported by the *chan_iax2.so* module.

History

The IAX protocol was developed by Digium for the purpose of communicating with other Asterisk servers (hence "the Inter-Asterisk eXchange protocol"). IAX is a transport protocol (much like SIP) that uses a single UDP port (4569) for both the channel signaling and Realtime Transport Protocol (RTP) streams. As discussed below, this makes it easier to firewall and more likely to work behind NAT.

IAX also has the unique ability to trunk multiple sessions into one dataflow, which can be a tremendous bandwidth advantage when sending a lot of simultaneous channels to a remote box. Trunking allows multiple data streams to be represented with a single datagram header, to lower the overhead associated with individual channels. This helps to lower latency and reduce the processing power and bandwidth required, allowing the protocol to scale much more easily with a large number of active channels between endpoints.

Future

Since IAX was optimized for voice, it has received some criticism for not better supporting video—but in fact, IAX holds the potential to carry pretty much any media

---

* Officially, the current version is IAX2, but all support for IAX1 has been dropped, so whether you say "IAX" or "IAX2," it is expected that you are talking about the Version 2.

stream desired. Because it is an open protocol, future media types are certain to be incorporated as the community desires them.

Security considerations

IAX includes the ability to authenticate in three ways: plain text, MD5 hashing, and RSA key exchange. This, of course, does nothing to encrypt the media path or headers between endpoints. Many solutions include using a Virtual Private Network (VPN) appliance or software to encrypt the stream in another layer of technology, which requires the endpoints to pre-establish a method of having these tunnels configured and operational. In the future, IAX may be able to encrypt the streams between endpoints with the use of an exchanged RSA key, or dynamic key exchange at call setup, allowing the use of automatic key rollover. This would be very attractive for creating a secure link with an institution such as your bank. The various law enforcement agencies, however, are going to want some level of access to such connections.

IAX and NAT

The IAX2 protocol was deliberately designed to work from behind devices performing NAT. The use of a single UDP port for both signaling and transmission of media also keeps the number of holes required in your firewall to a minimum. These considerations have helped make IAX one of the easiest protocols (if not the easiest) to implement in secure networks.

SIP

The Session Initiation Protocol (SIP) has taken the world of VoIP by storm. Originally considered little more than an interesting idea, SIP now seems poised to dethrone the mighty H.323 as the VoIP protocol of choice—certainly at the endpoints of the network. The premise of SIP is that each end of a connection is a peer, and the protocol negotiates capabilities between them. What makes SIP compelling is that it is a relatively simple protocol, with a syntax similar to that of other familiar protocols such as HTTP and SMTP.

SIP is supported in Asterisk with the *chan_sip.so* module.

History

SIP was originally submitted to the Internet Engineering Task Force (IETF) in February of 1996 as "draft-ietf-mmusic-sip-00." The initial draft looked nothing like the SIP we know today and contained only a single request type: a call setup request. In March of 1999, after 11 revisions, SIP RFC 2543 was born.

At first, SIP was all but ignored, as H.323 was considered the protocol of choice for VoIP transport negotiation. However, as the buzz grew, SIP began to gain popularity, and while there may be a lot of different factors that accelerated its growth, we'd like to think that a large part of its success is due to its freely available specification.

Future

SIP has earned its place as the protocol that justified VoIP. All new user and enterprise products are expected to support SIP, and any existing products will now be a tough sell unless a migration path to SIP is offered. SIP is widely expected to deliver far more than VoIP capabilities, including the ability to transmit video, music, and any type of real-time multimedia. SIP is poised to deliver the majority of new applications over the next few years.

Security considerations

SIP uses a challenge/response system to authenticate users. An initial INVITE is sent to the proxy with which the end device wishes to communicate. The proxy then sends back a 407 Proxy Authorization Request message, which contains a random set of characters referred to as a "nonce." This nonce is used along with the password to generate an MD5 hash, which is then sent back in the subsequent INVITE. Assuming the MD5 hash matches the one that the proxy generated, the client is then authenticated.

Denial of Service (DoS) attacks are probably the most common type of attack on VoIP communications. A DoS attack can occur when a large number of invalid INVITE requests are sent to a proxy server in an attempt to overwhelm the system. These attacks are relatively simple to implement, and their effects on the users of the system are immediate. SIP has several methods of minimizing the effects of DoS attacks, but ultimately they are impossible to prevent.

SIP implements a scheme to guarantee that a secure, encrypted transport mechanism (namely Transport Layer Security, or TLS) is used to establish communication between the caller and the domain of the callee. Beyond that, the request is sent securely to the end device, based upon the local security policies of the network. Note that the encryption of the media (that is, the RTP stream) is beyond the scope of SIP itself and must be dealt with separately.

More information regarding SIP security considerations, including registration hijacking, server impersonation, and session teardown, can be found in Section 26 of SIP RFC 3261.

SIP and NAT

Probably the biggest technical hurdle SIP has to conquer is the challenge of carrying out transactions across a NAT layer. Because SIP encapsulates addressing information in its data frames, and NAT happens at a lower network layer, the addressing information is not modified, and thus the media streams will not have the correct addressing information needed to complete the connection when NAT is in place. In addition to this, the firewalls normally integrated with NAT will not consider the incoming media stream to be part of the SIP transaction, and will block the connection.

H.323

This International Telecommunication Union (ITU) protocol was originally designed to provide an IP transport mechanism for video-conferencing. It has become the standard in IP-based video-conferencing equipment, and it briefly enjoyed fame as a VoIP protocol as well. While there is much heated debate over whether SIP or H.323 (or IAX) will dominate the VoIP protocol world, in Asterisk, H.323 has largely been deprecated in favour of IAX and SIP. H.323 has not enjoyed much success among users and enterprises, although it is still the most widely used VoIP protocol among carriers.

The two versions of H.323 supported in Asterisk are handled by the modules *chan_h323.so* (supplied with Asterisk) and *chan_oh323.so* (available as a free add-on).

 You have probably used H.323 without even knowing it—Microsoft's NetMeeting client is arguably the most widely deployed H.323 client.

History

H.323 was developed by the ITU in May of 1996 as a means to transmit voice, video, data, and fax communications across an IP-based network while maintaining connectivity with the PSTN. Since that time, H.323 has gone through several versions and annexes (which add functionality to the protocol), allowing it to operate in pure VoIP networks and more widely distributed networks.

Future

The future of H.323 is a subject of hot debate. If the media is any measure, it doesn't look good for H.323; it hardly ever gets mentioned (certainly not with the regularity of SIP). H.323 is commonly regarded as technically superior to SIP, but, as with so many other technologies, that ultimately might not matter. One of the factors that makes H.323 unpopular is its complexity—although many argue that the once-simple SIP is starting to suffer from the same problem.

H.323 still carries by far the majority of worldwide carrier VoIP traffic, but as people become less and less dependent on traditional carriers for their telecom needs, the future of H.323 becomes more difficult to predict with any certainty. While H.323 may not be the protocol of choice for new implementations, we can certainly expect to have to deal with H.323 interoperability issues for some time to come.

Security considerations

H.323 is a relatively secure protocol and does not require many security consider-ations beyond those that are common to any network communicating with the Inter-net. Since H.323 uses the RTP protocol for media communications, it does not natively support encrypted media paths. The use of a VPN or other encrypted tunnel between endpoints is the most common way of securely encapsulating communica-tions. Of course, this has the disadvantage of requiring the establishment of these secure tunnels between endpoints, which may not always be convenient (or even possible). As VoIP becomes used more often to communicate with financial institu-tions such as banks, we're likely to require extensions to the most commonly used VoIP protocols to natively support strong encryption methods.

H.323 and NAT

The H.323 standard uses the Internet Engineering Task Force (IETF) RTP protocol to transport media between endpoints. Because of this, H.323 has the same issues as SIP when dealing with network topologies involving NAT. The easiest method is to simply forward the appropriate ports through your NAT device to the internal client.

To receive calls, you will always need to forward TCP port 1720 to the client. In addition, you will need to forward the UDP ports for the RTP media and RTCP con-trol streams (see the manual for your device for the port range it requires). Older cli-ents, such as MS Netmeeting, will also require TCP ports forwarded for H.245 tunneling (again, see your client's manual for the port number range).

If you have a number of clients behind the NAT device, you will need to use a *gate-keeper* running in proxy mode. The gatekeeper will require an interface attached to the private IP subnet and the public Internet. Your H.323 client on the private IP subnet will then register to the gatekeeper, which will proxy calls on the clients' behalf. Note that any external clients that wish to call you will also be required to register with the proxy server.

At this time, Asterisk can't act as an H.323 gatekeeper. You'll have to use a separate application, such as the open source OpenH323 Gatekeeper (*http://www.gnugk.org*).

MGCP

The Media Gateway Control Protocol (MGCP) also comes to us from the IETF. While MGCP deployment is more widespread than one might think, it is quickly los-ing ground to protocols such as SIP and IAX. Still, Asterisk loves protocols, so natu-rally it has rudimentary support for it.

MGCP is defined in RFC 3435.* It was designed to make the end devices (such as phones) as simple as possible, and have all the call logic and processing handled by media gateways and call agents. Unlike SIP, MGCP uses a centralized model. MGCP phones cannot directly call other MGCP phones; they must always go through some type of controller.

Asterisk supports MGCP through the *chan_mgcp.so* module, and the endpoints are defined in the configuration file *mgcp.conf*. Since Asterisk provides only basic call agent services, it cannot emulate an MGCP phone (to register to another MGCP controller as a user agent, for example).

If you have some MGCP phones lying around, you will be able to use them with Asterisk. If you are planning to put MGCP phones into production on an Asterisk system, keep in mind that the community has moved on to more popular protocols, and you will therefore need to budget your software support needs accordingly. If possible (for example, with Cisco phones), you should upgrade MGCP phones to SIP.

Proprietary Protocols

Finally, let's take a look at two proprietary protocols that are supported in Asterisk.

Skinny/SCCP

The Skinny Client Control Protocol (SCCP) is proprietary to Cisco VoIP equipment. It is the default protocol for endpoints on a Cisco Call Manager PBX. Skinny is supported in Asterisk, but if you are connecting Cisco phones to Asterisk, it is generally recommended that you obtain SIP images for any phones that support it and connect via SIP instead.

UNISTIM

Support for Nortel's proprietary VoIP protocol, UNISTIM, has recently been added to Asterisk. This remarkable milestone means that Asterisk is the first PBX in history to natively support proprietary IP terminals from the two biggest players in VoIP, Nortel and Cisco.

Codecs

Codecs are generally understood to be various mathematical models used to digitally encode (and compress) analog audio information. Many of these models take into account the human brain's ability to form an impression from incomplete information. We've all seen optical illusions; likewise, voice-compression algorithms take

---

* RFC 3435 obsoletes RFC 2705.

advantage of our tendency to interpret what we *believe* we should hear, rather than what we *actually* hear.* The purpose of the various encoding algorithms is to strike a balance between efficiency and quality.†

Originally, the term CODEC referred to a COder/DECoder: a device that converts between analog and digital. Now, the term seems to relate more to COmpression/DECompression.

Before we dig into the individual codecs, take a look at Table 8-1—it's a quick reference that you may want to refer back to.

Table 8-1. Codec quick reference

Codec	Data bitrate (kbps)	Licence required?
G.711	64 kbps	No
G.726	16, 24, or 32 kbps	No
G.723.1	5.3 or 6.3 kbps	Yes (no for passthrough)
G.729A	8 kbps	Yes (no for passthrough)
GSM	13 kbps	No
iLBC	13.3 kbps (30-ms frames) or 15.2 kbps (20-ms frames)	No
Speex	Variable (between 2.15 and 22.4 kbps)	No

G.711

G.711 is the fundamental codec of the PSTN. In fact, if someone refers to PCM (discussed in the previous chapter) with respect to a telephone network, you are allowed to think of G.711. Two companding methods are used: μ-law in North America and A-law in the rest of the world. Either one delivers an 8-bit word transmitted 8,000 times per second. If you do the math, you will see that this requires 64,000 bits to be transmitted per second.

Many people will tell you that G.711 is an uncompressed codec. This is not exactly true, as companding is considered a form of compression. What is true is that G.711 is the base codec from which all of the others are derived.

---

* "Aoccdrnig to rsereach at an Elingsh uinervtisy, it deosn't mttaer in waht oredr the ltteers in a wrod are, the olny iprmoetnt tihng is taht frist and lsat ltteres are in the rghit pclae. The rset can be a toatl mses and you can sitll raed it wouthit a porbelm. Tihs is bcuseae we do not raed ervey lteter by istlef, but the wrod as a wlohe." (The source of this quote is unknown—see *http://www.bisso.com/ujg_archives/000228.html*.) Tihs is ture with snoud, too.

† On an audio CD, quality is far more important than bandwidth, so the audio is quantized at 16 bits (times 2, as it's stereo), with a sampling rate of 44,100 Hz. Considering that the CD was invented in the late 1970s, this was quite impressive stuff. The telephone network does not require this level of quality (and needs to optimize bandwidth), so telephone signals are encoded using 8 bits, at a sampling frequency of 8,000 Hz.

G.726

This codec has been around for some time (it used to be G.721, which is now obsolete), and it is one of the original compressed codecs. It is also known as Adaptive Differential Pulse-Code Modulation (ADPCM), and it can run at several bitrates. The most common rates are 16 kbps, 24 kbps, and 32 kbps. As of this writing, Asterisk currently supports only the ADPCM-32 rate, which is far and away the most popular rate for this codec.

G.726 offers quality nearly identical to G.711, but it uses only half the bandwidth. This is possible because rather than sending the result of the quantization measurement, it sends only enough information to describe the difference between the current sample and the previous one. G.726 fell from favor in the 1990s due to its inability to carry modem and fax signals, but because of its bandwidth/CPU performance ratio it is now making a comeback. G.726 is especially attractive because it does not require a lot of computational work from the system.

G.723.1

Not to be confused with G.723 (which is another obsolete version of ADPCM), this codec is designed for low-bitrate speech. It has two data bitrate settings: 5.3 kbps and 6.3 kbps. G.723.1 is one of the codecs required for compliance with the H.323 protocol (although other codecs may be employed by H.323). It is currently encumbered by patents and thus requires licensing if used in commercial applications. What this means is that while you can switch two G.723.1 calls through your Asterisk system, you are not allowed to decode them without a license.

G.729A

Considering how little bandwidth it uses, G.729A delivers impressive sound quality. It does this through the use of Conjugate-Structure Algebraic-Code-Excited Linear Prediction (CS-ACELP).[*] Because of patents, you can't use G729A without paying a licensing fee; however, it is extremely popular and is thus well supported on many different phones and systems.

To achieve its impressive compression ratio, this codec requires an equally impressive amount of effort from the CPU. In an Asterisk system, the use of heavily compressed codecs will quickly bog down the CPU.

G.729A uses 8 kbps of bandwidth.

---

[*] CELP is a popular method of compressing speech. By mathematically modeling the various ways humans make sounds, a codebook of sounds can be built. Rather than sending an actual sampled sound, a code corresponding to the sound is then sent. (Of course, there is much more to it than that.) Jason Woodward's Speech Coding page (*http://www-mobile.ecs.soton.ac.uk/speech_codecs/*) is a source of helpful information for the non-mathematically inclined. This is fairly heavy stuff, though, so wear your thinking cap.

GSM

GSM is the darling codec of Asterisk. This codec does not come encumbered with a licensing requirement the way that G.723.1 and G.729A do, and it offers outstanding performance with respect to the demand it places on the CPU. The sound quality is generally considered to be of a lesser grade than that produced by G.729A, but as much of this comes down to personal opinion, be sure to try it out.

GSM operates at 13 kbps.

iLBC

The Internet Low Bitrate Codec (iLBC) provides an attractive mix of low bandwidth usage and quality, and it is especially well suited to sustaining reasonable quality on lossy network links.

Naturally, Asterisk supports it (and support elsewhere is growing), but it is not as popular as the ITU codecs and thus may not be compatible with common IP telephones and commercial VoIP systems. IETF RFCs 3951 and 3952 have been published in support of iLBC, and iLBC is on the IETF standards track.

Because iLBC uses complex algorithms to achieve its high levels of compression, it has a fairly high CPU cost in Asterisk.

While you are allowed to use iLBC without paying royalty fees, the holder of the iLBC patent, Global IP Sound (GIPS), wants to know whenever you use it in a commercial application. The way you do that is by downloading and printing a copy of the iLBC license, signing it, and returning it to them. If you want to read about iLBC and its license, you can do so at *http://www.ilbcfreeware.org*.

iLBC operates at 13.3 kbps (30-ms frames) and 15.2 kbps (20-ms frames).

Speex

Speex is a Variable Bitrate (VBR) codec, which means that it is able to dynamically modify its bitrate to respond to changing network conditions. It is offered in both narrowband and wideband versions, depending on whether you want telephone quality or better.

Speex is a totally free codec, licensed under the Xiph.org variant of the BSD license.

An Internet draft for Speex is available, and more information about Speex can be found at its home page (*http://www.speex.org*).

Speex can operate at anywhere from 2.15 to 22.4 kbps, due to its variable bitrate

MP3

Sure thing, MP3 is a codec. Specifically, it's the Moving Picture Experts Group Audio Layer 3 Encoding Standard.* With a name like that, it's no wonder we call it MP3! In Asterisk, the MP3 codec is typically used for Music on Hold (MoH). MP3 is not a telephony codec, as it is optimized for music, not voice; nevertheless, it's very popular with VoIP telephony systems as a method of delivering Music on Hold.

 Be aware that music cannot usually be broadcast without a license. Many people assume that there is no legal problem with connecting a radio station or CD as a Music on Hold source, but this is very rarely true.

Quality of Service

Quality of Service, or *QoS* as it's more popularly termed, refers to the challenge of delivering a time-sensitive stream of data across a network that was designed to deliver data in an ad hoc, best-effort sort of way. Although there is no hard rule, it is generally accepted that if you can deliver the sound produced by the speaker to the listener's ear within 300 milliseconds, a normal flow of conversation is possible. When delay exceeds 500 milliseconds, it becomes difficult to avoid interrupting each other. Beyond one second, normal conversation becomes extremely awkward.

In addition to getting it there on time, it is also essential to ensure that the transmitted information arrives intact. Too many lost packets will prevent the far end from completely reproducing the sampled audio, and gaps in the data will be heard as static or, in severe cases, entire missed words or sentences.

TCP, UDP, and SCTP

If you're going to send data on an IP-based network, it will be transported using one of the three transport protocols discussed here.

Transmission Control Protocol

The Transmission Control Protocol (TCP) is almost never used for VoIP, for while it does have mechanisms in place to ensure delivery, it is not inherently in any hurry to do so. Unless you have an extremely low-latency interconnection between the two endpoints, TCP is going to tend to cause more problems than it solves.

The purpose of TCP is to guarantee the delivery of packets. In order to do this, several mechanisms are implemented, such as packet numbering (for reconstructing

---

* If you want to learn all about MPEG audio, do a web search for Davis Pan's paper entitled "A Tutorial on MPEG/Audio Compression."

blocks of data), delivery acknowledgment, and re-requesting lost packets. In the world of VoIP, getting the packets to the endpoint quickly is paramount—but 20 years of cellular telephony has trained us to tolerate a few lost packets.*

TCP's high processing overhead, state management, and acknowledgment of arrival work well for transmitting large amounts of data, but it simply isn't efficient enough for real-time media communications.

User Datagram Protocol

Unlike TCP, the User Datagram Protocol (UDP) does not offer any sort of delivery guarantee. Packets are placed on the wire as quickly as possible and released into the world to find their way to their final destinations, with no word back as to whether they get there or not. Since UDP itself does not offer any kind of guarantee that the data will arrive,† it achieves its efficiency by spending very little effort on what it is transporting.

 TCP is a more "socially responsible" protocol, because the bandwidth is more evenly distributed to clients connecting to a server. As the percentage of UDP traffic increases, it is possible that a network could become overwhelmed.

Stream Control Transmission Protocol

Approved by the IETF as a proposed standard in RFC 2960, SCTP is a relatively new transport protocol. From the ground up, it was designed to address the shortcomings of both TCP and UDP, especially as related to the types of services that used to be delivered over circuit-switched telephony networks.

Some of the goals of SCTP were:

- Better congestion-avoidance techniques (specifically, avoiding Denial of Service attacks)
- Strict sequencing of data delivery
- Lower latency for improved real-time transmissions

By overcoming the major shortcomings of TCP and UDP, the SCTP developers hoped to create a robust protocol for the transmission of SS7 and other types of PSTN signaling over an IP-based network.

---

* The order of arrival is important in voice communication, because the audio will be processed and sent to the caller ASAP. However, with a jitter buffer the order of arrival isn't as important, as it provides a small window of time in which the packets can be reordered before being passed on to the caller.

† Keep in mind that the upper-layer protocols or applications can implement their own packet acknowledgment systems.

Differentiated Service

Differentiated service, or DiffServ, is not so much a QoS mechanism as a method by which traffic can be flagged and given specific treatment. Obviously, DiffServ can help to provide QoS by allowing certain types of packets to take precedence over others. While this will certainly increase the chance of a VoIP packet passing quickly through each link, it does not guarantee anything.

Guaranteed Service

The ultimate guarantee of QoS is provided by the PSTN. For each conversation, a 64-kbps channel is completely dedicated to the call—the bandwidth is guaranteed. Similarly, protocols that offer guaranteed service can ensure that a required amount of bandwidth is dedicated to the connection being served. As with any packetized networking technology, these mechanisms generally operate best when traffic is below maximum levels. When a connection approaches its limits, it is next to impossible to eliminate degradation.

MPLS

Multiprotocol Label Switching (MPLS) is a method for engineering network traffic patterns independent of layer-3 routing tables. The protocol works by assigning short labels (MPLS frames) to network packets, which routers then use to forward the packets to the MPLS egress router, and ultimately to their final destinations. Traditionally, routers make an independent forwarding decision based on an IP table lookup at each hop in the network. In an MPLS network, this lookup is performed only once, when the packet enters the MPLS cloud at the ingress router. The packet is then assigned to a stream, referred to as a Label Switched Path (LSP), and identified by a label. The label is used as a lookup index in the MPLS forwarding table, and the packet traverses the LSP independent of layer-3 routing decisions. This allows the administrators of large networks to fine-tune routing decisions and to make the best use of network resources. Additionally, information can be associated with a label to prioritize packet forwarding.

RSVP

MPLS contains no method to dynamically establish LSPs, but you can use the Reservation protocol (RSVP) with MPLS. RSVP is a signaling protocol used to simplify the establishment of LSPs and to report problems to the MPLS ingress router. The advantage of using RSVP in conjunction with MPLS is the reduction in administrative overhead. If you don't use RSVP with MPLS, you'll have to go to every single router and configure the labels and each path manually. Using RSVP makes the network more dynamic by distributing control of labels to the routers. This enables the network to become more responsive to changing conditions, because it can be set up to change the paths based on certain conditions, such as a certain path going down (perhaps due to a faulty router). The configuration within the router will then be able to use RSVP to distribute new labels to the routers in the MPLS network, with no (or minimal) human intervention.

Best Effort

The simplest, least expensive approach to QoS is not to provide it at all—the "best effort" method. While this might sound like a bad idea, it can in fact work very well. Any VoIP call that traverses the public Internet is almost certain to be best effort, as QoS mechanisms are not yet common in this environment.

Echo

You may not realize it, but echo has been a problem in the PSTN for as long as there have been telephones. You probably haven't often experienced it, because the telecom industry has spent large sums of money designing expensive echo cancellation devices. Also, when the endpoints are physically close—e.g., when you phone your neighbor down the street—the delay is so minimal that anything you transmit will be returned back so quickly that it will be indistinguishable from the sidetone* normally occurring in your telephone.

Why Echo Occurs

Before we discuss measures to deal with echo, let's first take a look at why echo occurs in the analog world.

If you hear echo, it's not your phone that's causing the problem; it's the far end of the circuit. Conversely, echo heard on the far end is being generated at your end. Echo is caused by the fact that an analog local loop circuit has to transmit and receive on the same pair of wires. If this circuit is not electrically balanced, or if a low-quality telephone is connected to the end of the circuit, signals it receives can be reflected back, becoming part of the return transmission. When this reflected circuit gets back to you, you will hear the words you spoke just moments before. The human ear will perceive an echo after a delay of roughly 40 milliseconds.

In a cheap telephone, it is possible for echo to be generated in the body of the handset. This is why some cheap IP phones can cause echo even when the entire end-to-end connection does not contain an analog circuit.† In the VoIP world, echo is usually introduced either by an analog circuit somewhere in the connection, or by a

---

* As discussed in Chapter 7, sidetone is a function in your telephone that returns part of what you say back to your own ear, to provide a more natural-sounding conversation.

† Actually, the handset in any phone, be it traditional or VoIP, is an analog connection.

cheap endpoint reflecting back some of the signal (e.g., feedback through a hands-free or poorly designed handset). A good rule of thumb is to keep latency to less than 250 milliseconds.

Managing Echo

In the *zconfig.h* configuration file, you can choose from one of several echo canceller algorithms, with the default being MARK2. Experiment with the various echo cancellers on your network to determine the best one for your environment. Asterisk also has an option in the *zconfig.h* file to make the echo cancellation more aggressive. You can enable it by uncommenting the following line:

```
#define AGGRESSIVE_SUPPRESSOR
```

Note that aggressive echo cancellation can create a walkie-talkie, half-duplex effect. This should be enabled only if all other methods of reducing echo have failed.

Enable echo cancellation for Zaptel interfaces in the *zapata.conf* file. The default configuration enables echo cancellation with echocancel=yes. echocancelwhenbridged=yes will enable echo cancellation for TDM bridged calls. While bridged calls should not require echo cancellation, this may improve call quality.

When echo cancellation is enabled, the echo canceller learns of echo on the line by listening for it for the duration of the call. Consequently, echo may be heard at the beginning of a call and eventually lessen after period of time. To avoid this situation, you can employ a method called "echo training," which will mute the line briefly at the beginning of a call, and then send a tone from which the amount of echo on the line can be determined. This allows Asterisk to deal with the echo more quickly. Echo training can be enabled with echotraining=yes.

Asterisk and VoIP

It should come as no surprise that Asterisk loves to talk VoIP. But in order to do so, Asterisk needs to know which function it is to perform: that of client, server, or both. One of the most complex and often confusing concepts in Asterisk is the naming scheme of inbound and outbound authentication.

Users and Peers and Friends—Oh My!

Connections that authenticate to us, or that we authenticate, are defined in the *iax. conf* and *sip.conf* files as *users* and *peers*. Connections that do both may be defined as *friends*. When determining which way the authentication is occurring, it is always important to view the direction of the channels from Asterisk's viewpoint, as connections are being accepted and created by the Asterisk server.

Users

A connection defined as a user is any system/user/endpoint that we allow to connect to us. Keep in mind that a user definition does not provide a method with which to call that user—the user type is used simply to create a channel for incoming calls.* A user definition will require a context name to be defined to indicate where the incoming authenticated call will be placed in the dialplan (in *extensions.conf*).

Peers

A connection defined as a peer type is an outgoing connection. Think of it this way: *users* place calls to us, while we place calls to our *peers*. Since peers do not place calls to us, a peer definition does not typically require the configuration of a context name. However, there is one exception: if calls that originate from your system are returned to your system in a loopback, the incoming calls (which originate from a SIP proxy, not a user agent) will be matched on the peer definition. The default context should handle these incoming calls appropriately, although it's preferable for contexts to be defined for them on a per-peer basis.†

In order to know where to send a call to a host, we must know its location in relation to the Internet (that is, its IP address). The location of a peer may be defined either statically or dynamically. A dynamic peer is configured with host=dynamic under the peer definition heading. Because the IP address of a dynamic peer may change constantly, it must register with the Asterisk box to let it know what its IP address is, so calls can successfully be routed to it. If the remote end is another Asterisk box, the use of a register statement is required, as discussed below.

Friends

Defining a type as a friend is a shortcut for defining it as both a user and a peer. However, connections that are both a user and a peer aren't always defined this way, because defining each direction of call creation individually (using both a user and a peer definition) allows more granularity and control over the individual connections.

Figure 8-1 shows the flow of authentication control in relation to Asterisk.

---

* In SIP, this is not *always* the case. If the endpoint is a SIP proxy service (as opposed to a user agent), Asterisk will authenticate based on the peer definition, matching the IP address and port in the Contact field of the SIP header against the hostname (and port, if specified) defined for the peer (if the port is not specified, the one defined in the [general] section will be used). See the discussion of the SIP insecure option in Appendix A for more on this subject.

† For more information on this topic, see the discussion of the SIP context option in Appendix A.

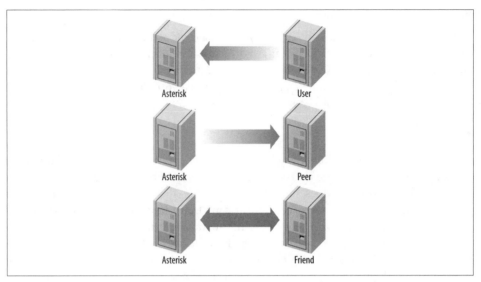

Figure 8-1. Call origination relationships of users, peers, and friends to Asterisk

register Statements

A register statement is a way of telling a remote peer where your Asterisk box is in relation to the Internet. Asterisk uses register statements to authenticate to remote providers when you are employing a dynamic IP address, or when the provider does not have your IP address on record. There are situations when a register statement is not required, but to demonstrate when a register statement *is* required, let's look at an example.

Say you have a remote peer that is providing DID services to you. When someone calls the number +1-800-555-1212, the call goes over the physical PSTN network to your service provider and into their Asterisk server, possibly over their T-1 connection. This call is then routed to your Asterisk server via the Internet.

Your service provider will have a definition in either their *sip.conf* or *iax.conf* configuration file (depending on whether you are connecting with the SIP or IAX protocol, respectively) for your Asterisk server. If you receive calls only from this provider, you would define them as a user (if they were another Asterisk system, you might be defined in their system as a peer).

Now let's say that your box is on your home Internet connection, with a dynamic IP address. Your service provider has a static IP address (or perhaps a fully qualified domain name), which you place in your configuration file. Since you have a dynamic address, your service provider specifies host=dynamic in its configuration file. In order to know where to route your +1-800-555-1212 call, your service provider needs to know where you are located in relation to the Internet. This is where the register statement comes into use.

The register statement is a way of authenticating and telling your peer where you are. In the [general] section of your configuration file, you would place a statement similar to this:

```
register => username:secret@my_remote_peer
```

You can verify a successful register with the use of the iax2 show registry and sip show registry commands at the Asterisk console.

Conclusion

If you listen to the buzz in the telecom industry, you might think that VoIP is the future of telephony. But to Asterisk, VoIP is more a case of "been there, done that." For Asterisk, the future of telephony is much more exciting. We'll take a look at that vision a bit later, in Chapter 11. In the next chapter, we are going to delve into one of the more revolutionary and powerful concepts of Asterisk: AGI, the Asterisk Gateway Interface.

## CHAPTER 9

The Asterisk Gateway Interface (AGI)

Even he, to whom most things that most people
would think were pretty smart were pretty dumb,
thought it was pretty smart.
—Douglas Adams, *The Salmon of Doubt*

The Asterisk Gateway Interface, or AGI, provides a standard interface by which external programs may control the Asterisk dialplan. Usually, AGI scripts are used to do advanced logic, communicate with relational databases (such as PostgreSQL or MySQL), and access other external resources. Turning over control of the dialplan to an external AGI script enables Asterisk to easily perform tasks that would otherwise be difficult or impossible.

This chapter covers the fundamentals of AGI communication. It will not teach you how to be a programmer—rather, we'll assume that you're already a competent programmer, so that we can show you how to write AGI programs. If you don't know how to do computer programming, this chapter probably isn't for you, and you should skip ahead to the next chapter.

Over the course of this chapter, we'll write a sample AGI program in each of the Perl, PHP, and Python programming languages. Note, however, that because Asterisk provides a standard interface for AGI scripts, these scripts can be written in almost any modern programming language. We've chosen to highlight Perl, PHP, and Python because they're the languages most commonly used for AGI programming.

Fundamentals of AGI Communication

Instead of releasing an API for programming, AGI scripts communicate with Asterisk over communications channels (file pointers, in programming parlance) known as STDIN, STDOUT, and STDERR. Most computer programmers will recognize these channels, but just in case you're not familiar with them we'll cover them here.

What Are STDIN, STDOUT, and STDERR?

STDIN, STDOUT, and STDERR are channels by which programs in Unix-like environments receive information from and send information to external programs. STDIN, or "standard input," is the information that is sent to the program, either from the keyboard or from another program. In our case, information coming from Asterisk itself comes in on the program's STDIN file handle. STDOUT, or "standard output," is the file handle that the AGI script uses to pass information back to Asterisk. Finally, the AGI script can use the STDERR ("standard error") file handle to write error messages to the Asterisk console.

Let's sum up these three communications concepts:

- An AGI script reads from STDIN to get information from Asterisk.
- An AGI script writes data to STDOUT to send information to Asterisk.
- An AGI script may write data to STDERR to send debug information to the Asterisk console.

At this time, writing to STDERR from within your AGI script writes the information only to the *first* Asterisk console—that is, the first Asterisk console started with the -c or -r parameters.

This is rather unfortunate, and will hopefully be remedied soon by the Asterisk developers.

If you're using the *safe_asterisk* program to start Asterisk (which you probably are), it starts a remote console on TTY9. (Try pressing Ctrl-Alt-F9, and see if you get an Asterisk command-line interface.) This means that all the AGI debug information will print on only that remote console. You may want to disable this console in *safe_asterisk* to allow you to see the debug information in another console. (You may also want to disable that console for security reasons, as you might not want just anyone to be able to walk up to your Asterisk server and have access to a console without any kind of authentication.)

The Standard Pattern of AGI Communication

The communication between Asterisk and an AGI script follows a predefined pattern. Let's enumerate the steps, and then we'll walk through one of the sample AGI scripts that come with Asterisk.

When an AGI script starts, Asterisk sends a list of variables and their values to the AGI script. The variables might look something like this:

```
agi_request: test.py
agi_channel: Zap/1-1
agi_language: en
agi_callerid:
agi_context: default
agi_extension: 123
agi_priority: 2
```

After sending these variables, Asterisk sends a blank line. This is the signal that Asterisk is done sending the variables and it is time for the AGI script to control the dialplan.

At this point, the AGI script sends commands to Asterisk by writing to STDOUT. After the script sends each command, Asterisk sends a response that the AGI script should read. This action (sending commands to Asterisk and reading the responses) can continue for the duration of the AGI script.

You may be asking yourself what commands you can use from within your AGI script. Good question—we'll cover the basic commands shortly.*

Calling an AGI Script from the Dialplan

In order to work properly, your AGI script must be executable. To use an AGI script inside your dialplan, simply call the AGI( ) application, with the name of the AGI script as the argument, like this:

```
exten => 123,1,Answer( )
exten => 123,2,AGI(agi-test.agi)
```

AGI scripts often reside in the AGI directory (usually located in */var/lib/asterisk/agi-bin*), but you can specify the complete path to the AGI script.

AGI( ), EAGI( ), DeadAGI( ), and FastAGI( )

In addition to the AGI( ) application, there are several other AGI applications suited to different circumstances. While they won't be covered in this chapter, they should be quite simple to figure out once you understand the basics of AGI scripting.

The EAGI( ) (enhanced AGI) application acts just like AGI( ), but allows your AGI script to read the inbound audio stream on file descriptor number three.

The DeadAGI( ) application is also just like AGI( ), but it works correctly on a channel that is dead (i.e., a channel that has been hung up). As this implies, the regular AGI( ) application doesn't work on dead channels.

The FastAGI( ) application allows the AGI script to be called across the network, so that multiple Asterisk servers can call AGI scripts from a central location.

In this chapter, we'll first cover the sample *agi-test.agi* script that comes with Asterisk (which was written in Perl), then write a weather report AGI program in PHP, and then finish up by writing an AGI program in Python to play a math game.

---

* To get a list of available AGI commands, type **show agi** at the Asterisk command-line interface. You can also refer to Appendix C for an AGI command reference.

Writing AGI Scripts in Perl

Asterisk comes with a sample AGI script called *agi-test.agi*. Let's step through the file while we cover the core concepts of AGI programming. While this particular script is written in Perl, please remember that your own AGI programs may be written in almost any programming language. Just to prove it, we're going to cover AGI programming in a couple of other languages later in the chapter.

Let's get started! We'll look at each section of the code in turn, and describe what it does.

```
#!/usr/bin/perl
```

This line tells the system that this particular script is written in Perl, so it should use the Perl interpreter to execute the script. If you've done much Linux or Unix scripting, this line should be familiar to you. This line assumes, of course, that your Perl binary is located in the */usr/bin/* directory. Change this to match the location of your Perl interpreter.

```
use strict;
```

use strict tells Perl to act, well, strict about possible programming errors, such as undeclared variables. While not absolutely necessary, enabling this will help you avoid common programming pitfalls.

```
$|=1;
```

This line tells Perl not to buffer its output—in other words, that it should write any data immediately, instead of waiting for a block of data before outputting it. You'll see this as a recurrent theme throughout the chapter.

> You should *always* use unbuffered output when writing AGI scripts. Otherwise, your AGI may not work as expected, because Asterisk may be waiting for the output of your program, while your program thinks it has sent the output to Asterisk and is waiting for a response.

```
# Set up some variables
my %AGI; my $tests = 0; my $fail = 0; my $pass = 0;
```

Here, we set up four variables. The first is a hash called AGI, which is used to store the variables that Asterisk passes to our script at the beginning of the AGI session. The next three are scalar values, used to count the total number of tests, the number of failed tests, and the number of passed tests, respectively.

```
while(<STDIN>) {
        chomp;
        last unless length($_);
        if (/^agi_(\w+)\:\s+(.*)$/) {
                $AGI{$1} = $2;
        }
}
```

As we explained earlier, Asterisk sends a group of variables to the AGI program at startup. This loop simply takes all of these variables and stores them in the hash named AGI. They can be used later in the program or simply ignored, but they should always be read from STDIN before continuing on with the logic of the program.

```
print STDERR "AGI Environment Dump:\n";
foreach my $i (sort keys %AGI) {
        print STDERR " -- $i = $AGI{$i}\n";
}
```

This loop simply writes each of the values that we stored in the AGI hash to STDERR. This is useful for debugging the AGI script, as STDERR is printed to the Asterisk console.*

```
sub checkresult {
        my ($res) = @_;
        my $retval;
        $tests++;
        chomp $res;
        if ($res =~ /^200/) {
                $res =~ /result=(-?\d+)/;
                if (!length($1)) {
                        print STDERR "FAIL ($res)\n";
                        $fail++;
                } else {
                        print STDERR "PASS ($1)\n";
                        $pass++;
                }
        } else {
                print STDERR "FAIL (unexpected result '$res')\n";
                $fail++;
        }
}
```

This subroutine reads in the result of an AGI command from Asterisk and decodes the result to determine whether the command passes or fails.

Now that the preliminaries are out of the way, we can get to the core logic of the AGI script.

```
print STDERR "1.  Testing 'sendfile'...";
print "STREAM FILE beep \"\"\n";
my $result = <STDIN>;
&checkresult($result);
```

This first test shows how to use the STREAM FILE command. The STREAM FILE command tells Asterisk to play a sound file to the caller, just as the Background( ) application does. In this case, we're telling Asterisk to play a file called *beep.gsm*.†

---

* Actually, to the first spawned Asterisk console (i.e., the first instance of Asterisk called with the –c or –r option). If *safe_asterisk* was used to start Asterisk, the first Asterisk console will be on TTY9, which means that you will not be able to view AGI errors remotely.

† Asterisk automatically selects the best format, based on translation cost and availability, so the file extension is never used in the function.

You will notice that the second argument is passed by putting in a set of double quotes, escaped by backslashes. Without the double quotes to indicate the second argument, this command does not work correctly.

You must pass *all required arguments* to the AGI commands. If you want to skip a required argument, you must send empty quotes (properly escaped in your particular programming language), as shown above. If you don't pass the required number of arguments, your AGI script will not work.

You should also make sure you pass a line feed (the \n on the end of the print statement) at the end of the command.

After sending the STREAM FILE command, this test reads the result from STDIN and calls the checkresult subroutine to determine if Asterisk was able to play the file. The STREAM FILE command takes three arguments, two of which are required:

- The name of the sound file to play back
- The digits that may interrupt the playback
- The position at which to start playing the sound, specified in number of samples (optional)

In short, this test told Asterisk to play back the file named *beep.gsm*, and then checked the result to make sure the command was successfully executed by Asterisk.

```
print STDERR "2.  Testing 'sendtext'...";
print "SEND TEXT \"hello world\"\n";
my $result = <STDIN>;
&checkresult($result);
```

This test shows us how to call the SEND TEXT command, which is similar to the SendText( ) application. This command will send the specified text to the caller, if the caller's channel type supports the sending of text.

The SEND TEXT command takes one argument: the text to send to the channel. If the text contains spaces (as in the example above), the argument should be encapsulated with quotes, so that Asterisk will know that the entire text string is a single argument to the command. Again, notice that the quotation marks are escaped, as they must be sent to Asterisk, not used to terminate the string in Perl.

```
print STDERR "3.  Testing 'sendimage'...";
print "SEND IMAGE asterisk-image\n";
my $result = <STDIN>;
&checkresult($result);
```

This test calls the SEND IMAGE command, which is similar to the SendImage( ) application. Its single argument is the name of an image file to send to the caller. As with the SEND TEXT command, this command works only if the calling channel supports the reception of images.

```
print STDERR "4.  Testing 'saynumber'...";
print "SAY NUMBER 192837465 \"\"\n";
my $result = <STDIN>;
&checkresult($result);
```

This test sends Asterisk the SAY NUMBER command. This command behaves identically to the SayNumber( ) dialplan application. It takes two arguments:

- The number to say
- The digits that may interrupt the command

Again, since we're not passing in any digits as the second argument, we need to pass in an empty set of quotes.

```
print STDERR "5.  Testing 'waitdtmf'...";
print "WAIT FOR DIGIT 1000\n";
my $result = <STDIN>;
&checkresult($result);
```

This test shows the WAIT FOR DIGIT command. This command waits the specified number of milliseconds for the caller to enter a DTMF digit. If you want the command to wait indefinitely for a digit, use -1 as the timeout. This application returns the decimal ASCII value of the digit that was pressed.

```
print STDERR "6.  Testing 'record'...";
print "RECORD FILE testagi gsm 1234 3000\n";
my $result = <STDIN>;
&checkresult($result);
```

This section of code shows us the RECORD FILE command. This command is used to record the call audio, similar to the Record( ) dialplan application. RECORD FILE takes seven arguments, the last three of which are optional:

- The filename of the recorded file.
- The format in which to record the audio.
- The digits that may interrupt the recording.
- The timeout (maximum recording time) in milliseconds, or -1 for no timeout.
- The number of samples to skip before starting the recording (optional).
- The word BEEP, if you'd like Asterisk to beep before the recording starts (optional).
- The number of seconds before Asterisk decides that the user is done with the recording and returns, even though the timeout hasn't been reached and no DTMF digits have been entered (optional). This argument must be preceded by s=.

In this particular case, we're recording a file called *testagi* (in the GSM format), with any of the DTMF digits 1 through 4 terminating the recording, and a maximum recording time of 3,000 milliseconds.

```
print STDERR "6a.  Testing 'record' playback...";
print "STREAM FILE testagi \"\"\n";
my $result = <STDIN>;
&checkresult($result);
```

The second part of this test plays back the audio that was recorded earlier, using the STREAM FILE command. We've already covered STREAM FILE, so this section of code needs no further explanation.

```
print STDERR "================== Complete =====================\n";
print STDERR "$tests tests completed, $pass passed, $fail failed\n";
print STDERR "================================================\n";
```

At the end of the AGI script, a summary of the tests is printed to STDERR, which should end up on the Asterisk console.

In summary, you should remember the following when writing AGI programs in Perl:

* Turn on strict language checking with the use strict command.*
* Turn off output buffering by setting $|=1.
* Data from Asterisk is received using a while(<STDIN>) loop.
* Write values with the print command.
* Use the print STDERR command to write debug information to the Asterisk console.

The Perl AGI Library

If you are interesting in building your own AGI scripts in Perl, you may want to check out the *Asterisk::AGI* Perl module written by James Golovich, which is located at *http://asterisk.gnuinter.net*. The *Asterisk::AGI* module makes it even easier to write AGI scripts in Perl.

Creating AGI Scripts in PHP

We promised we'd cover several languages, so let's go ahead and see what an AGI script in PHP looks like. The fundamentals of AGI programming still apply; only the programming language has changed. In this example, we'll write an AGI script to download a weather report from the Internet and deliver the temperature, wind direction, and wind speed back to the caller.

```
#!/usr/bin/php -q
<?php
```

The first line tells the system to use the PHP interpreter to run this script. The –q option turns off HTML error messages. You should ensure that there aren't any extra lines between the first line and the opening PHP tag, as they'll confuse Asterisk.

```
# change this to match the code of your particular city
# for a complete list of US cities, go to
# http://www.nws.noaa.gov/data/current_obs/
$weatherURL="http://www.nws.noaa.gov/data/current_obs/KMDQ.xml";
```

---

* This advice probably applies to any Perl program you might write, especially if you're new to Perl.

This tells our AGI script where to go to get the current weather conditions. In this example, we're getting the weather for Huntsville, Alabama. Feel free to visit the web site listed above for a complete list of stations throughout the United States of America.[*]

```
# don't let this script run for more than 60 seconds
set_time_limit(60);
```

Here, we tell PHP not to let this program run for more than 60 seconds. This is a safety net, which will end the script if for some reason it takes more than 60 seconds to run.

```
# turn off output buffering
ob_implicit_flush(false);
```

This command turns off output buffering, meaning that all data will be sent immediately to the AGI interface and will not be buffered.

```
# turn off error reporting, as it will most likely interfere with
# the AGI interface
error_reporting(0);
```

This command turns off all error reporting, as it can interfere with the AGI interface. (You might find it helpful to comment out this line during testing.)

```
# create file handles if needed
if (!defined('STDIN'))
{
    define('STDIN', fopen('php://stdin', 'r'));
}
if (!defined('STDOUT'))
{
    define('STDOUT', fopen('php://stdout', 'w'));
}
if (!defined('STDERR'))
{
    define('STDERR', fopen('php://stderr', 'w'));
}
```

This section of code ensures that we have open file handles for STDIN, STDOUT, and STDERR, which will handle all communication between Asterisk and our script.

```
# retrieve all AGI variables from Asterisk
while (!feof(STDIN))
{
    $temp = trim(fgets(STDIN,4096));
    if (($temp == "") || ($temp == "\n"))
    {
        break;
    }
}
```

[*] We apologize to our readers outside of the United States for using a weather service that only works for U.S. cities. If you can find a good international weather service that provides its data in XML, it shouldn't be too hard to change this AGI script to work with that particular service. Once we find one, we'll update this script for future editions.

```
    $s = split(":",$temp);
    $name = str_replace("agi_","",$s[0]);
    $agi[$name] = trim($s[1]);
}
```

Next, we'll read in all of the AGI variables passed to us by Asterisk. Using the fgets command in PHP to read the data from STDIN, we'll save each variable in the hash called $agi. Remember that we could use these variables in the logic of our AGI script, although we won't in this example.

```
# print all AGI variables for debugging purposes
foreach($agi as $key=>$value)
{
    fwrite(STDERR,"-- $key = $value\n");
    fflush(STDERR);
}
```

Here, we print the variables back out to STDERR for debugging purposes.

```
#retrieve this web page
$weatherPage=file_get_contents($weatherURL);
```

This line of code retrieves the XML file from the National Weather Service and puts the contents into the variable called $weatherPage. This variable will be used later on to extract out the pieces of the weather report that we want.

```
#grab temperature in Fahrenheit
if (preg_match("/<temp_f>([0-9]+)<\/temp_f>/i",$weatherPage,$matches))
{
    $currentTemp=$matches[1];
}
```

This section of code extracts the temperature (in Fahrenheit) from the weather report, using the preg_match command. This command uses Perl-compatible regular expressions[*] to extract out the needed data.

```
#grab wind direction
if (preg_match("/<wind_dir>North<\/wind_dir>/i",$weatherPage))
{
    $currentWindDirection='northerly';
}
elseif (preg_match("/<wind_dir>South<\/wind_dir>/i",$weatherPage))
{
    $currentWindDirection='southerly';
}
elseif (preg_match("/<wind_dir>East<\/wind_dir>/i",$weatherPage))
{
    $currentWindDirection='easterly';
}
elseif (preg_match("/<wind_dir>West<\/wind_dir>/i",$weatherPage))
{
    $currentWindDirection='westerly';
```

---

[*] The ultimate guide to regular expressions is O'Reilly's *Mastering Regular Expressions*, by Jeffrey Friedl.

```
    }
    elseif (preg_match("/<wind_dir>Northwest<\/wind_dir>/i",$weatherPage))
    {
        $currentWindDirection='northwesterly';
    }
    elseif (preg_match("/<wind_dir>Northeast<\/wind_dir>/i",$weatherPage))
    {
        $currentWindDirection='northeasterly';
    }
    elseif (preg_match("/<wind_dir>Southwest<\/wind_dir>/i",$weatherPage))
    {
        $currentWindDirection='southwesterly';
    }
    elseif (preg_match("/<wind_dir>Southeast<\/wind_dir>/i",$weatherPage))
    {
        $currentWindDirection='southeasterly';
    }
```

The wind direction is found through the use of preg_match (located in the wind_dir tags) and is assigned to the variable $currentWindDirection.

```
#grab wind speed
if (preg_match("/<wind_mph>([0-9.]+)<\/wind_mph>/i",$weatherPage,$matches))
{
    $currentWindSpeed = $matches[1];
}
```

Finally, we'll grab the current wind speed and assign it to the $currentWindSpeed variable.

```
# tell the caller the current conditions
if ($currentTemp)
{
    fwrite(STDOUT,"STREAM FILE temperature \"\"\n");
    fflush(STDOUT);
    $result = trim(fgets(STDIN,4096));
    checkresult($result);
    fwrite(STDOUT,"STREAM FILE is \"\"\n");
    fflush(STDOUT);
    $result = trim(fgets(STDIN,4096));
    checkresult($result);
    fwrite(STDOUT,"SAY NUMBER $currentTemp \"\"\n");
    fflush(STDOUT);
    $result = trim(fgets(STDIN,4096));
    checkresult($result);
    fwrite(STDOUT,"STREAM FILE degrees \"\"\n");
    fflush(STDOUT);
    $result = trim(fgets(STDIN,4096));
    checkresult($result);
    fwrite(STDOUT,"STREAM FILE fahrenheit \"\"\n");
    fflush(STDOUT);
    $result = trim(fgets(STDIN,4096));
    checkresult($result);
}
```

```
if ($currentWindDirection && $currentWindSpeed)
{
    fwrite(STDOUT,"STREAM FILE with \"\"\n");
    fflush(STDOUT);
    $result = trim(fgets(STDIN,4096));
    checkresult($result);
    fwrite(STDOUT,"STREAM FILE $currentWindDirection \"\"\n");
    fflush(STDOUT);
    $result = trim(fgets(STDIN,4096));
    checkresult($result);
    fwrite(STDOUT,"STREAM FILE wx/winds \"\"\n");
    fflush(STDOUT);
    $result = trim(fgets(STDIN,4096));
    checkresult($result);
    fwrite(STDOUT,"STREAM FILE at \"\"\n";)
    fflush(STDOUT);
    $result = trim(fgets(STDIN,4096));
    checkresult($result);
    fwrite(STDOUT,"SAY NUMBER $currentWindSpeed \"\"\n");
    fflush(STDOUT);
    $result = trim(fgets(STDIN,4096));
    checkresult($result);
    fwrite($STDOUT,"STREAM FILE miles-per-hour \"\"\n");
    fflush(STDOUT);
    $result = trim(fgets(STDIN,4096));
    checkresult($result);
}
```

Now that we've collected our data, we can send AGI commands to Asterisk (checking the results as we go) that will deliver the current weather conditions to the caller. This will be achieved through the use of the STREAM FILE and SAY NUMBER AGI commands.

We've said it before, and we'll say it again: when calling AGI commands, you must pass in all of the required arguments. In this case, both STREAM FILE and SAY NUMBER commands require a second argument; we'll pass empty quotes escaped by the backslash character.

You should also notice that we call the fflush command each time we write to STDOUT. While this is arguably redundant, there's no harm in ensuring that the AGI command is not buffered and is sent immediately to Asterisk.

```
function checkresult($res)
{
    trim($res);
    if (preg_match('/^200/',$res))
    {
        if (! preg_match('/result=(-?\d+)/',$res,$matches))
        {
            fwrite(STDERR,"FAIL ($res)\n");
            fflush(STDERR);
            return 0;
        }
        else
```

```
    {
        fwrite(STDERR,"PASS (".$matches[1].")\n");
        fflush(STDERR);
        return $matches[1];
    }
}
else
{
    fwrite(STDERR,"FAIL (unexpected result '$res')\n");
    fflush(STDERR);
    return -1;
}
}
```

The checkresult function is identical in purpose to the checkresult subroutine we saw in our Perl example. As its name suggests, it checks the result that Asterisk returns whenever we call an AGI command.

```
?>
```

At the end of the file, we have our closing PHP tag. Don't place any whitespace after the closing PHP tag, as it can confuse the AGI interface.

We've now covered two different languages, in order to demonstrate the similarities and differences of programming an AGI script in PHP as opposed to Perl. The following things should be remembered when writing an AGI script in PHP:

- Invoke PHP with the –q switch; it turns off HTML in error messages.
- Turn off the time limit, or set it to a reasonable value (newer versions of PHP automatically disable the time limit when PHP is invoked from the command line).
- Turn off output buffering with the ob_implicit_flush(false) command.
- Open file handles to STDIN, STDOUT, and STDERR (newer versions of PHP may have one or more of these file handles already opened—see the code above for a slick way of making this work across most versions of PHP).
- Read variables from STDIN using the fgets function.
- Use the fwrite function to write to STDOUT and STDERR.
- Always call the fflush function after writing to either STDOUT or STDERR.

The PHP AGI Library

For advanced AGI programming in PHP, you may want to check out the PHPAGI project at *http://phpagi.sourceforge.net*. It was originally written by Matthew Asham and is being developed by several other members of the Asterisk community.

Writing AGI Scripts in Python

The AGI script we'll be writing in Python, called "The Subtraction Game," was inspired by a Perl program written by Ed Guy and discussed by him at the 2004 AstriCon conference. Ed described his enthusiasm for the power and simplicity of Asterisk when he found he could write a quick Perl script to help his young daughter improve her math skills.

Since we've already written a Perl program using AGI, and Ed has already written the math program in Perl, we figured we'd take a stab at it in Python!

Let's go through our Python script.

```
#!/usr/bin/python
```

This line tells the system to run this script in the Python interpreter. For small scripts, you may consider adding the –u option to this line, which will run Python in unbuffered mode. This is not recommended, however, for larger or frequently used AGI scripts, as it can affect system performance.

```
import sys
import re
import time
import random
```

Here, we import several libraries that we'll be using in our AGI script.

```
# Read and ignore AGI environment (read until blank line)

env = {}
tests = 0;

while 1:
    line = sys.stdin.readline().strip()

    if line == '':
        break
    key,data = line.split(':')
    if key[:4] <> 'agi_':
        #skip input that doesn't begin with agi_
        sys.stderr.write("Did not work!\n");
        sys.stderr.flush()
        continue
    key = key.strip()
    data = data.strip()
    if key <> '':
        env[key] = data

sys.stderr.write("AGI Environment Dump:\n");
sys.stderr.flush()
for key in env.keys():
    sys.stderr.write(" -- %s = %s\n" % (key, env[key]))
    sys.stderr.flush()
```

This section of code reads in the variables that are passed to our script from Asterisk, and saves them into a dictionary named env. These values are then written to STDERR for debugging purposes.

```python
def checkresult (params):
    params = params.rstrip( )
    if re.search('^200',params):
        result = re.search('result=(\d+)',params)
        if (not result):
            sys.stderr.write("FAIL ('%s')\n" % params)
            sys.stderr.flush( )
            return -1
        else:
            result = result.group(1)
            #debug("Result:%s Params:%s" % (result, params))
            sys.stderr.write("PASS (%s)\n" % result)
            sys.stderr.flush( )
            return result
    else:
        sys.stderr.write("FAIL (unexpected result '%s')\n" % params)
        sys.stderr.flush( )
        return -2
```

The checkresult function is almost identical in purpose to the checkresult subroutine in the sample Perl AGI script we covered earlier in the chapter. It reads in the result of an Asterisk command, parses the answer, and reports whether or not the command was successful.

```python
def sayit (params):
    sys.stderr.write("STREAM FILE %s \"\"\n" % str(params))
    sys.stderr.flush( )
    sys.stdout.write("STREAM FILE %s \"\"\n" % str(params))
    sys.stdout.flush( )
    result = sys.stdin.readline( ).strip( )
    checkresult(result)
```

The sayit function is a simple wrapper around the STREAM FILE command.

```python
def saynumber (params):
    sys.stderr.write("SAY NUMBER %s \"\"\n" % params)
    sys.stderr.flush( )
    sys.stdout.write("SAY NUMBER %s \"\"\n" % params)
    sys.stdout.flush( )
    result = sys.stdin.readline( ).strip( )
    checkresult(result)
```

The saynumber function is a simple wrapper around the SAY NUMBER command.

```python
def getnumber (prompt, timelimit, digcount):
    sys.stderr.write("GET DATA %s %d %d\n" % (prompt, timelimit, digcount))
    sys.stderr.flush( )
    sys.stdout.write("GET DATA %s %d %d\n" % (prompt, timelimit, digcount))
    sys.stdout.flush( )
    result = sys.stdin.readline( ).strip( )
    result = checkresult(result)
```

```
sys.stderr.write("digits are %s\n" % result)
sys.stderr.flush( )
if result:
    return result
else:
    result = -1
```

The getnumber function calls the GET DATA command to get DTMF input from the caller. It is used in our program to get the caller's answers to the subtraction problems.

```
limit=20
digitcount=2
score=0
count=0
ttanswer=5000
```

Here, we initialize a few variables that we'll be using in our program.

```
starttime = time.time( )
t = time.time( ) - starttime
```

In these lines we set the starttime variable to the current time and initialize t to zero. We'll use the t variable to keep track of the number of seconds that have elapsed since the AGI script was started.

```
sayit("subtraction-game-welcome")
```

Next, we welcome the caller to the subtraction game.

```
while ( t < 180 ):

    big = random.randint(0,limit+1)
    big += 10
    subt= random.randint(0,big)
    ans = big - subt
    count += 1

    #give problem:
    sayit("subtraction-game-next");
    saynumber(big);
    sayit("minus");
    saynumber(subt);
    res = getnumber("equals",ttanswer,digitcount);

    if (int(res) == ans) :
        score+=1
        sayit("subtraction-game-good");
    else :
        sayit("subtraction-game-wrong");
        saynumber(ans);

    t = time.time( ) - starttime
```

This is the heart of the AGI script. We loop through this section of code and give subtraction problems to the caller until 180 seconds have elapsed. Near the top of the loop, we calculate two random numbers and their difference. We then present the problem to the caller, and read in the caller's response. If the caller answers incorrectly, we give the correct answer.

```
pct = float(score)/float(count)*100;
sys.stderr.write("Percentage correct is %d\n" % pct)
sys.stderr.flush( )

sayit("subtraction-game-timesup")
saynumber(score)
sayit("subtraction-game-right")
saynumber(count)
sayit("subtraction-game-pct")
saynumber(pct)
```

After the users are done answering the subtraction problems, they are given their scores.

As you have seen, the basics you should remember when writing AGI scripts in Python are:

- Flush the output buffer after every write. This will ensure that your AGI program won't hang while Asterisk is waiting for the buffer to fill and Python is waiting for the response from Asterisk.
- Read data from Asterisk with the sys.stdin.readline command.
- Write commands to Asterisk with the sys.stdout.write command. Don't forget to call sys.stdout.flush after writing.

The Python AGI Library

If you are planning on writing lot of Python AGI code, you may want to check out Karl Putland's Python module, *Pyst*. You can find it at *http://www.sourceforge.net/projects/pyst/*.

Debugging in AGI

Debugging AGI programs, as with any other type of program, can be frustrating. Luckily, there are two advantages to debugging AGI scripts. First, since all the communications between Asterisk and the AGI program happen over STDIN and STDOUT (and, of course, STDERR), you should be able to run your AGI script directly from the operating system. Second, Asterisk has a handy command for showing all the communications between itself and the AGI script: agi debug.

Debugging from the Operating System

As mentioned above, you should be able to run your program directly from the operating system to see how it behaves. The secret here is to act just like Asterisk does, providing your script with the following:

- A list of variables and their values, such as agi_test:1.
- A blank line feed (/n) to indicate that you're done passing variables.
- Responses to each of the AGI commands from your AGI script. Usually, typing **200 response=1** is sufficient.

Trying your program directly from the operating system may help you to more easily spot bugs in your program.

Using Asterisk's agi debug Command

The Asterisk command-line interface has a very useful command for debugging AGI scripts, which is called (appropriately enough) agi debug. If you type **agi debug** at an Asterisk console and then run an AGI, you'll see something like the following:

```
   -- Executing AGI("Zap/1-1", "temperature.php") in new stack
   -- Launched AGI Script /var/lib/asterisk/agi-bin/temperature.php
AGI Tx >> agi_request: temperature.php
AGI Tx >> agi_channel: Zap/1-1
AGI Tx >> agi_language: en
AGI Tx >> agi_type: Zap
AGI Tx >> agi_uniqueid: 1116732890.8
AGI Tx >> agi_callerid: 101
AGI Tx >> agi_calleridname: Tom Jones
AGI Tx >> agi_callingpres: 0
AGI Tx >> agi_callingani2: 0
AGI Tx >> agi_callington: 0
AGI Tx >> agi_callingtns: 0
AGI Tx >> agi_dnid: unknown
AGI Tx >> agi_rdnis: unknown
AGI Tx >> agi_context: incoming
AGI Tx >> agi_extension: 141
AGI Tx >> agi_priority: 2
AGI Tx >> agi_enhanced: 0.0
AGI Tx >> agi_accountcode:
AGI Tx >>
AGI Rx << STREAM FILE temperature ""
AGI Tx >> 200 result=0 endpos=6400
AGI Rx << STREAM FILE is ""
AGI Tx >> 200 result=0 endpos=5440
AGI Rx << SAY NUMBER 67 ""
   -- Playing 'digits/60' (language 'en')
   -- Playing 'digits/7' (language 'en')
AGI Tx >> 200 result=0
AGI Rx << STREAM FILE degrees ""
```

```
AGI Tx >> 200 result=0 endpos=6720
AGI Rx << STREAM FILE fahrenheit ""
AGI Tx >> 200 result=0 endpos=8000
    -- AGI Script temperature.php completed, returning 0
```

You'll see three types of lines while your AGI script is running. The first type, prefaced with AGI TX >>, are the lines that Asterisk transmits to your program's STDIN. The second type, prefaced with AGI RX <<, are the commands your AGI program writes back to Asterisk over STDOUT. The third type, prefaced by --, are the standard Asterisk messages presented as it executes certain commands.

To disable AGI debugging after it has been started, simply type **agi no debug** at an Asterisk console.

Using the agi debug command will enable you to see the communication between Asterisk and your program, which can be very useful when debugging. Hopefully, these two tips will greatly improve your ability to write and debug powerful AGI programs.

Conclusion

For a developer, AGI is one of the more revolutionary and compelling reasons to choose Asterisk over a closed, proprietary PBX. Still, AGI is only part of the picture. For those of us who are less developers and more systems integrators or power users, Chapter 10 will explore the wealth of accoutrements that make Asterisk compelling to so many people.

Asterisk for the Über-Geek

*The first ninety percent of the task takes ninety
percent of the time, and the last ten percent takes the
other ninety percent.*
—The Ninety:Ten Rule

The toughest part of writing this book was not finding things to write about, but rather deciding what we would not be able to write about. Now that we've covered the basics, you are ready to be told the truth: we have not taught you anywhere near all that there is to know about Asterisk. Well, okay, perhaps five percent, but likely less.

Now please understand, this is not because we didn't want to give you our very best; it's merely because Asterisk is, well, limitless (or so we believe).

In this chapter, we want to give you a taste of some of the wonders Asterisk holds in store for you. Pretty nearly every section in this chapter could become a book in itself (and they *will* become books, if Asterisk succeeds in the way we think it is going to).

Festival

Festival is a popular open source text-to-speech engine. The basic premise of using Festival with Asterisk is that your dialplan can pass a body of text to Festival, which will then "speak" the text to the caller. Probably the most obvious use for Festival would be to have it read your email to you when you are on the road.

Getting Festival Set Up and Ready for Asterisk

There are currently two ways to use Festival with Asterisk. The first (and easiest) method—without having to patch and recompile Festival—is to add the following text to Festival's configuration file (*festival.scm*, usually located in */etc/* or */usr/share/festival/*):

```
(define (tts_textasterisk string mode)
"(tts_textasterisk STRING MODE)
```

```
Apply tts to STRING. This function is specifically designed for use in server mode so
a single function call may synthesize the string. This function name may be added to
the server safe functions."
(let ((wholeutt (utt.synth (eval (list 'Utterance 'Text string)))))
(utt.wave.resample wholeutt 8000)
(utt.wave.rescale wholeutt 5)
(utt.send.wave.client wholeutt)))
```

You may place this text anywhere in the file, as long as it is not between any other parentheses.

The second (and more traditional) way is to compile Festival with an Asterisk-specific patch (located in the *contrib/* directory of the Asterisk source).

Information on both of these methods is contained in the *README.festival* file, located in the *contrib/* directory of the Asterisk source.

For either method, you'll need to modify the Festival access list in the *festival.scm* file. Simply search for the word "localhost," and replace it with the fully qualified domain name of your server.

Both of these methods set up Festival to be able to correctly communicate with Asterisk. After setting up Festival, you should start the Festival server. You can then call the Festival( ) application from within your dialplan.

Configuring Asterisk for Festival

The Asterisk configuration file that deals with Festival is aptly called *festival.conf*. Inside this file, you specify the hostname and port of your Festival server, as well some settings for the caching of Festival speech. For most installations (if you're going to run Festival on your Asterisk server), the defaults will work just fine.

Starting the Festival Server

To start the Festival server for debugging purposes, simply run `festival` with the `--server` argument, like this:

```
[root@asterisk ~]# festival --server
```

Once you're sure that the Festival server is running and not rejecting your connections, you can start Festival by typing:

```
[root@asterisk ~]# festival_server 2>&1 >/dev/null &
```

Calling Festival from the Dialplan

Now that Festival is configured and the Festival server is started, let's call it from within a simple dialplan:

```
exten => 123,1,Answer( )
exten => 123,2,Festival(Asterisk and Festival are working together)
```

You should always call the `Answer( )` application before calling `Festival( )`, to ensure that a channel is established.

As Asterisk connects to Festival, you should see output like this in the terminal where you started the Festival server:

```
[root@asterisk ~]# festival --server
server     Sun May  1 18:38:51 2005 : Festival server started on port 1314
client(1) Sun May  1 18:39:20 2005 : accepted from asterisk.localdomain
client(1) Sun May  1 18:39:21 2005 : disconnected
```

If you see output like the following, it means you didn't add the host to the access list in *festival.scm*:

```
[root@asterisk ~]# festival --server
server     Sun May  1 18:30:52 2005 : Festival server started on port 1314
client(1) Sun May  1 18:32:32 2005 : rejected from asterisk.localdomain not in access
list
```

Yet Another Way to Use Festival with Asterisk

Some people in the Asterisk community have reported good success by passing text to Festival's *text2wave* utility and then having Asterisk play back the resulting *.wav* file. For example, you might do something like this:

```
exten => 124,1,Answer( )
exten => 124,2,System(echo "This is a test of Festival" | /usr/bin/text2wave
-scale 1.5 -F 8000 -o /tmp/festival.wav)
exten => 124,3,Playback(/tmp/festival)
exten => 124,4,System(rm /tmp/festival.wav)
exten => 124,5,Hangup( )
```

This method also allows you to call other text-to-speech engines, such as the popular speech engine from Cepstral.[a] For this example, we'll assume that Cepstral is installed in */usr/local/cepstral/*:

```
exten => 125,1,Answer( )
exten => 125,2,System(/usr/local/cepstral/bin/swift -o /tmp/swift.wav
"This is a test of Cepstral")
exten => 125,3,Playback(/tmp/swift)
exten => 125,4,System(rm /tmp/swift.wav)
exten => 125,5,Hangup( )
```

a. Cepstral can be evaluated at *http://www.cepstral.com*. Cepstral is an inexpensive commercial derivative of Festival with very good-sounding voices.

Call Detail Recording

Without even being told, Asterisk assumes that you want to store CDR information. Quite a smart machine, yes?

By default, Asterisk will create a CSV* file and place it in the folder */var/log/asterisk/cdr-csv/*. To the naked eye, this file looks like a bit of a mess. If, however, you separate each line according to the commas, you will find that each line contains information about a particular call, and that the commas separate the following values:

accountcode
> Assigned if the application SetAccount( ) is used, or if configured for the channel in the channel configuration file (i.e., *sip.conf*). The account code is assigned on a per-channel basis.

src
> Received Caller*ID (string, 80 characters).

dst
> Destination extension.

dcontext
> Destination context.

clid
> Caller*ID with text (80 characters).

channel
> Channel used (80 characters).

dstchannel
> Destination channel, if appropriate (80 characters).

lastapp
> Last application, if appropriate (80 characters).

lastdata
> Last application data (arguments, 80 characters).

start
> Start of call (date/time).

answer
> Answer of call (date/time).

end
> End of call (date/time).

---

* A Comma Separated Values (CSV) file is a common method of formatting database-type information in a text file. You can open CSV files with a text editor, but most spreadsheet and database programs will also read them and properly parse them into rows and columns.

duration

> Total time in system, in seconds (integer), from dial to hangup.

billsec

> Total time call is up, in seconds (integer), from answer to hangup.

disposition

> What happened to the call (ANSWERED, NO ANSWER, BUSY).

amaflags

> What flags to use (DOCUMENTATION, BILL, IGNORE, etc.), specified on a per-channel basis, like *accountcode*. AMA flags stand for Automated Message Accounting flags, which are somewhat standard (supposedly) in the industry.

userfield

> A user-defined field, maximum 255 characters.

Storing CDRs in a Database

CDRs can also be stored in a database. Asterisk currently supports SQLite, PostGreSQL, MySQL, and unixODBC. The configuration details for these databases will not be covered in this book, but they are outlined in the Asterisk source code, under the *doc/* subdirectory. (For licensing reasons, *cdr_mysql* is in *asterisk-addons*.) Many people prefer to store their CDRs in a database because this makes it easier to query them for specific information, such as billing or toll fraud. We can use the CDR applications to manipulate the current CDR from the dialplan (adding information to the custom field, for example).

CDR Challenges

While Asterisk will happily store information about any calls that pass through it, it cannot store information it is not given. For example, if you have SIP devices that are allowed to reinvite, once Asterisk has finished setting up the calls, the devices will no longer need its assistance. Whether or not those devices subsequently report call detail information back to it is something Asterisk is unable to control. If CDRs are important, make sure your IP devices are not allowed to reinvite.[*]

Customizing System Prompts

In keeping with the seemingly limitless flexibility of Asterisk, you can also modify the system prompts. This is very simple to explain, but generally difficult to do well.

---

[*] Reinvites can be turned off in *sip.conf* with canreinvite=no. Similar functionality is controlled in *iax.conf* with notransfer=yes.

With over three hundred system prompts in the main distribution, and over six hundred more in the *asterisk-sounds* add on, if you're contemplating customizing all of them you'd better have either a lot of money or a lot of time on your hands.

An audio engineer is also recommended, to ensure that the recordings are normalized to −3 dB and that all prompts start and end at a zero-crossing point (with just the right amount of silence prepended and appended).

The Voice

If you are interested in The Voice of Asterisk, she is Allison Smith, and she can deliver customized recordings for you to use on your own system.

This is an extremely cool concept, as very few PBXs allow you to use the same voice in your custom recordings as is used by the system prompts.

To make use of Allison's talents, sign up at *http://thevoice.digium.com*.

Once you have the recordings, the actual implementation is easy—simply replace the files in */var/lib/asterisk/sounds/* with the ones you have created.

Alternatively, you can opt to record your own prompts and place them in a folder of your choosing. When you refer to sound files with the Playback( ) or Background( ) applications, you can refer to the full pathname of the sound file, or to any subdirectory of */var/lib/asterisk/sounds/*.

A useful way to convert your WAV files to GSM format is with the use of the *sox* application. To convert your files with *sox*, use:

```
# sox foo.wav -r 8000 foo.gsm resample -ql
```

If your WAV files are recorded in stereo, be sure to add the −c1 flag to write the files in mono. These recordings are often made through a PC, but check out the following sidebar—some people have had better luck recording from the dialplan.

Manager

Asterisk Manager provides an API that allows external programs the ability to create, monitor and manage Asterisk.[*] The Manager interface is a powerful mechanism for integrating external programs of all kinds into Asterisk.

To use the Manager, you must define an account in the file */etc/asterisk/manager.conf*. This file will look something like this:

```
[general]
enabled = yes
```

---

[*] An Application Program Interface (API) is a mechanism by which a program allows other programs to take control of it. Contrast this with AGI, which allows external programs to be called from the dialplan.

Sound Recording from the Dialplan

Surprisingly, one of the easiest ways to get respectable-quality recordings is not through a PC with fancy editing software, but rather through a telephone set. There are many reasons for this, but the most important is that the telephone will tend to filter out background noise (such as white noise caused by HVAC equipment) and will record at a consistent audio level.

This little addition to your dialplan will allow you to easily create recordings, which will be placed in your system's */tmp/* folder (from there, you can rename them and move them wherever you'd like):

```
exten => _66XX,1,Wait(2)
exten => _66XX,2,Record(/tmp/prompt${EXTEN:2}:wav)
exten => _66XX,3,Wait(1)
exten => _66XX,4,Playback(/tmp/prompt${EXTEN:2})
exten => _66XX,5,Wait(2)
exten => _66XX,6,Hangup( )
```

This little snippet will allow you to dial from 6600 to 6699, and it will record prompts in the */tmp/* folder using the names *prompt00.wav* to *prompt99.wav*. After you complete recording (by pressing the # key), it will play your prompt back to you and hang up.

Be sure to move your prompts out of the */tmp/* dir to the Asterisk sounds directory. To keep the dialplan readable, rename your *promptXX* files to a meaningful names—e.g., **mv /tmp/prompt00.wav /var/lib/asterisk/sounds/custom/welcome-message.wav**.

```
port = 5038
bindaddr = 0.0.0.0

[oreilly]
secret = notvery
;deny=0.0.0.0/0.0.0.0
;permit=209.16.236.73/255.255.255.0
read = system,call,log,verbose,command,agent,user
write = system,call,log,verbose,command,agent,user
```

In the [general] section, you have to enable the service by setting the parameter enabled = yes. The TCP port to use will default to 5038.

For each user, you will specify the username in square brackets ([]), followed by the password for that user (secret), any IP addresses you wish to deny access to, any IP addresses you wish to permit access to, and the read and write permissions for that user.

Manager Commands

It is important to keep in mind that the Manager interface is designed to be used by programs, not fingers. That's not to say that you can't issue commands to it directly—just don't expect a typical console interface, because that's not what Manager is for.

Commands to Manager are delivered in packages with the following syntax (lines are terminated with CRLF):

```
Action: <action type>
<Key 1>: <Value 1>
<Key 2>: <Value 2>
etc ...
<Variable>: <Value>
<Variable>: <Value>
etc...
```

For example, to authenticate with Manager (which is required if you expect to have any interaction whatsoever), you would send the following:

```
Action: login
Username: oreilly
Secret: notvery
<CRLF>
```

An extra CRLF on a blank line will send the entire package to Manager.

Once authenticated, you will be able to initiate actions, as well as see events generated by Asterisk. On a busy system, this can get quite complicated and become totally impossible to keep track of with the unaided eye.

The Flash Operator Panel

The Flash Operator Panel (FOP) is far and away the most popular example of the power of the Manager interface. FOP creates a web-based visual view of your system and allows you control of calls.

FOP is most commonly used to enable a live attendant to view the users in the system and connect calls between them. It can also be used in a call-center environment to provide CRM-triggered screen pops.[*]

The FOP management interface is shown in Figure 10-1. To grab a copy of FOP, head to *http://www.asternic.org*.

Call Files

Call files allow you to create calls through the Linux shell. These powerful events are triggered by depositing a *.call* file in the directory */var/spool/asterisk/outgoing/*. The actual name of the file does not matter, but it's good form to give the file a meaningful name and to end the filename with *.call*.

---

[*] Customer Relationship Management (CRM) is an interface companies use to help manage customer information and interaction.

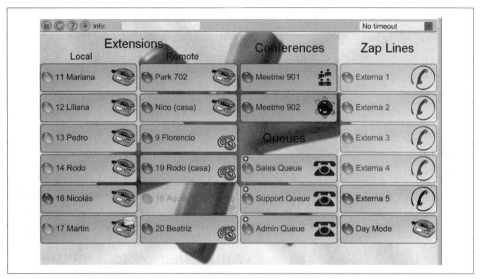

Figure 10-1. The Flash Operator Panel management interface

When a call file appears in the outgoing folder, Asterisk will almost immediately* act on the instructions contained therein.

Call files are formatted in the following manner. First, we define where we want to call:

```
Channel: <channel>
```

We can control how long to wait for a call to be answered (the default is 45 seconds), how long to wait between call retries, and the maximum number of retries. If `MaxRetries` is omitted, the call will be attempted only once:

```
WaitTime: <number>
RetryTime: <number>
MaxRetries: <number>
```

If the call is answered, we specify where to connect it here:

```
Context: <context-name>
Extension: <ext>
Priority: <priority>
```

Alternatively, we can specify a single application and pass arguments to it:

```
Application: Playback()
Data: hello-world
```

Next, we set the Caller*ID of the outgoing call:

```
CallerID: Asterisk <800-555-1212>
```

Then we set channel variables, as follows:

```
SetVar: john=Zap/1/5551212
SetVar: sally=SIP/1000
```

---

* We're talking seconds or less.

and add a CDR account code:

```
Account: documentation
```

 When you create a call file, do *not* do so from the spool directory. Asterisk monitors the spool aggressively and will try to grab your file before you've even finished writing it. Create call files in some other folder, and then mv them into the spool directory.

DUNDi

If there were any concerns that Mark Spencer was in danger of running out of good ideas, Distributed Universal Number Discovery (DUNDi) ought to lay such thoughts to rest. DUNDi is poised to be as revolutionary as Asterisk. The DUNDi web site (*http://www.dundi.com*) says it best: "DUNDi™ is a peer to peer system for locating Internet gateways to telephony services. Unlike traditional centralized services (such as the remarkably simple and concise *ENUM** standard), DUNDi is fully distributed with no centralized authority whatsoever."

How Does DUNDi Work?

Think of DUNDi as a large phone book that allows you to ask peers if they know of an alternative VoIP route to an extension number or PSTN telephone number.

For example, assume you are connected to the *DUNDi-test* network (a free and open network that terminates calls to traditional PSTN numbers). You ask your friend Bob if he knows how to reach 1-800-555-1212, a number for which you have no direct access. Bob replies, "I don't know how to reach that number, but let me ask my peer Sally."

Bob asks Sally if she knows how to reach the requested number, and she responds with, "You can reach that number at *IAX2/dundi:very_long_password@hostname/extension*." Bob then stores the address in his database and passes on to you the information about how to reach 1-800-555-1212 via VoIP, allowing you an alternative method of reaching the same destination through a different network.

Because Bob has stored the information he found, he'll be able to provide it to any peers who later request the same number from him, so the lookup won't have to go any further. This helps reduce the load on the network and increases response times for numbers that are looked up often. (However, it should be noted that DUNDi creates a rotating key, and thus stored information is valid for a limited period of time.)

DUNDi performs lookups dynamically, either with a switch => statement in your *extensions.conf* file or with the use of the DUNDiLookup( ) application. DUNDi is available only in Asterisk Version 1.2 or higher.

* *http://www.faqs.org/rfc/rfc2916.txt.*

You can use the DUNDi protocol in a private network as well. Say you're the Asterisk administrator of a very large enterprise installation and you wish to simplify the administration of extension numbers. You could use DUNDi in this situation, allowing multiple Asterisk boxes (presumably located at each of the company's locations and peered with one another) to perform dynamic lookups for the VoIP addresses of extensions on the network.

Configuring Asterisk for Use with DUNDi

There are three files that need to be configured for DUNDi: *dundi.conf*, *extensions.conf*, and *iax.conf*.* The *dundi.conf* file controls the authentication of peers who we allow to perform lookups through our system. This file also manages the list of peers to whom we might submit our own lookup requests. Since it is possible to run several different networks on the same box, it is necessary to define a different section for each peer, and then configure the networks in which that peer is allowed to perform lookups. Additionally, we need to define which peers we wish to use to perform lookups.

The General Peering Agreement

The General Peering Agreement (GPA) is a legally binding license agreement that is designed to prevent abuse of the DUNDi protocol. Before connecting to the *DUNDi-test* group, you are required to sign a GPA. The GPA is used to protect the members of the group and to create a "trust" between the members. It is a requirement of the *DUNDi-test* group that your complete and accurate contact information be configured in *dundi.conf*, so that members of your peer group can contact you. The GPA can be found in the *doc/* subdirectory of the Asterisk source.

General configuration

The [general] section of *dundi.conf* contains parameters relating to the overall operation of the DUNDi client and server:

```
; DUNDi configuration file
;
[general]
;
department=IT
organization= toronto.example.com
locality=Toronto
stateprov=ON
country=CA
email=support@toronto.example.com
phone=+19055551212
;
```

---

* The *dundi.conf* and *extensions.conf* files must be configured. We have chosen to configure *iax.conf* for our address advertisement on the network, but DUNDi is protocol-agnostic—thus *sip.conf*, *h323.conf*, or *mgcp.conf* could be used instead.

```
; Specify bind address and port number.  Default is 4520
;bindaddr=0.0.0.0
port=4520
entityid=FF:FF:FF:FF:FF:FF
ttl=32
autokill=yes
;secretpath=dundi
```

The entity identifier defined by entityid should generally be the Media Access Control (MAC) address of an interface in the machine. The entity ID defaults to the first Ethernet address of the server, but you can override this with entityid, as long as it is set to the MAC address of *something* you own. The MAC address of the primary external interface is recommended. This is the address that other peers will use to identify you.

The Time To Live (ttl) field defines how many peers away we wish to receive replies from and is used to break loops. Each time a request is passed on down the line because the requested number is not known, the value in the TTL field is decreased by one, much like the TTL field of an ICMP packet. The TTL field also defines the maximum number of seconds we are willing to wait for a reply.

When you request a number lookup, an initial query (called a DPDISCOVER) is sent to your peers requesting that number. If you do not receive an acknowledgment (ACK) of your query (DPDISCOVER) within 2,000 ms (enough time for a single transmission only), and autokill is equal to yes, Asterisk will send a CANCEL to the peers. (Note that an acknowledgment is not necessarily a reply to the query; it is just an acknowledgment that the peer has received the request.) The purpose of autokill is to keep the lookup from stalling due to hosts with high latency. In addition to the yes and no options, you may also specify the number of milliseconds to wait.

The *pbx_dundi* module creates a rotating key and stores it in the local Asterisk database (AstDB). The key name secret is stored in the dundi family. The value of the key can be viewed with the database show command at the Asterisk console. The database family can be overridden with the secretpath option.

Creating mapping contexts

The *dundi.conf* file defines DUNDi contexts that are mapped to dialplan contexts in your *extensions.conf* file. DUNDi contexts are a way of defining distinct and separate directory service groups. The contexts in the mapping section point to contexts in the *extensions.conf* file, which control the numbers that you advertise. When you create a peer, you need to define which mapping contexts you will allow this peer to search. You do this with the permit statement (each peer may contain multiple permit statements). Mapping contexts are related to dialplan contexts in the sense that they are a security boundary for your peers.

 Phone numbers must be advertised in the following format:

```
<country_code><area_code><prefix><number>
```

For example, a complete North American number could be advertised as 14165551212.

All DUNDi mapping contexts take the form of:

```
dundi_context => local_context,weight,technology,destination[,options]]
```

The following configuration creates a DUNDi mapping context that we will use to advertise our local phone numbers to the *DUNDi-test* group. Note that this should all appear on one line:

```
dundi-test => dundi-local,0,IAX2,dundi:${SECRET}@toronto.example.com/${NUMBER},
nounsolicited,nocomunsolicit,nopartial
```

In this example, the mapping context is `dundi-test`, which points to the `dundi-local` context within *extensions.conf* (providing a listing of phone numbers to reply to). Numbers that resolve to the PBX should be advertised with a *weight* of zero (directly connected). Numbers higher than 0 indicate an increased number of hops or paths to reach the final destination. This is useful when multiple replies for the same lookup are received at the end that initially requested the number—a *weight* with a lower number will be the preferred path.

If we can reply to a lookup, our response will contain the method by which the other end can connect to the system. This includes the technology to use (such as IAX2, SIP, H323, and so on), the username and password with which to authenticate, which host to send the authentication to, and finally the extension number.

Asterisk provides some shortcuts to allow us to create a "template" with which we can build our responses. The following channel variables can be used to construct the template:

${SECRET}
: Replaced with the password stored in the local AstDB

${NUMBER}
: The number being requested

${IPADDR}
: The IP address to connect to

 It is generally safest to statically configure the hostname, rather than making use of the ${IPADDR} variable. The ${IPADDR} variable will sometimes reply with an address in the private IP space, which is unreachable from the Internet.

Defining DUNDi peers

DUNDi peers are defined in the *dundi.conf* file. Peers are identified by the unique layer-two MAC address of an interface on the remote system. The *dundi.conf* file is where we define what context to search for peers requesting a lookup and which peers we want to use when doing a lookup for a particular network.

```
[00:00:00:00:00:00] ; Remote Office
model = symmetric
host = montreal.example.com
inkey = montreal
outkey = toronto
include = dundi-test
permit = dundi-test
qualify = yes
dynamic=yes
```

The remote peer's identifier (MAC address) is enclosed in square brackets ([ ]). The inkey and outkey are the public/private key pairs that we use for authentication. Key pairs are generated with the *astgenkey* script, located in the *./asterisk/contrib/scripts/* source directory. Be sure to use the –n flag so that you don't have to initialize passwords every time you start Asterisk:

```
# cd /var/lib/asterisk/keys
# /usr/src/asterisk/contrib/scripts/astgenkey -n toronto
```

The resulting keys, *toronto.pub* and *toronto.key*, will be placed in your */var/lib/asterisk/keys/* directory. The *toronto.pub* file is the public key, which you should post to a web server so that it is easily accessible for anyone with whom you wish to peer. When you peer, you can give your peers the HTTP-accessible public key, which they can then place in their */var/lib/asterisk/keys/* directories.

After you have downloaded the keys, you must reload the *res_crypto.so* and *pbx_dundi.so* modules in Asterisk:

```
*CLI> reload res_crypto.so
    -- Reloading module 'res_crypto.so' (Cryptographic Digital Signatures)
    -- Loaded PRIVATE key 'toronto'
    -- Loaded PUBLIC key 'toronto'

*CLI> reload pbx_dundi.so
    -- Reloading module 'pbx_dundi.so' (Distributed Universal Number Discovery
(DUNDi))
  == Parsing '/etc/asterisk/dundi.conf': Found
```

Then, create the dundi user in the *iax.conf* file to allow connections into your Asterisk system. When a call is authenticated, the extension number being requested is passed to the dundi-local context in the *extensions.conf* file, where the call is then handled by Asterisk.

Allowing remote connections

Here is the user definition for the dundi user:

```
[dundi]
type=user
dbsecret=dundi/secret
context=dundi-local
disallow=all
allow=ulaw
allow=g726
```

Instead of using a static password, Asterisk regenerates passwords every 3,600 seconds (1 hour). The value is stored in */dundi/secret* of the Asterisk database and advertised using the ${SECRET} variable defined within the mapping context lines in *dundi.conf*. You can see the current keys for all peers, including your local public and private keys, by performing a **show keys** at the Asterisk CLI.

The context entry dundi-local is where authorized callers are sent in *extensions.conf*. From there, we can manipulate the call just as we would in the dialplan of any other incoming connection.

Configuring the dialplan

The *extensions.conf* file handles what numbers you advertise and what you do with the calls that connect to them. The dundi-local context performs double duty:

- It controls the numbers we advertise, referenced by the dundi mapping context in *dundi.conf*.

- It controls what to do with the call, referenced by the dundi user in *iax.conf*.

You have the power of dialplan pattern matching to advertise ranges of numbers and to control the incoming calls. In the following dialplan, we are only advertising the number +1-416-555-1212, but pattern matching could just as easily have been employed to advertise a range of numbers or extensions:

```
[dundi-local]
exten => 14165551212,1,NoOp(dundi-local: Number advertisement and incoming)
exten => 14165551212,n,Answer()
exten => 14165551212,n(call),Dial(SIP/1000)
exten => 14165551212,n,Voicemail(u1000)
exten => 14165551212,n,Hangup()
exten => 14165551212,n(call)+101,Voicemail(b1000)
exten => 14165551212,n,Hangup()
```

Conclusion

That's pretty much all this chapter is going to teach you, but it's nowhere near all there is to learn. Hopefully, you are starting to get an idea of how big this Asterisk thing really is.

In the next chapter, we're going to try and predict the future of telecom, and we'll discuss how (and why) we believe that Asterisk is well positioned to play a starring role.

## CHAPTER 11

Asterisk: The Future of Telephony

First they ignore you, then they laugh at you,
then they fight you, then you win.
—Mahatma Gandhi

We have arrived at the final chapter of this book. We've covered a lot, but we hope that you now realize that we have barely begun to scratch the surface of this phenomenon called Asterisk. To wrap things up, we want to spend some time exploring what we might see from Asterisk and open source telephony in the near future.

While prognostication is always a thankless task, we are confident in asserting that open source communications engines such as Asterisk herald a shift in thinking that will transform the telecommunications industry. In this chapter, we will discuss some of our reasons for this belief.

The Problems with Traditional Telephony

Although Alexander Graham Bell is most famously remembered as the father of the telephone, the reality is that during the latter half of the 1800s, dozens of minds were at work on the project of carrying voice over telegraph lines. These people were mostly business-minded folks, looking to create a product through which they might make their fortunes.

We have come to think of traditional telephone companies as monopolies, but this was not true in their early days. The early history of telephone service took place in a very competitive environment, with new companies springing up all over the world, often with little or no respect for the patents they might be violating. Some of the monopolies got their start through the waging (and winning) of patent wars.

It's interesting to contrast the history of the telephone with the history of Linux and the Internet. While the telephone was created as a commercial exercise, and the telecom industry was forged through lawsuits and corporate takeovers, Linux and the

Internet arose out of the academic community, which has always valued the sharing of knowledge over profit.

The cultural differences are obvious. Telecommunications technologies tend to be closed, confusing, and expensive, while networking technologies are generally open, well documented, and competitive.

Closed Thinking

If one compares the culture of the telecommunications industry to that of the Internet, it is sometimes difficult to believe the two are related. The Internet was designed by enthusiasts, whereas contributing to the development of the PSTN is impossible for any individual to contemplate. This is an exclusive club; membership is not open to just anyone.*

The International Telecommunication Union (ITU) clearly exhibits this type of closed thinking. If you want access to their knowledge, you have to be prepared to pay for it. Membership requires proof of your qualifications, and you will be expected to pay tens of thousands of dollars in annual dues.

Although the ITU is the United Nations's sanctioned body responsible for international telecommunications, many of the VoIP protocols (SIP, MGCP, RTP, STUN) come not from the hallowed halls of the ITU, but rather from the IETF (which publishes all of its standards free to all, and allows anyone to submit an Internet Draft for consideration).

Open protocols such as SIP may have a tactical advantage over ITU protocols such as H.323 due to the ease with which one can obtain them. Although H.323 is widely deployed by carriers as a VoIP protocol in the backbone, it is much more difficult to find H.323-based endpoints; newer products are far more likely to support SIP.

The success of the IETF's open approach has not gone unnoticed by the mighty ITU. It has recently become possible to download up to three documents free of charge from the ITU web site.† Openness is clearly on their minds. Recent statements by the ITU suggest that there is a desire to achieve "Greater participation in ITU by civil society and the academic world." Mr. Houlin Zhao, the ITU's Director of the Telecommunication Standardization Bureau (TSB), believes that "ITU should take some steps to encourage this."‡

---

* Contrast this with the IETF's membership page, which states: "The IETF is not a membership organization (no cards, no dues, no secret handshakes :-)... It is open to any interested individual... Welcome to the IETF." Talk about community!

† Considering the thousands of documents available, and the fact that each document generally contains references to dozens more, the value of this free information is difficult to judge.

‡ *http://www.itu.int/ITU-T/tsb-director/itut-wsis/files/wg-wsis-Zhao-rev1.pdf.*

The roadmap to achieving this openness is unclear, but they are beginning to realize the inevitable.

As for Asterisk, it embraces both the past and the future—H.323 support is available, although the community has for the most part shunned H.323 in favor of the IETF protocol SIP and the darling of the Asterisk community, IAX.

Limited Standards Compliancy

One of the oddest things about all the standards that exist in the world of legacy tele-communications is the various manufacturers' seeming inability to implement them consistently. Each manufacturer desires a total monopoly, so the concept of interoperability tends to take a back seat to being first to market with a creative new idea.

The ISDN protocols are a classic example of this. Deployment of ISDN was (and in many ways still is) a painful and expensive proposition, as each manufacturer decided to implement it in a slightly different way. ISDN could very well have helped to usher in a massive public data network, 10 years before the Internet. Unfortunately, due to its cost, complexity, and compatibility issues, ISDN never delivered much more than voice, with the occasional video or data connection for those willing to pay. ISDN is quite common (especially in Europe, and in North America in larger PBX implementations), but it is not delivering anywhere near the capabilities that were envisioned for it.

As VoIP becomes more and more ubiquitous, the need for ISDN will disappear.

Slow Release Cycles

It can take months, or sometimes years, for the big guys to admit to a trend, let alone release a product that is compatible with it. It seems that before a new technology can be embraced, it must be analyzed to death, and then it must pass successfully through various layers of bureaucracy before it is even scheduled into the development cycle. Months or even years must pass before any useful product can be expected. When those products are finally released, they are often based on hardware that is obsolete; they also tend to be expensive and to offer no more than a minimal feature set.

These slow release cycles simply don't work in today's world of business communications. On the Internet, new ideas can take root in a matter of weeks and become viable in extremely short periods of time. Since every other technology must adapt to these changes, so too must telecommunications.

Open source development is inherently better able to adapt to rapid technological change, which gives it an enormous competitive advantage.

The spectacular crash of the telecom industry may have been caused in large part by an inability to change. Perhaps that continued inability is why recovery has been so slow. Now, there is no choice: change, or cease to be. Community-driven technologies such as Asterisk will see to that.

Refusing to Let Go of the Past and Embrace the Future

Traditional telecommunications companies have lost touch with their customers. While the concept of adding functionality beyond the basic telephone is well understood, the idea that the user should be the one defining this functionality is not.

Nowadays, people have nearly limitless flexibility in every other form of communication. They simply cannot understand why telecommunications cannot be delivered as flexibly as the industry has been promising for so many years. The concept of flexibility is not familiar to the telecom industry, and very well might not be until open source products such as Asterisk begin to transform the fundamental nature of the industry. This is a revolution similar to the one Linux and the Internet willingly started over 10 years ago (and IBM unwittingly started with the PC, 15 years before that). What is this revolution? The commoditization of telephony hardware and software, enabling a proliferation of tailor-made telecommunications systems.

Paradigm Shift

In his article "Paradigm Shift" (*http://tim.oreilly.com/articles/paradigmshift_0504. html*), Tim O'Reilly talks about a paradigm shift that is occurring in the way technology (both hardware and software) is delivered.[*] O'Reilly identifies three trends: *the commoditization of software*, *network-enabled collaboration*, and *software customizability (software as a service)*. These three concepts provide evidence to suggest that open source telephony is an idea whose time has come.

The Promise of Open Source Telephony

> Every good work of software starts by scratching a developer's personal itch.
> —Eric S. Raymond, *The Cathedral and the Bazaar*

In his book *The Cathedral and the Bazaar* (O'Reilly), Eric S. Raymond explains that "Given enough eyeballs, all bugs are shallow." The reason open source software development produces such consistent quality is simple: crap can't hide.

---

[*] Much of the following section is merely our interpretation of O'Reilly's article. To get the full gist of these ideas, the full read is highly recommended.

The Itch That Asterisk Scratches

In this era of custom database and web site development, people are not only tired of hearing that their telephone system "can't do that," they quite frankly just don't believe it. The creative needs of the customers, coupled with the limitations of the technology, have spawned a type of creativity born of necessity: telecom engineers are like contestants in an episode of "Junkyard Wars," trying to create functional devices out of a pile of mismatched components.

The development methodology of a proprietary telephone system dictates that it will have a huge number of features, and that the number of features will in large part determine the price. Manufacturers will tell you that their products give you hundreds of features, but if you only need five of them, who cares? Worse, if there's one missing feature you really can't do without, the value of that system will be diluted by the fact that it can't completely address your needs.

The fact that a customer might only need five out of five hundred features is ignored, and that customer's desire to have five unavailable features that address the needs of his business is dismissed as unreasonable.* Until flexibility becomes standard, telecom will remain stuck in the last century—all the VoIP in the world notwithstanding.

Asterisk addresses that problem directly, and solves it in a way that few other telecom systems can. This is extremely disruptive technology, in large part because it is based on concepts that have been proven time and time again: "the closed-source world cannot win an evolutionary arms race with open-source communities that can put orders of magnitude more skilled time into a problem."†

Open Architecture

One of the stumbling blocks of the traditional telecommunications industry has been its apparent refusal to cooperate with itself. The big telecommunications giants have all been around for over a hundred years. The concept of closed, proprietary systems is so ingrained in their culture that even their attempts at standards compliancy are tainted by their desire to get the jump on the competition, by adding that one feature that no one else supports. For an example of this thinking, one simply has to look at the VoIP products being offered by the telecom industry today. While they claim standards compliance, the thought that you would actually expect to be able to

---

* From the perspective of the closed-source industry, their attitude is understandable. In his book *The Mythical Man-Month: Essays on Software Engineering* (Addison-Wesley), Fred Brooks opined that "the complexity and communication costs of a project rise with the square of the number of developers, while work done only rises linearly." Without a community-based development methodology, it is very difficult to deliver products that at best are little more than incremental improvements over their predecessors, and at worst are merely collections of patches.

† Eric S. Raymond, *The Cathedral and the Bazaar.*

---

connect a Cisco phone to a Nortel switch, or that an Avaya voicemail system could be integrated via IP to a Siemens PBX, is not one that bears discussing.

In the computer industry, things are different. Twenty years ago, if you bought an IBM server, you needed an IBM network and IBM terminals to talk to it. Now, that IBM server is likely to interconnect to Dell terminals though a Cisco network (and run Linux, of all things). Anyone can easily think of thousands of variations on this theme. If any one of these companies were to suggest that we could only use their products with whatever they told us, they would be laughed out of business.

The telecommunications industry is facing the same changes, but it's in no hurry to accept them. Asterisk, on the other hand, is in a big hurry to not only accept change, but embrace it.

Cisco, Nortel, Avaya, and Polycom IP phones (to name just a few) have all been successfully connected to Asterisk systems. There is no other PBX in the world today that can make this claim. None.

Openness is the power of Asterisk.

Standards Compliance

In the past few years, it has become clear that standards evolve at such a rapid pace that to keep up with them requires an ability to quickly respond to emerging technology trends. Asterisk, by virtue of being an open source, community-driven development effort, is uniquely suited to the kind of rapid development that standards compliance demands.

Asterisk does not focus on cost-benefit analysis or market research. It evolves in response to whatever the community finds exciting—or necessary.

Lightning-Fast Response to New Technologies

After Mark Spencer attended his first SIP Interoperability Test (SIPIT) event, he had a rudimentary but working SIP stack for Asterisk coded within a few days. This was before SIP had emerged as the protocol of choice in the VoIP world, but he saw its value and momentum and ensured that Asterisk would be ready.

This kind of foresight and flexibility is typical in an open-source development community (and very unusual in a large corporation).

Passionate Community

The *Asterisk-users* list receives over three hundred email messages per day. Over ten thousand people are subscribed to it. This kind of community support is unheard of in the world of proprietary telecommunications, while in the open source world it is commonplace.

The very first AstriCon event was expected to attract one hundred participants. Nearly five hundred showed up (far more wanted to but couldn't attend). This kind of community support virtually guarantees the success of an open source effort.

Some Things That Are Now Possible

So what sorts of things can be built using Asterisk? Let's look at some of the things we've come up with.

Legacy PBX migration gateway

Asterisk can be used as a fantastic bridge between an old PBX and the future. You can place it in front of the PBX as a gateway (and migrate users off the PBX as needs dictate), or you can put it behind the PBX as a peripheral application server. You can even do both at the same time, as shown in Figure 11-1.

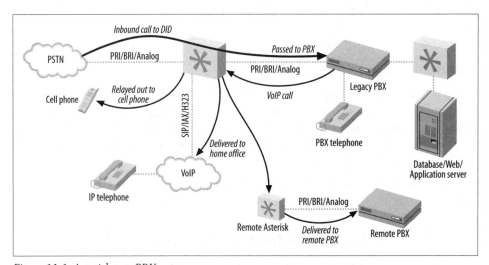

Figure 11-1. Asterisk as a PBX gateway

Here are some of the options you can implement:

Keep your old PBX, but evolve to IP
Companies that have spent vast sums of money in the past few years buying proprietary PBX equipment want a way out of proprietary jail, but they can't stomach the thought of throwing away all of their otherwise functioning equipment. No problem—Asterisk can solve all kinds of problems, from replacing a voicemail system to providing a way to add IP-based users beyond the nominal capacity of the system.

Find-me-follow-me

Provide the PBX a list of numbers where you can be reached, and it will ring them all whenever a call to your DID (Direct Inward Dialing, a.k.a. phone number) arrives. Figure 11-2 illustrates this technology.

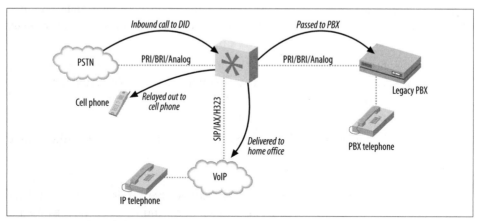

Figure 11-2. Find-me-follow-me

VoIP calling

If a legacy telephony connection from an Asterisk PBX to an old PBX can be established, Asterisk can provide access to VoIP services, while the old PBX continues to connect to the outside world as it always has. As a gateway, Asterisk simply needs to emulate the functions of the PSTN, and the old PBX won't know that anything has changed. Figure 11-3 shows how you can use Asterisk to VoIP-enable a legacy PBX.

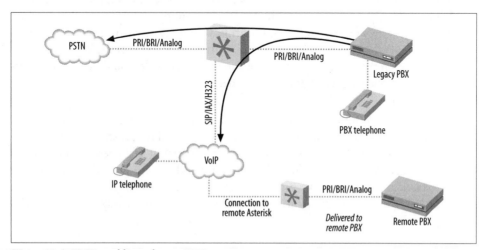

Figure 11-3. VoIP-enabling a legacy PBX

Low-barrier IVR

Many people confuse the term "Interactive Voice Response," or IVR, with the Automated Attendant (AA). Since the Automated Attendant was the very first thing IVR was used for, this is understandable. Nevertheless, to the telecom industry, the term IVR represents far more than an AA. An AA generally does little more than present a way for callers to be transferred to extensions, and it is built into most proprietary voicemail systems—but IVR can be so much more.

IVR systems are generally very expensive, not only to purchase, but also to configure. A custom IVR system will usually require connectivity to an external database or application. Asterisk is arguably the perfect IVR, as it embraces the concepts of connectivity to databases and applications at its deepest level.

Here are a few examples of relatively simple IVRs an Asterisk system could be used to create:

Weather reporting
> Using the Internet, you can obtain text-based weather reports from around the world in a myriad of ways. Capturing these reports and running them through a purpose-built parser (Perl would probably eat this up) would allow the information to be available to the dialplan. Asterisk's sound library already contains all the required prompts, so it would not be an onerous task to produce an interactive menu to play current forecasts for anywhere in the world.

Math programs
> Ed Guy of Pulver.com did a presentation at Astricon 2004 in which he talked about a little math program he'd cooked up for his daughter to use. The program took him no more than an hour to write. What it did was present her with a number of math questions, the answers to which she keyed into the telephone. When all the questions were tabulated, the system presented her with her score. This extremely simple Asterisk application would cost tens of thousands of dollars to implement on any closed PBX platform, assuming it could be done at all.* As is so often the case, things that are simple for Asterisk would be either impossible or massively expensive with any other IVR system.

Distributed IVR
> The cost of a proprietary IVR system is such that when a company with many small retail locations wants to provide IVR, it is forced to transfer callers to a central server to process the transactions. With Asterisk, it becomes possible to distribute the application to each node, and thus handle the requests locally. Literally thousands of little Asterisk systems deployed at retail locations across the world could serve up IVR functionality in a way that would be impossible to achieve with

---

* See Chapter 9 for further details.

any other system. No more long-distance transfers to a central IVR server, no more huge trunking facility dedicated to the task—more power with less expense.

These are three rather simple examples of the potential of Asterisk.

Conference rooms

This little gem is going to end up being one of the killer functions of Asterisk. In the Asterisk community, everyone finds themselves using conference rooms more and more, for purposes such as these:

- Small companies need an easy way for business partners to get together for a chat
- Sales teams have a meeting once per week where everyone can dial in from wherever they are
- Development teams designate a common place and time to update each other on progress

Home automation

Asterisk is still too much of an über-geek's tool to be able to serve in the average home, but with no more than average Linux and Asterisk skills, the following things become plausible:

Monitoring the kids
> Parents who want to check up on the babysitter (or the kids home alone) could dial an extension context protected by a password. Once authenticated, a two-way audio connection would be created to all the IP phones in the house, allowing mom and dad to listen for trouble. Creepy? Yes. But an interesting concept nonetheless.

Locking down your phones
> Going out for the night? Don't want the babysitter tying up the phone? No problem! A simple tweak to the dialplan, and the only calls that can be made are to 911, your cell phone, and the pizza parlor. Any other call attempt will get the recording "We are paying you to babysit our kids, not make personal calls."
> Pretty evil, huh?

Controlling the alarm system
> You get a call while on vacation that your mom wants to borrow some cooking utensils. She forgot her key, and is standing in front of the house shivering. Piece of cake; a call to your Asterisk system, a quick digit string into the context you created for the purpose, and your alarm system is instructed to disable the alarm for 15 minutes. Mom better get her stuff and get out quick, though, or the cops'll be showing up!

Managing teenagers' calls

How about allocating a specific phone-time limit to your teenagers? To use the phone, they have to enter their access codes. They can earn extra minutes by doing chores, scoring all As, dumping that annoying bum with the bad haircut— you get the idea. Once they've used up their minutes... click... you get your phone back.

Incoming calls can be managed as well, via Caller ID. "Donny, this is Suzy's father. She is no longer interested in seeing you, as she has decided to raise her standards a bit. Also, you should consider getting a haircut."

The Future of Asterisk

We've come to love the Internet, both because it is so rich in content and inexpensive and, perhaps more importantly, because it allows us to define how we communicate. As its ability to carry richer forms of media advances, we'll find ourselves using it more and more. Once Internet voice delivers quality that rivals (or betters) the capabilities of the PSTN, the phone company had better look for another line of business. The PSTN will cease to exist and become little more than one more communications protocol the Internet happily carries for us. As with most of the rest of the Internet, open source technologies will lead this transformation.

Speech Processing

The dream of having our technical inventions talk to us is older than the telephone itself. Each new advance in technology spurs a new wave of eager experimentation. Generally, results never quite meet expectations, possibly because as soon as a machine says something that *sounds* intelligent, most people assume that it *is* intelligent.

People who program and maintain computers realize their limitations, and thus tend to allow for their weaknesses. Everybody else just expects their computers and software to work. The amount of thinking a user must do to interact with a computer is often inversely proportional to the amount of thinking the design team did. Simple interfaces belie complex design decisions.

The challenge, therefore, is to design a system that has anticipated the most common desires of its users, and can also adroitly handle unexpected challenges.

Festival

The Festival text-to-speech server can transform text into spoken words. While this is a whole lot of fun to play with, there are many challenges to overcome.

For Asterisk, an obvious value of text to speech might be the ability to have your telephone system read your emails back to you. Of course, if you've noticed the somewhat poor grammar, punctuation, and spelling typically found in email messages these days, you can perhaps appreciate the challenges this poses.

One cannot help but wonder if the emergence of text to speech will inspire a new generation of people dedicated to proper writing. Seeing spelling and punctuation errors on the screen is frustrating enough—having to hear a computer speak such things will require a level of Zazen that few possess.

Sphinx

If text to speech is rocket science, speech recognition is science fiction.

Speech recognition can actually work very well, but unfortunately this is generally true only if you provide it with the right conditions—and the right conditions are not those found on a telephone network. Even a perfect PSTN connection is considered to be at the lowest acceptable limit for accurate speech recognition. Add in compressed and lossy VoIP connections, or a cell phone, and you will discover far more limitations than uses.

Asterisk has the potential to be a fantastic system for speech recognition, due to its flexibility. Unfortunately, speech recognition itself is not yet mature enough to be put to the kinds of uses we want of it. As this technology ripens, the open source community is the most likely to embrace it and provide flexible, powerful platforms on which to run it.

High-Fidelity Voice

As we gain access to more and more bandwidth, it becomes less and less easy to understand why we still use low-fidelity codecs. Many people do not realize that Skype uses a higher fidelity than a telephone; it's a large part of the reason why Skype has a reputation for sounding so good.

If you were ever to phone CNN, wouldn't you love to hear James Earl Jones's mellifluous voice saying "This is CNN," instead of some tinny electronic recording? And if you think Allison Smith* sounds good through the phone, you should hear her in person!

In the future, we will expect, and get, high-fidelity voice though our communications equipment.

---

* Allison Smith is The Voice of Asterisk—it is her voice in all of the system prompts. To have Allison produce your own prompt, simply visit *http://thevoice.digium.com*.

Video

Video is in some ways already compatible with Asterisk. The problem is not so much one of functionality as one of bandwidth and processing power. More significantly, it is not yet important enough to the community to merit the attention it needs.

The challenge of video-conferencing

The concept of video-conferencing has been around since the invention of the cathode ray tube. The telecom industry has been promising a video-conferencing device in every home for decades.

As with so many other communications technologies, if you have video-conferencing in your house, you are probably running it over the Internet, with a simple, inexpensive webcam. Still, it seems that people see video-conferencing as a bit gimmicky. Yes, you can see the person you're talking to, but there's something missing.

Why we love video-conferencing

Video-conferencing promises a richer communications experience than the telephone. Rather than hearing a disembodied voice, the nuances of speech that come from eye-to-eye communication are possible.

Why video-conferencing may never totally replace voice

There are some challenges to overcome, though, and not all of them are technical.

Consider this: using a plain telephone, people working from their home offices can have business conversations, un-showered, in their underwear, feet on the desk, coffee in hand—if they use a telephone. A similar video conversation would require half an hour of grooming to prepare for, and couldn't happen in the kitchen, on the patio, or... well, you get the idea.

Also, the promise of eye-to-eye communication over video will never happen as long as the focal points of the participants are not in line with the cameras. If you look at the camera, your audience will see you looking at them, but you won't see them. If you look at your screen to see whom you are talking to, the camera will show you looking down at something—not at your audience. That looks impersonal. Perhaps if a videophone could be designed like a Tele-Prompt-R, where the camera was behind the screen, it wouldn't feel so unnatural. As it stands, there's something psychological that's missing. Video ends up being a gimmick.

Wireless

Since Asterisk is fully VoIP-enabled, wireless is all part of the package.

Wi-Fi

Wi-Fi is going to be the office mobility solution for VoIP phones. This technology is already quite mature. The biggest hurdle is the cost of handsets, which can be expected to improve as competitive pressure from around the world drives down prices.

Wi-MAX

Since we are so bravely predicting so many things, it's not hard to predict that Wi-MAX spells the beginning of the end for traditional cellular telephone networks.

With wireless Internet access within the reach of most communities, what value will there be in expensive cellular service?

Unified Messaging

This is a term that has been hyped by the telecom industry for years, but adoption has been far slower than predicted.

Unified Messaging is the concept of tying voice and text-messaging systems into one. With Asterisk, the two don't need to be artificially combined, as Asterisk already treats them the same way.

Just by examining the terms, *unified* and *messaging*, we can see that the integration of email and voicemail must be merely the beginning—Unified Messaging needs to do a lot more than just that if it is to deserve its name.

Perhaps we need to define "messaging" as communication that does not occur in real time. In other words, when you send a message, you expect that the reply may take moments, minutes, hours, or even days to arrive. You compose what you wish to say, and your audience is expected to compose a reply.

Contrast this with conversing, which happens in real time. When you talk to someone on a telephone connection, you expect no more than a few seconds' delay before the response arrives.

Tim O'Reilly delivered a speech entitled "Watching the Alpha Geeks: OS X and the Next Big Thing" (*http://www.macdevcenter.com/pub/a/mac/2002/05/14/oreilly_wwdc_keynote.html*), in which he talked about someone piping IRC through a text-to-speech engine. One could imagine doing the reverse as well, allowing us to join an IRC or Instant Messaging chat over our Wi-Fi phone, our Asterisk PBX providing the speech-to-text-to-speech translations.

Peering

As monopoly networks such as the PSTN give way to community-based networks like the Internet, there will be a period of time where it is necessary to interconnect the two. While the traditional providers would prefer that the existing model be carried into the new paradigm, it is increasingly likely that telephone calls will become little more than another application the Internet happily carries.

But a challenge remains: how to manage the telephone numbering plan with which we are all familiar and comfortable?

E.164

The ITU defined a numbering plan in their E.164 specification. If you've used a telephone to make a call across the PSTN, you can confidently state that you are familiar with the concept of E.164 numbering. Prior to the advent of publicly available VoIP, nobody cared about E.164 except the telephone companies—nobody needed to.

Now that calls are hopping from PSTN to Internet to who-knows-what, some consideration must be given to E.164.

ENUM

In response to this challenge, the IETF has sponsored the Telephone Number Mapping (ENUM) working group, the purpose of which is to map E.164 numbers into the Domain Name System (DNS).

While the concept of ENUM is sound, it requires cooperation from the telecom industry to achieve success. However, cooperation is not what the telecom industry is famous for, and thus far ENUM has foundered.

e164.org

The folks at e164.org are trying to contribute to the success of ENUM. You can log on to this site, register your phone number, and inform the system of alternative methods of communicating with you. This means that someone who knows your phone number can connect a VoIP call to you, as the e164.org DNS zone will provide the IP addressing and protocol information needed to connect to your location.

As more and more people publish VoIP connectivity information, fewer and fewer calls will be connected through the PSTN.

DUNDi

Distributed Universal Number Discovery (DUNDi) is an open routing protocol designed to maintain dynamic telecom routing tables between compatible systems.* While Asterisk is currently the only PBX to support DUNDi, the openness of the standard ensures that anyone can implement it.

DUNDi has huge potential, but it is very much in its infancy. This is the one to watch.

---

* See the previous chapter for more information.

Challenges

As is true with any worthwhile thing, Asterisk will face challenges. Let's take a glance at what some of them may be.

Too much change, too few standards

These days, the Internet is changing so fast, and offers so much diverse content, that it is impossible for even the most attentive geek to keep on top of it all. While this is as it should be, it also means that an enormous amount of technology churn is an inevitable part of keeping any communications system current.

VoIP spam

Yes, it's coming. There will always be people who believe they have the right to inconvenience and harass others in their pursuit of money. Efforts are underway to try and address this, but only time will tell how efficacious they will be.

Fear, uncertainty, and doubt

The industry is making the transition from ignorance to laughter. If Gandhi is correct, we can expect the fight to begin soon.

As their revenue streams become increasingly threatened by open source telephony, the traditional industry players are certain to mount a fear campaign, in hopes of undermining the revolution.

Bottleneck engineering

There is a rumor making the rounds that the major network providers will begin to artificially cripple VoIP traffic by tagging and prioritizing the traffic of their premium VoIP services and, worse, detecting and bumping any VoIP traffic generated by services not approved by them.

Some of this is already taking place, with service providers blocking traffic of certain types through their networks, ostensibly due to some public service being rendered (such as blocking popular file-sharing services to protect us from piracy). In the United States, the FCC has taken a clear stand on the matter and fined companies that engage in such practices. In the rest of the world, regulatory bodies are not always as accepting of VoIP.

What seems clear is that the community and the network will find ways around blockages, just as they always have.

Regulatory wars

The recently departed Chairman of the United States Federal Communications Commission, Michael Powell, delivered a gift that may well have altered the path of the

VoIP revolution. Rather than attempting to regulate VoIP as a telecom service, he has championed the concept that VoIP represents an entirely new way of communicating and requires its own regulatory space in which to evolve.

VoIP will become regulated, but not everywhere as a telephony service. Some of the regulations that may be created include:

Presence information for emergency services
> One of the characteristics of a traditional PSTN circuit is that it is always in the same location. This is very helpful to emergency services, as they can pinpoint the location of a caller by identifying the address of the circuit from which the call was placed. The proliferation of cell phones has made this much more difficult to achieve, since a cell phone does not have a known address. A cell phone can be plugged into any network and can register to any server. If the phone does not identify its physical location, an emergency call from it will provide no clue as to the where the caller is. VoIP creates similar challenges.

Call monitoring for law enforcement agencies
> Law enforcement agencies have always been able to obtain wiretaps on traditional circuit-switched telephone lines. While regulations are being enacted that are designed to achieve the same end on the network, the technical challenge of delivering this functionality will probably never be completely solved. People value their privacy, and the more governments want to stifle it, the more effort will be put toward maintaining it.

Anti-monopolistic practices
> These practices are already being seen in the U.S., with fines being levied against network providers who attempt to filter traffic based on content.

When it comes to regulation, Asterisk is both a saint and a devil: a saint because it feeds the poor, and a devil because it empowers the phrackers and spammers like nothing ever has. The regulation of open source telephony may in part be determined by how well the community regulates itself. Concepts such as DUNDi, which incorporate anti-spam processes, are an excellent start. On the other hand, concepts such as Caller ID-spoofing are ripe with opportunities for abuse.

Quality of Service

Due to the best-effort reality of the TCP/IP-based Internet, it is not yet known how well increasing real-time VoIP traffic will affect overall network performance. Currently, there is so much excess bandwidth in the backbone that best-effort delivery is generally quite good indeed. Still, it has been proven time and time again that whenever we are provided with more bandwidth, we figure out a way to use it up. The 1-MB DSL connection undreamt of 5 years ago is now barely adequate.

Perhaps a corollary of Moore's Law* will apply to network bandwidth. QoS may become moot, due to the network's ability to deliver adequate performance without

---

* Gordon Moore wrote a paper in 1965 that predicted the doubling of transistors on a processor every few years.

any special processing. Organizations that require higher levels of reliability may elect to pay a premium for a higher grade of service. Perhaps the era of paying by the minute for long-distance connections will give way to paying by the millisecond for guaranteed low latency, or by the percentage point for reduced packet loss. Premium services will offer the five-nines* reliability the traditional telecom companies have always touted as their advantage over VoIP.

Complexity

Open systems require new approaches toward solution design. Just because the hardware and software are cheap doesn't mean the solution will be. Asterisk does not come out of the box ready to run; it has to be designed and built, and then maintained. While the base software is free, and the hardware costs will be based on commodity pricing, it is fair to say that the configuration costs for a highly customized system will be a sizeable part of the solution costs—in many cases, because of its high degree of complexity and configurability, more than would be expected with a traditional PBX.

The rule of thumb is generally considered to be something like this: if it can be done in the dialplan, the system design will be roughly the same as for any similarly featured traditional PBX. Beyond that, only experience will allow one to accurately estimate the time required to build a system.

There is much to learn.

Opportunities

Open source telephony creates limitless opportunities. Here are some of the more compelling ones.

Tailor-made private telecommunications networks

Some people would tell you that price is the key, but we believe that the real reason Asterisk will succeed is because it is now possible to build a telephone system as one would a web site: with complete, total customization of each and every facet of the system. Customers have wanted this for years. Only Asterisk can deliver.

---

* This term refers to 99.999%, which is touted as the reliability of traditional telecom networks. Achieving five nines requires that service interruptions for an entire year total no more than 5 minutes and 15 seconds. Many people believe that VoIP will need to achieve this level of reliability before it can be expected to fully replace the PSTN. Many other people believe that the PSTN doesn't even come close to five-nines reliability. We believe that this could have been an excellent term to describe high reliability, but marketing departments abuse it far too frequently.

Low barrier to entry

Anyone can contribute to the future of communicating. It is now possible for someone with an old $200 PC to develop a communications system that has intelligence to rival the most expensive proprietary systems. Granted, the hardware would not be production-ready, but there is no reason the software couldn't be. This is one of the reasons why closed systems will have a hard time competing. The sheer number of people who have access to the required equipment is impossible to equal in a closed shop.

Hosted solutions of similar complexity to corporate web sites

The design of a PBX was always a kind of art form, but before Asterisk, the art lay in finding creative ways to overcome the limitations of the technology. With limitless technology, those same creative skills can now be properly applied to the task of completely answering the needs of the customer. Open source telephony engines such as Asterisk will enable this. Telecom designers will dance for joy, as their considerable creative skills will now actually serve the needs of their customers, rather than be focused on managing kludge.

Proper integration of communications technologies

Ultimately, the promise of open source comes to nothing if it cannot fulfill the need people have to solve problems. The closed industries lost sight of the customer, and tried to fit the customer to the product.

Open source telephony brings voice communications in line with other information technologies. It is finally possible to properly begin the task of integrating email, voice, video, and anything else we might conceive of over flexible transport networks (whether wired or wireless), in response to the needs of the user, not the whims of monopolies.

Welcome to the future of telecom!

VoIP Channels

VoIP channels in Asterisk represent connections to the protocols they support. Each protocol you wish to use requires a configuration file, containing general parameters defining how your system handles the protocol as well as specific parameters for each channel (or device) you will want to reference in your dialplan. In this appendix, we'll take an in-depth look at the IAX and SIP configuration files.

IAX

The IAX configuration file (*iax.conf*) contains all of the configuration information Asterisk needs to create and manage IAX protocol channels. The sections in the file are separated by headings, which are formed by a word framed in square brackets ([ ]). The name in the brackets will be the name of the channel, with one notable exception: the [general] section, which is not a channel, is the area where global protocol parameters are defined.

This section examines the various general and channel-specific settings for *iax.conf*. We will define each parameter, and then give an example of its use. Certain options may have several valid arguments. These arguments are listed beside the option, separated with the pipe symbol (|). For example, bandwidth=low|medium|high means that the bandwidth option accepts one of the values low, medium, or high as its argument.

You can insert comments anywhere in the *iax.conf* file, by preceding the comment text with the semicolon character (;). Everything to the right of the semicolon will be ignored. Feel free to use comments liberally.

General IAX Settings

The first non-comment line in your *iax.conf* file must be the heading [general]. The parameters in this section will apply to all connections using this protocol, unless defined differently in a specific channel's definition. Since some of these settings can be defined on a per-channel basis, we have identified settings that are always global with the tag "(global)" and those that can optionally be configured for individual channels with the tag "(channel)." If you define a channel parameter under the [general] section, you do not need to define it in each channel; its value becomes the default. Keep in mind that setting a parameter in the [general] section does not prevent you from setting it differently for specific channels; it merely makes this setting the default. Also keep in mind that not defining these parameters may, in some cases, cause a system default to be used instead.

Here are the parameters that you can configure:

accountcode *(channel)*

> The account code can be defined on a per-user basis. If defined, this account code will be assigned to a call record whenever no specific user account code is set. The accountcode name configured will be used as the *filename.csv* in the */var/log/asterisk/cdr-csv/* directory to store Call Detail Records (CDRs) for the user/peer/friend.
>
>   accountcode=iax-username

allow *and* disallow *(channel)*

> Specific codecs can be allowed or disallowed, limiting codec use to those preferred by the system designer. allow and disallow can also be defined on a per-channel basis. Keep in mind that allow statements in the [general] section will carry over to each of the channels, unless you reset with a disallow=all. Codec negotiation is attempted in the order in which the codecs are defined. Best practice suggests that you define disallow=all, followed by explicit allow statements for each codec you wish to use. If nothing is defined, allow=all is assumed.
>
>   disallow=all
>   allow=ulaw
>   allow=gsm
>   allow=ilbc

amaflags *(channel)*

> Automatic Message Accounting (AMA) is defined in the Telcordia Family of Documents listed under FR-AMA-1. These documents specify standard mechanisms for generation and transmission of CDRs. You can specify one of four AMA flags to apply to all IAX connections.
>
>   amaflags=default|omit|billing|documentation

authdebug *(global)*

> You can minimize the amount of authorization debugging by disabling it with authdebug=no. Authorization debugging is enabled by default if not explicitly disabled.
>
>   authdebug=no

autokill *(global)*

> To minimize the danger of stalling when a host is unreachable, you can set autokill to yes to specify that any new connection should be torn down if an ACK

is not received within 2,000 ms. (This is obviously not advised for hosts with high latency.) Alternatively, you can replace yes with the number of milliseconds to wait before considering a peer unreachable. autokill configures the wait for all IAX2 peers, but you can configure it differently for individual peers with the use of the qualify command.

```
autokill=1500
```

bandwidth *(channel)*

bandwidth is a shortcut that may help you get around using disallow=all and multiple allow statements to specify which codecs to use. The valid options are:

high

Allows all codecs (G.723.1, GSM, ulaw, alaw, G.726, ADPCM, slinear, LPC10, G.729, Speex, iLBC).

medium

Allows all codecs except slinear, ulaw, and alaw.

low

Allows all medium codecs except G.726 and ADPCM.

```
bandwidth=low|medium|high
```

bindport *and* bindaddr *(global)*

These optional parameters allow you to control the IP interface and port on which you wish to accept IAX connections. If omitted, the port will be set to 4569, and all IP addresses in your Asterisk system will accept incoming IAX connections. If multiple bind addresses are configured, only the defined interfaces will accept IAX connections. The address 0.0.0.0 tells Asterisk to listen on all interfaces.

```
bindport=4569
bindaddr=192.168.0.1
```

codecpriority *(channel)*

The codecpriority option controls which end of an inbound call leg will have priority over the negotiation of codecs. If set in the [general] section, the selected options will be inherited by all user entries in the channel configuration file; however, they can be defined in the individual user entries for more granular control. If set in both the [general] and user sections, the user entry will override that which is configured in the [general] section. If this parameter is not configured, the value defaults to host.

Valid options include:

caller

The inbound caller has priority over the host.

host

The host has priority over the inbound caller.

> Codec preferences are not considered—this is the default behavior before the implementation of codec preferences.

reqonly

> Codec preferences are ignored, and the call is accepted only if the requested codec is available.

> > codecpriority=caller|host|disabled|reqonly

delayreject *(global)*

> If an incorrect password is received on an IAX channel, this will delay the sending of the REGREQ or AUTHREP reject messages, which will help to secure against brute-force password attacks. The delay time is 1,000 ms.

> > delayreject=yes|no

forcejitterbuffer *(channel)*

> Since Asterisk attempts to bridge channels (endpoints) directly together, the endpoints are normally allowed to perform jitter buffering themselves. However, if the endpoints have a poor jitter buffer implementation, you may wish to force Asterisk to perform jitter buffering no matter what. You can force jitter buffering to be performed with forcejitterbuffer=yes.

> > forcejitterbuffer=yes

jitterbuffer *(channel)*

> *Jitter* refers to the varying latency between packets. When packets are sent from an end device, they are sent at a constant rate with very little latency variation. However, as the packets traverse the Internet, the latency between the packets may become varied; thus, they may arrive at the destination at different times, and possibly even out of order.

> The jitter buffer is, in a sense, a staging area where the packets can be reordered and delivered in a regulated stream. Without a jitter buffer, the user may perceive anomalies in the stream, experienced as static, strange sound effects, garbled words, or, in severe cases, missed words or syllables.

> The jitter buffer affects only data received from the far end. Any data you transmit will not be affected by your jitter buffer, as the far end will be responsible for the de-jittering of its incoming connections.

> The jitter buffer is enabled with the use of jitterbuffer=yes.

> > jitterbuffer=yes|no

language *(channel)*

> This sets the language flag to whatever you define. The global default language is English. The language that is set is sent by the channel as an information element. It is also used by applications such as SayNumber( ) that have different files for different languages. Keep in mind that languages other than English are not explicitly installed on the system, and it is up to you to configure the system to ensure that the language you specify is handled properly.

> > language=en

mailboxdetail *(global)*

If mailboxdetail is set to yes, the new/old message count is sent to the user, instead of a simple statement of whether new and old messages exist. mailboxdetail can also be set on a per-peer basis.

```
mailboxdetail=yes
```

maxjitterbuffer *(channel)*

This parameter is used to set the maximum size of the jitter buffer, in milliseconds. Be sure not to set maxjitterbuffer too high, or you will needlessly increase your latency.

```
maxjitterbuffer=500
```

regcontext *(channel)*

By specifying the context that contains the actions to perform, you can configure Asterisk to perform a number of actions when a peer registers to your server. This option works in conjunction with regexten, by specifying the extension to execute. If no regexten is configured, the peer name is used as the extension. Asterisk will dynamically create and destroy a NoOp at priority 1 for the extension. All actions to be performed upon registration should start at priority 2. More than one regexten may be supplied, if separated by an &. regcontext can be set on a per-peer basis or globally.

```
regcontext=registered-phones
```

regexten *(channel)*

The regexten option is used in conjunction with regcontext to specify the extension to be executed within the configured context. If regexten is not explicitly configured, the peer name is used as the extension to match.

```
regexten=myphone
```

resyncthreshold *(channel)*

The resynchronize threshold is used to resynchronize the jitter buffer if a significant change is detected over a few frames, assuming that the change was caused by a timestamp mixup. The resynchronization threshold is defined as the measured jitter plus the resyncthreshold value, defined in milliseconds.

```
resyncthreshold=1000
```

tos *(global)*

Asterisk can set the Type of Service (TOS) bits in the IP header to help improve performance on routers that respect TOS bits in their routing calculations. The following values are valid:

lowdelay

Minimize delay.

throughput

Maximize throughput.

reliability

Maximize reliability.

mincost

    Minimize cost.

    No bits set.

        `tos=lowdelay|throughput|reliability|mincost|none`

trunk *(channel)*

    IAX2 trunking enables Asterisk to send media (as mini-frames) from multiple channels using a single header. The reduction in overhead makes the IAX2 protocol more efficient when sending multiple streams to the same endpoint (usually another Asterisk server).

        `trunk=yes|no`

trunkfreq *(channel)*

    trunkfreq is used to control how frequently you send trunk messages, in milliseconds. Trunk messages are sent in conjunction with the trunk=yes command.

        `trunkfreq=20`

Retrieving Dialplan Information from a Remote Asterisk Box

Asterisk can retrieve dialplan information from another Asterisk box with the use of a `switch =>` statement. When this occurs, the Asterisk IAX channel driver must wait for a reply from the remote box before it can continue with other IAX-related processes. This is especially troubling when you have multiple `switch` statements nested throughout multiple boxes—if a `switch` statement has to traverse several boxes, there could be an appreciable delay before a result is returned.

When the global `iaxcompat` option is set to yes, Asterisk will spawn a separate thread when the `switch` lookup is being performed. The use of this thread allows the main IAX channel driver to continue on with other processes while the thread waits for the reply. A small performance hit is incurred with this option.

    `iaxcompat=yes|no`

---

register Statements

The register switch (`register =>`) is used to register your Asterisk box to a remote server—this lets the remote end know where you are, in case you are configured with a dynamic IP address. Note that `register` statements are used only when the remote end has you configured as a *peer*, and when host=dynamic.

The basic format for a `register` statement is:

    `register => `*`username`*`:`*`password`*`@`*`remote-host`*

The *password* is optional (if not configured on the remote system).

Alternatively, you can specify an RSA key by framing the appropriate RSA key name[*] in square brackets ([ ]):

```
register => username:[rsa-key-name]@remote-host
```

By default, register requests will be sent via port 4569. You can direct them to a different port by explicitly specifying it, as follows:

```
register => username:password@remote-host:1234
```

IAX Channel Definitions

With the general settings defined, we can now define our channels. Defining a guest channel is recommended whenever you want to accept anonymous IAX calls. This is a very common way for folks in the Asterisk community to contact one another. Before you decide that this is not for you, keep in mind that anyone whom you want to be able to connect to you via IAX (without you specifically configuring an account for them) will need to connect as a guest. This account, in effect, becomes your "IAX phone number." Your guest channel definition should look something like this:

```
[guest]
type=user
context=incoming
callerid="Incoming IAX Guest"
```

 No doubt the spammers will find a way to harass these addresses, but in the short term this has not proven to be a problem. In the long term, we'll probably use DUNDi.[†]

If you wish to accept calls from the Free World Dialup network, Asterisk comes with a predefined security key that ensures that anonymous connections cannot spoof an incoming Free World Dialup call. You'll want to set up an `iaxfwd` channel:

```
[iaxfwd]
type=user
context=incoming
auth=rsa
inkeys=freeworlddialup
```

If you have resources advertised on a DUNDi network, the associated user must be defined in *iax.conf*:

```
[dundi]
type=user
dbsecret=dundi/secret
context=dundi-incoming
```

---

[*] Asterisk RSA keys are typically located in */var/lib/asterisk/keys/*. You can generate your own keys using the *astkeygen* script.

[†] See Chapter 9 for more information regarding DUNDi.

IAX Authentication

IAX provides authentication mechanisms to allow for a reasonable level of security between endpoints. This does not mean that the audio information cannot be captured and decoded, but it does mean that you can more carefully control who is allowed to make connections to your system. Three levels of security are supported on IAX channels. The auth option defines which authentication method to use on the channel: plaintext, md5, or rsa.

plaintext, in IAX, offers very little security. While it will prevent connection to the channel unless a valid password is supplied, the fact that the password is stored in *iax. conf* in plain text and is transmitted and received as plain text makes this a very insecure authentication method.

md5 improves the security on the network connection; however, both ends still require a plain-text secret in the *iax.conf* file. Here's how it works: Box A requests a connection with Box B, which in turn replies with an authorization request including a randomly generated number. Box A then generates an MD5 hash using the value supplied in the secret field of *iax.conf* and the random number from Box B. The hash is returned in the authorization reply, and Box B compares it to the hash it generated locally. If the hashes match, authorization is granted.

rsa provides the most security. Before using RSA, each end must create a public and private key pair through the *astgenkey* script, typically located in */usr/src/asterisk/contrib/scripts/*. The public key must then be given to the far end. Each end of the circuit must include the public key of the far end in its channel definition, using the inkeys and outkey parameters.

RSA keys are stored in */var/lib/asterisk/keys/*. Public keys are named *name*.pub; private keys are named *name*.key. Private keys must be encrypted with 3DES.

If you have IAX-based devices (such as an IAXy), or IAX-based users at a remote node, you may want to provide them with their own channels with which to connect to the system.

Let's say you have a user on a remote node for whom you want to define an IAX channel. We'll call this hypothetical channel sushi. The channel definition might look something like this:

```
[sushi]
type=user
context=local_users
auth=md5,plaintext,rsa
secret=wasabi
notransfer=yes
jitterbuffer=yes
callerid="Happy Tempura" <(800) 555-1234>
accountcode=seaweed
deny=0.0.0.0/0.0.0.0
```

```
permit=192.168.1.100/255.255.255.0
language=en
```

Incoming calls for this channel will arrive in the context local_users and will ask the system to accept the Caller ID Happy Tempura <(800) 555-1234>. The system will be willing to accept MD5, plain-text, or RSA authentication from this user, so long as the password wasabi is provided and the call comes from the IP address 192.168.1.100. All calls related to this channel will be assigned the account code seaweed. Because we've defined notransfer, the media path for this channel will always pass through Asterisk; it cannot be redirected to another IAX node.

If you yourself are a remote node, and you need to connect into a remote node as a user, you would define that main node as your peer:

```
[sashimi_platter]
type=peer
username=sushi
secret=wasabi
host=192.168.1.101
qualify=yes
trunk=yes
```

A *peer* can be referenced from the dialplan with the name contained in square brackets but authenticate with a different username. The host is specified using either IP dotted notation or a fully qualified domain name (FQDN). You can determine the latency between you and the remote host, and whether the peer is alive, with qualify=yes. To minimize the amount of overhead for multiple calls going to the same peer, you can trunk them.

Trunking is unique to IAX and is designed to take advantage of the fact that two large sites may have multiple simultaneous VoIP connections between them. IAX trunking reduces overhead by loading several channels into each signaling packet. You can enable trunking for a channel with trunk=yes in *iax.conf*.

Figure A-1 shows a channel with trunking disabled, and Figure A-2 shows a channel with trunking enabled.

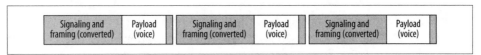

Figure A-1. Trunking disabled

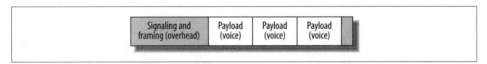

Figure A-2. Trunking enabled

Channel-specific parameters

Now, let's take a look at the channel-specific parameters:

callerid

> You can set a suggested Caller ID string for a user or peer with callerid. If you
> define a Caller ID field for a user, any calls that come in on that channel will
> have that Caller ID assigned to them, regardless of what the far end sends to
> you. If you define Caller ID for a *peer*, you are requesting that the far end use
> that to identify you (although you have no way of ensuring it will do so). If you
> want incoming users to be able to define their own Caller IDs (i.e., for guests),
> make sure you do not set the callerid field.
>
> ```
> callerid=John Smith <(800) 555-1234>
> ```

defaultip

> The defaultip setting complements host=dynamic. If a host has not yet regis-
> tered with your server, you'll attempt to send messages to the default IP address
> configured here.
>
> ```
> defaultip=192.168.1.101
> ```

inkeys

> You can use the inkeys option to authenticate a user with the use of an RSA key.
> To associate more than one RSA key with a user channel definition, separate the
> key names with a colon (:). Any one of those keys will be sufficient to validate a
> connection. The "inkey" is the public key you distribute to your users.
>
> ```
> inkeys=server_one:server_two
> ```

mailbox

> If you associate a mailbox with a peer within the channel definition, voicemail
> will send a message waiting indication (MWI) to the nodes on the end of that
> channel. If the mailbox number is in a voicemail context other than default, you
> can specify it as *mailbox@context*. To associate multiple mailboxes with a single
> peer, use multiple mailbox statements.
>
> ```
> mailbox=1000@internal
> ```

outkey

> You can use the outkey option to authenticate a peer with the use of an RSA key.
> Only one RSA key may be used for outgoing authentication. The "outkey" is not
> distributed; it is your private key.
>
> ```
> outkey=private_key
> ```

qualify

> You can set qualify to yes, no, or a time in milliseconds. If you set qualify=yes,
> PING messages will be sent periodically to the remote peers to determine whether
> they are available and what the latency between replies is. The peers will respond
> with PONG messages. A peer will be determined unreachable if no reply is received
> within 2,000 ms (to change this default, instead set qualify to the number of
> milliseconds to wait for the reply).

---

sendani

The SS7 PSTN network uses Automatic Number Identification (ANI) to identify a caller, and Caller ID is what is delivered to the user. The Caller ID is generated from the ANI, so it's easy to confuse the two. Blocking Caller ID sets a privacy flag on the ANI, but the backbone network still knows where the call is coming from.

    sendani=yes

 ANI has been around for a while. Its original purpose was to deliver the billing number of the originating party on a long-distance call to the terminating office. Unlike Caller ID, ANI does not require SS7, as it can be transmitted using DTMF. Also, ANI cannot be blocked.

SIP

Just as with IAX, the SIP configuration file (*sip.conf*) contains configuration information for SIP channels. The headings for the channel definitions are formed by a word framed in square brackets ([ ])—again, with the exception of the [general] section, where we define global SIP parameters. Don't forget to use comments generously in your *sip.conf* file. Precede the comment text with a semicolon; everything to the right will be ignored.

General SIP Parameters

The following options are to be used within the [general] section of *sip.conf*:

allowguest

If set to no, this disallows guest SIP connections. The default is to allow guest connections. SIP normally requires authentication, but you can accept calls from users who do not support authentication (i.e., do not have a secret field defined). Certain SIP appliances (such as the Cisco Call Manager v4.1) do not support authentication, so they will not be able to connect if you set allowguest=no.

    allowguest=no

bindaddr *and* bindport

These optional parameters allow you to control the IP interface and port on which you wish to accept SIP connections. If omitted, the port will be set to 5060, and all IP addresses in your Asterisk system will accept incoming SIP connections. If multiple bind addresses are configured, only those interfaces will listen for connections. The address 0.0.0.0 tells Asterisk to listen on all interfaces.

    bindaddr=0.0.0.0
    bindport=5060

callevents

Set this to yes when you want SIP to generate Manager events. This will be important if you have external programs that use the Asterisk Manager interface, such as the Flash Operator Panel.

```
callevents=yes
```

checkmwi

This option specifies the default amount of time, in seconds, between mailbox checks for peers.

```
checkmwi=30
```

compactheaders

You can set compactheaders to yes or no. If it's set to yes, the SIP headers will use a compact format, which may be required if the size of the SIP header is larger than the maximum transmission unit (MTU) of your IP headers, causing the IP packet to be fragmented. Do not use this option unless you know what you are doing.

```
compactheaders=no
```

defaultexpirey

This sets the default SIP registration expiration time, in seconds, for incoming and outgoing registrations. A client will normally define this value when it initially registers, so the default value you set here will be used only if the client does not specify a timeout when it registers. If you are registering to another user agent server (UAS), this is the registration timeout that it will send to the far end.

```
defaultexpirey=300
```

externhost

externhost takes a fully qualified domain name as its argument. If Asterisk is behind NAT, the SIP header will normally use the private IP address assigned to the server. If you set this option, Asterisk will perform periodic DNS lookups on the hostname and replace the private IP address with the IP address returned from the DNS lookup.

```
externhost=my.hostname.tld
```

 The use of externhost is not recommended in production systems, because if the IP address of the server changes, the wrong IP address will be set in the SIP headers until the next lookup is performed. The use of externip is recommended instead.

externip

externip takes an IP address as its argument. If Asterisk is behind NAT, the SIP header will normally use the private IP address assigned to the server. The remote server will not know how to route back to this address; thus, it must be replaced with a valid, routable address.

```
externip=216.239.39.104
```

externrefresh

> If externhost is used, externrefresh configures how long, in seconds, should pass between DNS lookups.
>
> ```
> externrefresh=30
> ```

localnet

> localnet is used to tell Asterisk which IP addresses are considered local, so that the address in the SIP header can be translated to that specified by externip or the IP address can be looked up with externhost.
>
> ```
> localnet=192.168.1.0/24
> localnet=172.16.0.0/16
> ```

maxexpirey

> This sets the maximum amount of time, in seconds, until a peer's registration expires.
>
> ```
> maxexpirey=3600
> ```

notifymimetype

> This takes as its argument a string specifying the MIME type used for the message waiting notification (MWI) in the SIP NOTIFY message. The most common setting for this field is text/plain, although it can be customized if need be.
>
> ```
> notifymimetype=text/plain
> ```

pedantic

> You can set pedantic to yes or no. Setting it to yes enables slow pedantic checking for phones that require it, such as the Pingtel, and enables more strict SIP RFC compliancy. In an effort to improve performance, SIP RFC compliance is not normally strictly adhered to.
>
> ```
> pedantic=yes
> ```

realm

> This option sets the realm for digest authentication. Set realm to your fully qualified domain name, which must be globally unique.
>
> ```
> realm=my.hostname.tld
> ```

recordhistory

> You can set recordhistory to yes or no to enable or disable SIP history recording for all channels. (See sip history and sip no history in Appendix E.)
>
> ```
> recordhistory=yes
> ```

relaxdtmf

> You can set relaxdtmf to yes or no. Setting it to yes will relax the DTMF detection handling. Use this if Asterisk is having a difficult time determining the DTMF on the SIP channel. Note that this may cause "talkoff," where Asterisk incorrectly detects DTMF when it should not.
>
> ```
> relaxdtmf=yes
> ```

srvlookup

> DNS SRV records are a way of setting up a logical, resolvable address where you can be reached. This allows calls to be forwarded to different locations without the need to change the logical address. By using SRV records, you gain many of the advantages of DNS, whereas disabling them removes the ability to place SIP calls based on domain names. (Note that if multiple records are returned, Asterisk will use only the first.) DNS SRV record lookups are recommended. To enable them, set srvlookup=yes in the [general] section of *sip.conf*.
>
>> srvlookup=yes

tos

> Asterisk can set the Type of Service (TOS) bits in the IP header to help improve performance on routers that respect TOS bits in their routing calculations. The following values are valid:

lowdelay

> Minimize delay.

throughput

> Maximize throughput.

reliability

> Maximize reliability.

mincost

> Minimize cost.

> No bits set.
>
>> tos=lowdelay|throughput|reliability|mincost|none

useragent

> useragent takes as its argument a string specifying the value for the useragent field in the SIP header. The default value is asterisk.
>
>> useragent=asterisk

videosupport

> You can set videosupport to yes or no. Setting it to yes will enable SIP video support. Video support works only between two endpoints—Asterisk does not support video conferencing at this time.
>
>> videosupport=yes

SIP Channel Definitions

Now that we've covered the global SIP parameters, we will discuss the channel-specific parameters. These parameters can be defined for a user, a peer, or both (as noted in parentheses):

accountcode *(both)*

> The account code can be defined on a per-user basis. If defined, this account code will be assigned to a call record whenever no specific user account code is set. The accountcode name configured will be used as the *filename.csv* in the */var/log/asterisk/cdr-csv/* directory to store CDRs for the user/peer/friend.
>
> ```
> accountcode=iax-username
> ```

allow *and* disallow *(both)*

> Specific codecs can be allowed or disallowed, limiting codec use to those preferred by the system designer. allow and disallow can also be defined on a per-channel basis. Keep in mind that allow statements in the [general] section will carry over to each of the channels, unless you reset with a disallow=all. Codec negotiation is attempted in the order in which the codecs are defined. Best practice suggests that you define disallow=all, followed by explicit allow statements for each codec you wish to use. If nothing is defined, allow=all is assumed.
>
> ```
> disallow=all
> allow=ulaw
> allow=gsm
> allow=ilbc
> ```

amaflags *(both)*

> Automatic Message Accounting (AMA) is defined in the Telcordia Family of Documents listed under FR-AMA-1. These documents specify standard mechanisms for generation and transmission of CDRs. You can specify one of four AMA flags (default, omit, billing, or documentation) to apply to all SIP connections.
>
> ```
> amaflags=documentation
> ```

callerid *(both)*

> You can set a suggested Caller ID string for a user or peer with callerid. If you define a Caller ID field for a user, any calls that come in on that channel will have that Caller ID assigned to them, regardless of what the far end sends to you. If Caller ID is defined for a peer, you are requesting that the far end use that to identify you (keep in mind, however, that you have no way to ensure that it will do so). If you want incoming callers to be able to define their own Caller IDs (i.e., for guests), make sure you do not set the callerid field.
>
> ```
> callerid=John Smith <(800) 555-1234>
> ```

callgroup *and* pickupgroup *(both)*

> You can use the callgroup parameter to assign a channel definition to one or more groups, and you can use the pickupgroup option in conjunction with this parameter to allow a ringing phone to be answered from another extension. The pickupgroup option is used to control which callgroups a channel may pick up—a channel is given authority to answer another ringing channel if it is assigned to the same pickupgroup as the ringing channel's callgroup. By default, remote ringing extensions can be answered with *8 (this is configurable in the *features.conf* file).
>
> ```
> callgroup=1,3-5
> pickupgroup=1,3-5
> ```

canreinvite *(both)*

The SIP protocol tries to connect endpoints directly. However, Asterisk must remain in the transmission path between the endpoints if it is required to detect DTMF. (For more information, see Chapter 4.)

```
canreinvite=no
```

context *(both)**

A context is assigned to a channel definition to direct incoming calls into the matching context in *extensions.conf*, where call handling is performed (see Chapters 4 and 5). Any channel connecting to an Asterisk machine has to have a context defined into which it will arrive. The context is essential for any user channel definition—if you do not define a context, incoming calls will be directed to the default context.

```
context=incoming
```

defaultip *(peer)*

The defaultip setting complements host=dynamic. If a host has not yet registered with your server, you'll attempt to send messages to the default IP address configured here.

```
defaultip=192.168.1.101
```

deny *(both)*

Specific IP addresses and ranges can be controlled with the deny option. To restrict access from a range of IP addresses, use a subnet mask—for example, deny=192.168.1.0/255.255.255.0. You can also deny all addresses with deny=0.0.0.0/0.0.0.0 and then allow only certain addresses with the permit command. Be aware of the security implications of this setting. (See also permit.)

```
deny=0.0.0.0/0.0.0.0
```

disallow *(both)*

See allow.

dtmfmode *(both)*

You can set dtmfmode to inband, rfc2833, or info. DTMF digits can be sent either in band (as part of the audio stream), or out of band (as signaling information), using the RFC 2833 or INFO methods. The inband method only works reliably when using an uncompressed codec such as G.711, ulaw, or alaw. The recommended method is to use rfc2833; however, some devices—such as those by Grandstream—support the info method.

```
dtmfmode=rfc2833
```

---

* You should be aware of an unusual scenario that will require a context definition for a peer. If a SIP call that originated from your system returns to your system for some reason (perhaps because a call you sent to a remote Asterisk box has been forwarded back to you), that call will authenticate not from a user definition (as would be expected), but rather from a peer definition. Since peers don't normally have contexts, this will cause such a call to arrive in the default context. While this will work, the default context shouldn't really be used to handle incoming calls. The solution is to define a context, on a per-peer basis, for any peers that may return calls to you (such as in a call-forwarding situation, or if the remote end is a SIP proxy server). To experiment with this, you can call your Free World Dialup number; the call will come right back to you.

fromdomain *(peer)*

This allows you to set the domain in the From: field of the SIP header. It may be required by some providers for authentication.

```
fromdomain=my.hostname.tld
```

fromuser *(peer)*

This allows you to set the username with which to authenticate. The name contained within the square brackets of the channel definition is usually used, but this can be overridden with the fromuser option. This allows a channel definition to be referenced with a name other than that used to authenticate.

```
fromuser=john_smith
```

host *(peer)*

This configures the host to which this peer is to connect. Use a fully qualified domain name.

```
host=remote.hostname.tld
```

incominglimit *(both)*

This option limits the total number of simultaneous calls for a peer or user. It sets the max number of simultaneous outgoing calls for a peer, or the max number of incoming calls for a user.

```
incominglimit=3
```

insecure *(both)*

When an INVITE is received from a remote location, Asterisk attempts to authenticate the string of characters before the @ sign on the INVITE line received in the SIP header with the name of a channel definition in *sip.conf*. If the remote end is a user agent, it will authenticate based on a user definition. However, if the remote end is a SIP proxy service, it will authenticate on the peer entry. When calls come from a provider such as Free World Dialup, which acts as a proxy for the true remote end who is calling you, that provider cannot authenticate the call on behalf of the endpoint. Since it would be impractical to have an authentication configured for every FWD user, and since FWD cannot respond to a 407 Proxy Authentication Required response, there must be an alternate way to allow calls from these callers.

If you set insecure=invite, you'll determine which peer to match on by comparing the IP address or hostname and port number to those provided in the Contact field of the SIP header with the host and port options in *sip.conf*. If a match is found, authentication will not be required on the initial INVITE, and the call will be allowed.

If you have multiple endpoints behind a NAT device, you need to enable insecure=port to match only against the IP address. To not require authentication on the incoming INVITE for the peer, set insecure=invite,port.

```
insecure=invite
```

language *(both)*

> This sets the language flag to whatever you define. The global default language is English. The language that is set is sent by the channel as an information element. It is also used by applications such as SayNumber( ) that have different files for different languages. Keep in mind that languages other than English are not explicitly installed on the system, and it is up to you to configure the system to ensure that the language you specify is handled properly.
>
> > `language=en`

mailbox *(peer)*

> If you associate a mailbox with a peer within the channel definition, voicemail will send a message waiting indication to the nodes on the end of that channel. If the mailbox number is in a voicemail context other than default, you can specify it as *mailbox@context*. To associate multiple mailboxes with a single peer, use multiple mailbox statements.
>
> > `mailbox=1000@internal`

md5secret *(both)*

> If you do not wish to have plain-text secrets in your *sip.conf* files, you can use md5secret to configure the MD5 hash that can be used for authentication. To generate the MD5 hash from the Linux console, use the following command:
>
> > `# echo -n "`*`username:realm:secret`*`" | md5sum`
>
> Be sure to use the –n flag, or echo will add a \n to the end of the string; the line feed will then be calculated into the MD5 hash, creating the incorrect hash. The *realm*, if not specified with the realm option (discussed in the list of general SIP parameters), defaults to asterisk. If both an md5secret and a secret are specified in the same channel definition, the secret will be ignored.
>
> > `md5secret=0bcbe762982374c276fb01af6d272dca`

musicclass *(both)*

> This option sets the default Music on Hold class.
>
> > `musicclass=classical`

nat *(both)*

> You can set nat to yes, no, or never. If you set it to yes, Asterisk ignores the IP address in the SIP and SDP headers and responds to the address and port in the IP header. The never option is for devices that cannot handle rport in the SIP header, such as the Uniden UIP200.
>
> > `nat=yes`

permit *(both)*

> See deny.

pickupgroup *(both)*

> See callgroup.

port *(peer)*

You can use this to define the port on which to listen for SIP signaling, if you want to listen on a nonstandard port. (The default port for SIP signaling is 5060.)

    port=5060

progressinband *(both)*

You can set progressinband to yes, no, or never, to configure whether or not to generate in-band ringing. Normally, Asterisk will send the progress of a call via a few methods, such as 183 Session Progress, 180 Ringing, 486 Busy, and so on. If you set progressinband=yes, Asterisk will indicate the call progress in band by generating tones.

    progressinband=yes

promiscredir *(both)*

You can set promiscredir to yes or no. Normally, when you perform call forwarding on a phone, Asterisk will use the Local channel (for example, *Local/18005551212@peer*). If you set promiscredir=yes, Asterisk will use the SIP channel instead, which enables you to forward the calls to remote boxes.

    promiscredir=yes

> Note that if Asterisk performs a redirect to itself when promiscredir=yes, the system will receive an INVITE with the same Caller ID and detect a loop to itself. SIP does not have the ability to perform a hairpin call, so the channel will then be destroyed.

qualify *(peer)*

You can set qualify to yes, no, or a time in milliseconds. If you set qualify=yes, NOTIFY messages will be sent periodically to the remote peers to determine whether they are available and what the latency between replies is. A peer is determined unreachable if no reply is received within 2,000 ms (to change this default, instead set qualify to the number of milliseconds to wait for the reply). Use this option in conjunction with nat=yes to keep the path through the NAT device alive.

    qualify=yes

regcontext *(peer)*

By specifying the context that contains the actions to perform, you can configure Asterisk to perform a number of actions when a peer registers to your server. This option works in conjunction with regexten, by specifying the extension to execute. If no regexten is configured, the peer name is used as the extension. Asterisk will dynamically create and destroy a NoOp at priority 1 for the extension. All actions to be performed upon registration should start at priority 2. More than one regexten may be supplied, if separated by an &. regcontext can be set on a per-peer basis or globally.

    regcontext=peer_registrations

regexten *(peer)*

> The regexten option is used in conjunction with `regcontext` to specify the extension that is executed within the configured context. If regexten is not explicitly configured, the peer name is used as the extension to match.
>
> ```
> regexten=1000
> ```

rtpholdtimeout *(peer)*

> This takes as its argument an integer, specified in seconds. It terminates a call if no RTP data is received while on hold. The value of `rtpholdtimeout` must be greater than that of `rtptimeout`. (See also `rtptimeout`.)
>
> ```
> rtpholdtimeout=120
> ```

rtptimeout *(peer)*

> This takes as its argument an integer, specified in seconds. It terminates a call if no RTP data is received within the time specified.
>
> ```
> rtptimeout=60
> ```

secret *(both)*

> This sets the password to use for authentication.
>
> ```
> secret=welcome
> ```

setvar *(both)*

> This sets a channel variable, which will be available when a channel to the peer or user is created and will be destroyed when the call is hung up. For example, to set the channel variable `foo` with a value of `bar`, use `setvar=foo=bar`.
>
> ```
> setvar=foo=bar
> ```

username *(peer)*

> The username field allows you to attempt contact with a peer before it has registered with you. At registration, a SIP device tells Asterisk which SIP URI to use to contact it. The username is used in conjunction with `defaultip` to create the SIP URI in the SIP INVITE header. This might be useful following a reboot, in order to place a call. The endpoints will not attempt to register with the server until their registration timeouts expire, so you will not know their locations. For non-dynamic hosts, you will require the username to be specified, as it is used to construct the authorization username.
>
> ```
> username=john_smith
> ```

Application Reference

AbsoluteTimeout( )

Sets the maximum number of seconds a call may last

AbsoluteTimeout(*length*)

Sets the absolute time limit of a call to *length* seconds. Calls lasting longer than *length* seconds will be sent to the T (absolute timeout) extension, if it exists. Otherwise, the channel will be hung up.

If *length* is set to zero (0), the timeout is disabled.

Each time AbsoluteTimeout( ) runs, it overrides the previous timeout setting. Asterisk starts the timeout countdown at the time the application is called, not at the time the call starts.

```
; limit calls to ex-girlfriend to 300 seconds
exten => 123,1,AbsoluteTimeout(300)
exten => 123,2,Dial(${EX-GIRLFRIEND})
exten => T,1,Playback(im-sorry)
exten => T,2,Playback(vm-goodbye)
exten => T,3,Hangup( )
```

See Also

DigitTimeout( ), ResponseTimeout( ), the T extension

AddQueueMember( )

Dynamically adds queue members to the specified call queue

AddQueueMember(*queuename*[,*interface*[,*penalty*]])

Dynamically adds the specified *interface* to an existing queue named *queuename*, as specified in *queues.conf*. If specified, *penalty* sets the penalty for queues to use this member. Members with a lower penalty are called before members with a higher penalty.

If *interface* is already a member of the queue and there exists an n+101 priority (where n is the number of the current priority), the call will continue at that priority. Otherwise, it will return an error.

Calling AddQueueMember( ) without an *interface* argument will use the interface that the caller is currently using.

```
; add SIP/3000 to the techsupport queue, with a penalty of 1
exten => 123,1,AddQueueMember(techsupport,SIP/3000,1)
```

See Also

RemoveQueueMember( ), *queues.conf*

ADSIProg( )   Loads an ADSI script into an ADSI-capable phone

ADSIProg(*script*)

Programs an Analog Display Services Interface (ADSI) phone with the given *script*. If none is specified, the default script, *asterisk.adsi*, is used. The path for the *script* is relative to the Asterisk configuration directory (usually */etc/asterisk/*). You may also provide the full path to the script.

To get the CPE ID and other information from your ADSI-capable phone, use the GetCPEID( ) application.

```
; program the ADSI phone with the telcordia-1.adsi script
exten => 123,1,ADSIProg(telcordia-1.adsi)
```

See Also

GetCPEID( ), *adsi.conf*

AgentCallbackLogin( )   Enables agent login with callback

AgentCallbackLogin([*AgentNo*][,[*options*][*exten*]@*context*])

Allows a call agent identified by *AgentNo* to log into the call queue system, to be called back when a call comes in for that agent.

When a call comes in for the agent, Asterisk calls the specified *exten* (with an optional *context*).

The *options* argument may contain the letter s, which causes the login to be silent.

```
; silently log in as agent number 42, and have Asterisk
; call SIP/400 when a call comes in for this agent
exten => 123,1,AgentCallbackLogin(42,s,SIP/400)
```

See Also

AgentLogin( )

AgentLogin( )   Allows a call agent to log into the system

AgentLogin([*AgentNo*][,*options*])

Logs the current caller into the call queue system as a call agent (optionally identified by *AgentNo*). While logged in, the agent can receive calls and will hear a beep on the line when a new call comes in. The agent can hang up the call by pressing the asterisk (*) key.

The *options* argument may contain the letter s, which causes the login to be silent.

```
; silently log in as agent number 42, as defined in agents.conf
exten => 123,1,AgentLogin(42,s)
```

See Also

AgentCallbackLogin( )

AgentMonitorOutgoing( )   Records an agent's outgoing calls

AgentMonitorOutgoing([*options*])

Records all outbound calls made by a call agent.

This application tries to figure out the ID of the agent who is placing outgoing call based on a comparison of the Caller ID of the current interface and the global variable set by the AgentCallbackLogin( ) application. As such, it should be used only in conjunction with (and after!) the AgentCallbackLogin( ) application. It uses the monitoring functions in the chan_agent module instead of the Monitor( ) application to record the calls. This means that call recording must be configured correctly in the *agents.conf* file.

By default, recorded calls are saved to the */var/spool/asterisk/monitor/* directory. This may be overridden by changing the savecallsin parameter in *agents.conf*.

If the Caller ID and/or agent ID are not found, this application will go to priority n+1, if it exists (where n is the current priority).

Returns 0 unless overridden by one of the options.

The *options* argument may include one or more of the following:

d   Make this application return -1 if there is an error condition and there is no extension n+101.

c   Change the Call Detail Record so that the source of the call is recorded as Agent/agent_id.

n   Don't generate warnings when there is no Caller ID or if the agent ID is not known. This option is useful if you want to have a shared context for agent and non-agent calls.

```
; record outbound calls for this agent, and change the CDR to reflect
; that the call is being made by an agent
exten => 123,1,AgentMonitorOutgoing(c)
```

See Also

AgentCallbackLogin( ), *agents.conf*

AGI( )   Executes an AGI-compliant application

[E]AGI(*program*[,*arguments*])

Executes an Asterisk Gateway Interface–compliant *program* on the current channel. AGI programs allow external programs (written in almost any language) to control the telephony channel by playing audio, reading DTMF digits, and so on. Asterisk communicates with the AGI program on STDIN and STDOUT. The specified *arguments* are passed to the AGI program.

The *program* must be set as executable in the underlying filesystem. The program path is relative to the Asterisk AGI directory, which by default is */var/lib/asterisk/agi-bin/*.

If you want to run an AGI when no channel exists (such as in an h extension), use the DeadAGI( ) application instead. You may want to use the FastAGI( ) application if you want to do AGI processing across the network.

If you want access to the inbound audio stream from within your AGI program, use EAGI( ) instead of AGI( ). Inbound audio can then be read in on file descriptor number three.

Returns -1 on hangup or if the program requested a hangup, or 0 on non-hangup exit.

```
; call the demo AGI program
exten => 123,1,AGI(agi-test)
exten => 123,2,EAGI(eagi-test)
```

See Also

DeadAGI( ), FastAGI( ), Chapter 9

AlarmReceiver( )   Provides support for receiving alarm reports from a burglar or fire alarm panel

AlarmReceiver( )

Emulates an alarm receiver, and allows Asterisk to receive and decode special data from fire and/or burglar alarm panels. At this time, only the Ademco Contact ID format is supported.

When called, AlarmReceiver( ) will handshake with the alarm panel, receive events, validate them, handshake them, and store them until the panel hangs up. Once the panel hangs up, the application will run the command line specified by the eventcmd setting in *alarmreceiver.conf* and pipe the events to the standard input of the application. *alarmreceiver.conf* also contains settings for DTMF timing and for the loudness of the acknowledgment tones.

 This application is not guaranteed to be reliable, so don't depend on it unless you have extensively tested it. If you use this application without extensive testing, you may be putting your life and property at great risk.

This application always returns 0.

```
; set up Asterisk to answer a call from a supported fire alarm panel
exten => s,1,AlarmReceiver( )
```

See Also

alarmreceiver.conf

Answer( )

```
Answer( )
```

Causes Asterisk to answer the channel if it is currently ringing. If the current channel is not ringing, this application does nothing.

It is usually a good idea to use Answer( ) on the channel before calling any other applications, unless you have a very good reason not to. Most applications require that the channel be answered before they are called, and may not work correctly otherwise.

Returns 0 unless it tries to answer the channel and fails.

```
exten => 123,1,Answer( )
exten => 123,2,Wait(1)
exten => 123,3,Playback(tt-weasels)
```

See Also

```
Hangup( )
```

AppendCDRUserField( )

```
AppendCDRUserField(value)
```

Appends *value* to the user field of the Call Detail Record (CDR). The user field is often used to store arbitrary data about the call, which may not be appropriate for any of the other fields.

Always returns 0.

```
; set the user field to 'abcde'
exten => 123,1,SetCDRUserField(abcde)
; now append 'xyz'
exten => 123,1,AppendCDRUserField(xyz)
```

See Also

```
SetCDRUserField( ), ForkCDR( ), NoCDR( ), ResetCDR( )
```

Authenticate( )

```
Authenticate(password[,options])
```

Requires a caller to enter a given *password* in order to continue execution of the next priority in the dialplan. Authenticate( ) gives the caller three chances to enter the password correctly. If the password is not correctly entered after three tries, the channel is hung up.

If *password* begins with the / character, it is interpreted as a file that contains a list of valid passwords (one per line). Passwords may also be stored in the Asterisk database (AstDB); see the d option below.

A set of *options* may be provided, consisting of one or more of the letters in the following list.

a   Sets the CDR field named accountcode and the channel variable ACCOUNTCODE to the password that is entered

d   Interprets the path as the database key from the Asterisk database in which to find the password, not a literal file. When using a database key, the value associated with the key can be anything.

r   Removes the database key upon successful entry (valid with d only).

Returns 0 if the user enters a valid password within three tries, or -1 otherwise (or on hangup).

```
; force the caller to enter the password before continuing, and set the CDR field
; named 'accountcode' to the entered password
exten => 123,1,Answer
exten => 123,2,Authenticate(1234,a)
exten => 123,3,Playback(pin-number-accepted)
exten => 123,4,SayDigits(${ACCOUNTCODE})
```

See Also

VMAuthenticate( ), Chapter 6

Background( )   Plays a file while accepting touch-tone (DTMF) digits

Background(*filename1*[&*filename2*...][,*options*[,*language*]])

Plays the specified audio file(s) while waiting for the user to begin entering an extension. Once the user begins to enter an extension, the playback is terminated. The *filename* should be specified without a file extension, as Asterisk will automatically find the file format with the lowest translation cost.

Valid *options* include one of the following:

skip
    Causes the playback of the message to be skipped if the channel is not in the "up" state (i.e., hasn't yet been answered). If skip is specified, the application will return immediately should the channel not be off-hook.

noanswer
    Does not answer the channel before playing the specified file. Without this option, the channel will automatically be answered before the sound is played. Not all channels support playing messages before being answered.

The *language* argument may be used to specify a language to use for playing the prompt, if it differs from the current language of the channel.

Returns -1 if the channel was hung up, or if the given filename does not exist; otherwise, returns 0.

```
exten => 123,1,Answer( )
exten => 123,2,Background('exter-ext-of-person');
```

See Also

Playback( ), BackgroundDetect( ), the show translation command

---

BackgroundDetect( )

BackgroundDetect(*filename*[,*sil*[,*min*[,*max*]]])

Similar to Background( ), but attempts to detect talking.

During the playback of the file, audio is monitored in the receive direction. If a period of non-silence that is greater than *min* milliseconds yet less than *max* milliseconds and is followed by silence for at least *sil* milliseconds occurs, the audio playback is aborted and processing jumps to the talk extension, if available.

If unspecified, *sil*, *min*, and *max* default to 1,000 ms, 100 ms, and infinity, respectively.

Returns -1 on hangup, and 0 on successful playback completion with no exit conditions.

```
exten => 123,1,BackgroundDetect(tt-monkeys)
exten => 123,2,Playback(im-sorry)
exten => talk,1,Playback(yes-dear)
```

See Also

Playback( ), Background( )

Busy( )

Busy([*timeout*])

Requests that the channel indicate the busy condition and then waits for the user to hang up or for the optional *timeout* (in seconds) to expire.

This application only signals a busy condition to the bridged channel. Each particular channel type has its own way of communicating the busy condition to the caller. You can use Playtones(busy) to play a busy tone to the caller.

Always returns -1.

```
exten => 123,1,Playback(im-sorry)
exten => 123,2,Playtones(busy)
exten => 123,3,Busy( )
```

See Also

Congestion( ), Progress( ), Playtones( )

CallingPres( )

CallingPres(*presentation*)

Changes the presentation parameters for the Caller ID on a Q931 PRI connection. These parameters should be set before placing an outgoing call. The argument *presentation* controls two things: whether or not the person being called can view the Caller ID information (known as *presentation*), and whether or not the Caller ID information has been verified by an authoritative source (known as *screening*).

This application has been replaced by the SetCallerPres() application, which is easier to use and less dependent on the internal Zaptel structures.

This application takes the call presentation setting and the screening setting and combines them into one number. The values themselves are defined in the ITU Q931 standard, as shown in Tables B-1 and B-2.

Table B-1. Screening is controlled by bits 2 and 1

Bit 2	Bit 1	Explanation
0	0	Caller ID information was provided by the user, and not screened.
0	1	Caller ID information was provided by the user, and successfully verified.
1	0	Caller ID information was provided by the user, and verification failed.
1	1	Caller ID information was provided by the network.

Table B-2. Presentation is controlled by bits 7 and 6

Bit 7	Bit 6	Explanation
0	0	Presentation of the Caller ID information is allowed.
0	1	Presentation of the Caller ID information is restricted.
1	0	The number is not available due to interworking.
1	1	Reserved.

Bits 3, 4, 5, and 8 should all be set to zero (0). Please note that the bits are numbered from most significant to least significant, like this: 87654321.

```
; set presentation to:
; Presentation Allowed            (00000000)
; Network Provided                (00000011)
; ------------------              ----------
; Result = 3 (bitwise AND)        (00000011)
exten => 123,1,CallingPres(3)
exten => 123,2,Dial(Zap/g1/8885551212)

; set presentation to:
; Presentation Restricted          (00100000)
; User-provided, verified, and passed (00000001)
; ------------------              ----------
; Result = 33 (bitwise AND)       (00100001)
exten => 124,1,CallingPres(33)
exten => 124,2,Dial(Zap/g1/8885551213)
```

See Also

SetCallerPres( ), SetCallerID( )

ChangeMonitor( )

ChangeMonitor(*filename_base*)

Changes the name of the recorded file created by monitoring a channel with the Monitor( ) application. This application has no effect if the channel is not monitored. The argument *filename_base* is the new filename base to use for monitoring the channel.

```
; start recording this channel with a basename of 'sample'
exten => 123,1,Monitor(sample)
; change the filename base to 'example'
exten => 123,2,ChangeMonitor(example)
```

See Also

Monitor( ), StopMonitor( )

ChanIsAvail( )

ChanIsAvail(*technology1/resource1*[&*technology2/resource2*...][,*option*])

Checks to see if any of the requested channels are available. If none of the requested channels are available, the new priority will be n+101 (where n is the current priority), unless that priority does not exist or an error occurs, in which case ChanIsAvail( ) will exit and return -1.

If any of the requested channels are available, the next priority will be n+1 and ChanIsAvail( ) will return 0.

ChanIsAvail( ) sets the following channel variables:

${AVAILCHAN}
> The name of the available channel, including the call session number used to perform the test.

${AVAILORIGCHAN}
> The canonical channel name that was used to create the channel—that is, the channel name without any session number.

${AVAILSTATUS}
> The status code for the channel.

If the option s (which stands for "state") is specified, Asterisk will consider the channel unavailable whenever it is in use, even if it can take another call.

 This application does not work correctly on MGCP channels.

```
; check both Zap/1 and Zap/2 to see if they're available
exten => 123,1,ChanIsAvail(Zap/1&Zap/2)
; if we go to priority 2, then one of the channels is available
; in priority 2, we'll dial our number on the available channel
exten => 123,2,NoOp(${AVAILORIGCHAN})
```

```
exten => 123,3,Dial(${AVAILORIGCHAN}/5551212)
; if we go to priority 101, then neither Zap/1 nor Zap/2 is available
exten => 123,3,Playback(all-circuits-busy-now)
```

CheckGroup( )

Checks the number of channels in a particular group

CheckGroup(*max*[*@category*])

Checks to see if the total number of channels in the current channel's group exceeds the *max* argument. If the number does not exceed *max*, the application continues to the next priority. If the number of channels in the group is higher than *max*, and priority n+101 exists (where n is the current priority), execution continues at that priority. Otherwise, the application terminates and -1 is returned.

When the optional *category* argument is passed, this application checks the total number of channels in the group category. See SetGroup( ) for more information about categories.

```
exten => 123,1,SetGroup(support)
exten => 123,2,CheckGroup(5)
; if there are less than five calls in the support group
exten => 123,3,Dial(${SUPPORT})
; if there are more than five calls in the support group
exten => 123,103,Playback(im-sorry)
```

See Also

SetGroup( ), GetGroupCount( ), GetGroupMatchCount( )

Congestion( )

Indicates congestion on the channel

Congestion([*timeout*])

Requests that the channel indicate congestion and then waits for the user to hang up or for the optional *timeout* (in seconds) to expire.

This application only signals congestion; it doesn't actually play a congestion tone to the user. You can use Playtones(congestion) to play a congestion tone to the caller.

Always returns -1.

```
; if the Caller ID is 555-1234, always play congestion
exten => 123,1,GotoIf($[${CALLERIDNUM} = 5551234]?5:2)
exten => 123,2,Playtones(congestion)
exten => 123,3,Congestion( )
exten => 123,4,Hangup( )
exten => 123,5,Dial(Zap/1)
```

See Also

Busy( ), Progress( ), Playtones( )

ControlPlayback( )   Plays a file, with the ability to fast forward and rewind the file

ControlPlayback(*filename*[,*skipms*[,*ffchar*[,*rewchar*[,*stopchar*[,*pausechr*]]]]])

Plays back a given filename (without the file extension), while allowing the caller to move forward and backward through the file by pressing *ffchar* and *rewchar*. By default, you can use * and # to rewind and fast-forward the playback of the file. If *stopchar* is specified, the application will stop playback when *stopchar* is pressed. If the file does not exist, the application jumps to priority n+101, if present (where n is the current priority number).

The *skipms* option specifies how far forward or backward to jump in the file with each press of *ffchar* or *rewchar*.

A *pausechr* option may also be specified, which will pause playback of the file. Pressing *pausechr* again will continue the playback of the file.

Returns -1 if the channel was hung up during playback.

```
; allow the caller to control the playback of this file
exten => 123,1,ControlPlayback(tt-monkeys|3000|#|*|5|0)
```

See Also

Playback( ), Background( )

Curl( )   Loads an external URL and assigns the result to a variable

Curl(*URL*[,*postdata*])

Downloads the given *URL* and assigns it to the channel variable named CURL. If specified, the *postdata* argument is passed to the URL as an HTTP POST. Curl( ) is often used to signal external applications of dialplan events.

Returns 0, or -1 on fatal errors.

```
; post the Caller ID number and unique call ID to a URL
exten => 123,1,Curl(http://localhost/test.
php,CallerID=${CALLERID}&UniqueCallID={$UNIQUEID})
; now use the NoOp( ) application to print the result to the Asterisk console
exten => 123,2,NoOp(${CURL})
```

Cut( )   Assigns part of one variable to another variable

Cut(*newvar=varname,delimiter,fieldspec*)

Cuts an existing variable named *varname* into several pieces, and assigns one or more of the pieces to a new variable named *newvar*.

The *delimiter* argument is the character on which to cut *varname*. It defaults to -.

fieldspec is the number of the field you want to assign to *newvar*. Fields are counted starting with 1. The *fieldspec* may be specified as a range (with -) or a group of ranges and fields (with &). If more than one field is selected, Cut( ) leaves the delimiter between the fields.

Returns 0, or -1 on hangup or error.

```
exten => 123,1,Set(TEST=123-456-7890)
exten => 123,2,Cut(FIRST=TEST,-,2)      ; gives us 456
exten => 123,3,Cut(SECOND=TEST,,1-2)    ; gives us 123-456
exten => 123,4,Cut(THIRD=TEST,-,1&3)    ; gives us 123-7890
```

DateTime( )
Says the specified time in a custom format

DateTime([*unixtime*][,*timezone*[,*format*]])

Says the time *unixtime*, in the time zone specified by *timezone*, according to the format specified in *format*.

The *unixtime* argument is the time, in seconds, since January 1, 1970. It may be negative for dates before 1970. *unixtime* defaults to the current time.

The *timezone* argument specifies the time zone of the specified time. See */usr/share/zoneinfo/* for a list of valid time zones. *timezone* defaults to the current time zone of the Asterisk server.

The *format* argument specifies which parts of the date and time should be read. See *voicemail.conf* for formatting options. *format* defaults to "ABdY 'digits/at' IMp".

Returns 0, or -1 on hangup.

```
; today's date and time
exten => 123,1,DateTime( )
; today's date
exten => 124,1,DateTime(,,BdY)
; A specified date
exten => 125,1,DateTime(871624800,,BdY)
```

DBdel( )
Deletes a key from the AstDB

DBdel(*family/key*)

Deletes the key specified by *key* from the key family named *family* in the AstDB.

Always returns 0.

```
exten => 123,1,DBput(test/name=John) ; add name to AstDB
exten => 123,2,DBget(NAME=test/name) ; retrieve name from AstDB
exten => 123,3,DBdel(test/name)      ; delete from AstDB
```

See Also

DBdeltree( ), DBput( ), DBget( )

DBdeltree( )
Deletes a family or key tree from the Asterisk database

DBdeltree(*family*[/*keytree*])

Deletes the specified *family* or *keytree* from the AstDB.

Always returns 0.

```
; create a couple of entries in the AstDB
exten => 123,1,DBput(test/blue)
exten => 123,2,DBput(test/green)
; now delete the key family named test
exten => 123,3,DBdeltree(test)
```

See Also

DBdel( ), DBput( ), DBget( )

DBget( )   Retrieves a key from the AstDB

DBget(*varname=family/key*)

Retrieves a key value from the Asterisk database and stores it in the variable specified by *varname*. If the requested key is not found, control jumps to priority n+101 (where n is the current priority), if it exists.

Always returns 0.

```
; put an entry in the AstDB
exten => 123,1,DBput(test/color=blue)
; now retrieve it and assign it to a variable
exten => 123,2,DBget(COLOR=test/color)
```

See Also

DBdel( ), DBdeltree, DBput( )

DBput( )   Stores a value in the AstDB

DBput(*family/key=value*)

Stores the given *value* in the corresponding *family* and *key* in the AstDB.

Always returns 0.

```
; put an entry in the AstDB
exten => 123,1,DBput(test/color=blue)
```

See Also

DBdel( ), DBdeltree, DBget( )

DeadAGI( )   Executes an AGI-compliant script on a dead (hung-up) channel

DeadAGI(*program,args*)

Executes an AGI-compliant *program* on a dead (hung-up) channel. AGI allows Asterisk to launch external programs written in almost any language to control a telephony channel, play audio, read DTMF digits, and so on by communicating with the AGI protocol on STDIN and STDOUT. The arguments specified by *args* will be passed to the program.

This application has been written specifically for dead channels, as the normal AGI interface doesn't work correctly if the channel has been hung up.

Use the show agi command on the command-line interface to list all of the available AGI commands.

Returns -1 if the application requested a hangup, or 0 on a non-hangup exit.

> exten => h,1,DeadAGI(agi-test)

See Also

AGI( ), FastAGI( )

Dial( )   Attempts to connect channels

Dial(*tech/username:password@hostname/extension,ring-timeout,flags*)

Allows you to connect together all of the various channel types.[*] Dial( ) is the most important application in Asterisk—you'll want to read through this section a few times.

Any valid channel type (such as SIP, IAX2, H.323, MGCP, Local, or Zap) is acceptable to Dial( ), but the parameters that need to be passed to each channel will depend on the information the channel type needs to do its job. For example, a SIP channel will need a network address and user to connect to, whereas a Zap channel is going to want some sort of phone number.

When you specify a channel type that is network-based, you can pass the destination host (name or IP address), username, password, and remote extension as part of the options to Dial( ), or you can refer to the name of a channel entry in the appropriate *.conf* file; all the required information will then need to be obtained from that file. The username and password can be replaced with the name contained within square brackets ([ ]) of the channel configuration file. The hostname is optional.

This is a valid Dial statement:

> exten => s,1,Dial(SIP/sake:arigato@thathostoverthere.tld)

This is effectively identical:

> exten => s,1,Dial(SIP/some_SIP_friend)

but will work only if there is a channel defined in *sip.conf* as [some_SIP_friend], whose channel definition contains fromuser=sake, password=arigato, and host=thathostoverthere.tld.

An extension number is often attached after the address information, like this:

> exten => s,1,Dial(IAX2/user:pass@otherend.com/**500**)

This asks the far end to connect the call to extension 500 in the context in which the channel arrived. The extension is not required by Dial( ), as the information in the remote

---

[*] The fact that Asterisk will happily connect IAX, SIP, H.323, Skinny, PRI, FX(O/S), and anything else is amazing, but possibly the most amazing of all is the Local channel. By allowing a single Dial( ) command to connect to multiple Local channels, one Dial( ) event can trigger a multitude of completely independent and unique actions in other parts of the dialplan. The power of this concept is truly revolutionary and has to be experienced to be believed.

end's channel configuration file may be used, or the remote server will pass the call to the s extension in the context in which the call came in. Ultimately, the far end controls what happens to the call—you can only request a specific treatment.

If no *ring-timeout* is specified, the channel will ring indefinitely. This is not always a bad thing, so don't feel you need to set it—just be aware that "indefinitely" could mean a very long time. *ring-timeout* is specified in seconds. The ring timeout always follows the addressing information, like this:

```
exten => s,1,Dial(IAX2/user:pass@otherend.com/500,ring-timeout)
```

Much of the power of the Dial( ) application is in the flags. These are assigned following the addressing and timeout information, like this:

```
exten => s,1,Dial(IAX2/user:pass@otherend.com/500,60,flags)
```

 Here's something important to note: if you don't have a timeout specified, and you want to assign flags, you must still assign a spot for the timeout. You do this by adding an extra comma in the spot where the timeout would normally go, like this:

```
exten => s,1,Dial(IAX2/user:pass@otherend.com/500,,flags)
```

The valid flags that may be used with the Dial( ) application are:

d   Allows the user to dial a one-digit extension while waiting for a call to be answered. The call will then exit to that extension (either in the current context, if it exists, or in the context specified by ${EXITCONTEXT}).

t   Permits the called party to transfer a call by pressing the # key. Please note that if this option is used, reinvites are disabled, as Asterisk needs to monitor the call to detect when the called party presses the # key.

T   Permits the caller to transfer a connected call by pressing the # key. Again, note that if this option is used, reinvites are disabled, as Asterisk needs to monitor the call to detect when the caller presses the # key.

w   Permits the called user to start and stop recording the call audio to disk by pressing the automon sequence (as configured in *features.conf*). If the variable TOUCH_MONITOR is set, its value will be passed as the arguments to the Monitor( ) application when recording is started. If it is not set, the default values of WAV||m are passed to Monitor( ).

W   Permits the calling user to record the call audio to disk by pressing the automon sequence (as configured in *features.conf*).

f   Forces the Caller ID to be set as the extension of the line making or redirecting the outgoing call. This is done because some PSTN providers will not allow the Caller ID to be set to anything other than that which is assigned to you. For example, if you had a PRI, you would use the f flag to override any Caller ID set locally on a SIP phone.

o   Uses the Caller ID received on the inbound leg of the call for the Caller ID on the outbound leg of the call. This is useful if you are accepting a call and then forwarding it to another destination, but you wish to pass the Caller ID from the inbound leg of the call instead of overwriting it with the local Caller ID settings. This is the default behavior on Asterisk versions prior to 1.2.

r Indicates ringing to the calling party, without passing any audio until the call is answered. This flag is not normally required to indicate ringing, as Asterisk will signal ringing if a channel is actually being called.

m[*class*]

 Provides music to the calling party until the call is answered. You may also optionally indicate the Music on Hold class.

M(*x*[^*arg*])

 Executes the macro *x* upon the connection of a call, optionally passing arguments delimited by ^. The macro can also set the MACRO_RESULT channel variable to one of the following:

 ABORT

  Hangs up both legs of the call

 CONGESTION

  Acts as if the line encountered congestion

 BUSY

  Acts as if the line was busy (goes to n+101, where n is the current priority)

 CONTINUE

  Hangs up the called party and continues on in the dialplan

GOTO:<context>^<extension>^<priority>

 Transfers the call to the specified destination

h Allows the called user to hang up the channel by pressing *.

H Allows the calling user to hang up the channel by pressing *.

C Resets the Call Detail Record for the call. Since the CDR time is set to when you Answer( ) the call, you may wish to reset the CDR so the end user is not billed for the time prior to the Dial( ) application being invoked.

P[(*x*)]

 Sets the privacy mode, optionally specifying *x* as the family/key value in the local AstDB. Useful for accepting calls based on a blacklist (explicitly denying calls from listed numbers) or whitelist (explicitly accepting calls from listed numbers). See also LookupBlacklist( ).

g Goes on in the context if the destination channel hangs up.

G(*context^extension^priority*)

 Transfers both parties to the specified destination, if the call is answered.

A(*x*)

 Plays an announcement to the called party; *x* is the filename of the sound file to play as the announcement.

D([*called*][:*calling*])

 Sends DTMF digits after the call has been answered, but before the call is bridged. The *called* parameter is passed to the called party, and the *calling* parameter is passed to the calling party. Either parameter may be used individually.

L(*x*[:*y*][:*z*])

 Limits the call to *x* milliseconds, warning when *y* milliseconds are left and repeating every *z* milliseconds until the limit is reached. The *x* parameter is required; the *y* and *z* parameters are optional. The following special variables may also be set to provide additional control:

LIMIT_PLAYAUDIO_CALLER=yes|no
> Specifies whether to play sounds to the caller

LIMIT_PLAYAUDIO_CALLEE=yes|no
> Specifies whether to play sounds to the callee

LIMIT_TIMEOUT_FILE=*filename*
> Specifies which file to play when time is up

LIMIT_CONNECT_FILE=*filename*
> Specifies which file to play when call begins

LIMIT_WARNING_FILE=*filename*
> Specifies the file to play if the argument *y* is defined

n
Prevents jumping to priority n+101 (where n is the number of the current priority) if all channels are deemed busy.

A call may also be parked instead of being transferred (which is done with the t or T flags). Calls are normally parked by transferring them to extension 700, but that's configurable in the *features.conf* file.

The Dial( ) application sets the following variables upon exiting:

DIALEDTIME
> The total time elapsed from execution of Dial( ) until completion.

ANSWEREDTIME
> The total time elapsed during the call.

DIALSTATUS
> The status of the call, set as one of the following values:

> CHANUNAVAIL
>> The channel is unavailable.

> CONGESTION
>> The channel returned a congestion signal, usually indicating that it was unable to complete the connection.

> NOANSWER
>> The channel did not answer in the time indicated by the ring-timeout option.

> BUSY
>> The dialed channel is currently busy.

> ANSWER
>> The channel answered the call.

> CANCEL
>> The call was cancelled.

```
; dial a seven-digit number on Zap channel 4
exten => 123,1,Dial(Zap/4/2317154)

; dial the same number, but this time only have it ring for 10 seconds
; before continuing on with the dialplan
```

```
exten => 124,1,Dial(Zap/4/2317154,10)
exten => 124,2,Playback(im-sorry)
exten => 124,3,Hangup( )

; dial the same number, but this time with no timeout, and using the
; t, T, and m flags
exten => 125,1,Dial(Zap/4/2317154,,tTm)

; dial extension 500 at a remote host (over the IAX protocol), using
; the specified username and password
exten => 126,1,Dial(IAX/username:password@remotehost/500)

; dial a number, but limit the call to 5 minutes (300,000 milliseconds)
; start warning the caller 4 minutes (240,000 milliseconds) into the call,
; and repeat the warning every 30 seconds (30,000 milliseconds)
exten => 127,1,Dial(Zap/4/2317154,,L[300000:240000:30000])
```

DigitTimeout( )   Sets the maximum timeout between digits

DigitTimeout(*seconds*)

Sets the maximum amount of time permitted between digit presses when the caller is entering an extension. If the time period specified by *seconds* elapses after the caller enters a digit, the extension will be considered complete and will be interpreted.

Note that if a valid extension is typed in it will not have to time out to be tested, so typically at the expiration of this timeout, the extension will be considered invalid (and thus control will be passed to the i extension, or, if it doesn't exist, the call will be terminated).

Always returns 0.

```
exten => 123,1,DigitTimeout(3)
exten => 123,2,Background(enter-ext-of-person)
exten => i,1,Playaback(im-sorry)
exten => i,2,Goto(123,1)
```

See Also

AbsoluteTimeout( )

Directory( )   Provides a dialable directory of extensions

Directory(*vm-context*[,*dial-context*[,*options*]])

Presents users with a directory of extensions from which they may select by name. The list of names and extensions is discovered from *voicemail.conf*. The *vm-context* argument is required; it specifies the context of *voicemail.conf* to use.

The *dial-context* argument is the context to use for dialing the users, and it defaults to *vm-context* if unspecified. Currently, the only option that can be specified in the *options* argument is f, which causes the directory to match based on the first name in *voicemail.conf* instead of the last name.

If the user enters 0 (zero) and there exists an extension o (the lowercase letter o) in the current context, the call control will go to that extension. Entering * will exit similarly, but to the a extension, much like Voicemail( )'s behavior.

Returns 0 unless the user hangs up.

```
exten => *,1,Directory(default,incoming)
exten => #,1,Directory(default,incoming,f)
```

See Also

voicemail.conf

DISA( )   Direct Inward System Access: allows inbound callers to make outbound calls

DISA(*password*[,*context*[,*callerid*[,*mailbox*[@*vmcontext*]]]])
DISA(*password-file*[,*callerid*[,*mailbox*[@*vmcontext*]]])

Allows outside callers to obtain an "internal" system dial tone and to place calls from it as if they were placing calls from within the switch. The user is given a dial tone, after which she should enter her passcode, followed by the pound sign (#). If the passcode is correct, the user is then given a system dial tone on which a call may be placed.

 Obviously, this type of access has serious security implications, and *extreme* care must be taken not to compromise the security of your phone system.

The *password* argument is a numeric passcode that the user must enter to be able to make outbound calls. Using this syntax, all callers to this extension will use the same password. To allow users to use DISA( ) without a password, put the string "no-password" instead of the password.

The *context* argument specifies the context in which the user will be dialing. If no context is specified, the DISA( ) application defaults the context to disa.

The *callerid* argument specifies a new Caller ID string that will be used on the outbound call.

The *mailbox* argument is the mailbox number (and optional voicemail context, *vmcontext*) of a voicemail box. The caller will hear a stuttered dial tone if there are any new messages in the specified voicemail box.

Additionally, you may use an alternate syntax and pass the name of a global password file instead of the *password* and *context* arguments. On each line, the file may contain either a passcode, or a passcode and context, separated by a pipe character (|). If a context is not specified, the application defaults to the context named disa.

If the user login is successful, the application parses the dialed number in the specified *context*.

```
; allow outside callers to call 1-800 numbers, as long
; as they know the passcode. Set their Caller IDs to make
; it appear that they are dialing from within the company
[incoming]
```

```
exten => 123,1,DISA(4569,disa,"Company ABC" <(234) 123-4567>)

[disa]
exten => _1800NXXXXXX,1,Dial(Zap/4/${EXTEN})
```

DumpChan( )

Dumps information about the calling channel to the console

DumpChan([*min_verbose_level*])

Displays information about the calling channel, as well as a listing of all channel variables. If *min_verbose_level* is specified, output is displayed only when the verbosity level is currently set to that number or greater.

Always returns 0.

```
exten => s,1,Answer( )
exten => s,2,DumpChan( )
exten => s,3,Background(enter-ext-of-person)
```

DUNDiLookup( )

Looks up a phone number using DUNDi

DUNDiLookup(*number*[,*context*[,*options*]])

Looks up the given phone *number* in the *context* specified, or in the reserved e164 context if not specified. On completion, the variables ${DUNDTECH} and ${DUNDDEST} will contain the appropriate technology and destination to access the number. If no answer was found, and the priority n+101 (where n is the current priority) exists, execution will continue at that priority.

The *options* argument is currently ignored.

Returns -1 if the channel is hung up during the lookup, or 0 otherwise.

```
; look up a number via DUNDi, and dial it
exten => 123,1,DUNDiLookup(8885551212)
exten => 123,2,Dial(${DUNDITECH}/${DUNDDEST})
; if DUNDi lookup fails, dial it on a Zap channel instead
exten => 123,102,Dial(Zap/4/1888551212)
```

See Also

ENUMLookup( )

EAGI( )

See AGI( ).

Echo( )

```
Echo( )
```

Echoes audio read from the channel back to the channel. This application is often used to test the latency and voice quality of a VoIP link. The caller may press the # key to exit.

Returns 0 if the user exits with the # key, or -1 if the user hangs up.

```
exten => 123,1,Echo( )
exten => 123,2,Playback(vm-goodbye)
```

See Also

```
Milliwatt( )
```

EndWhile( )

```
EndWhile( )
```

Returns to the previously called While( ) application. See While( ) for a complete description of how to use a while loop.

```
exten => 123,1,Set(COUNT=1)
exten => 123,2,While($[ ${COUNT} < 5 ])
exten => 123,3,SayNumber(${COUNT})
exten => 123,4,Set(COUNT=$[ ${COUNT} + 1 ])
exten => 123,5,EndWhile( )
```

See Also

```
While( ),GotoIf( )
```

ENUMLookup( )

```
ENUMLookup(number)
```

Looks up the telephone number specified by *number* via ENUM, and sets the variable ENUM with the result. For VoIP URIs, this variable will look like TECHNOLOGY/URI.

A good SIP, H.323, IAX, or IAX2 entry will result in normal-priority handling, whereas a good TEL entry will increase the priority by 51 (if the priority exists). If the lookup was *not* successful and there exists a priority n+101 (where n is the current priority), that priority will be taken next.

Currently, the only recognized ENUM services are SIP, H.323, IAX, IAX2, and TEL.

Returns -1 on hangup or 0 on completion, regardless of whether the lookup was successful.

```
; look up the phone number
exten => 123,1,ENUMLookup(8885551212)
; go to priority 2 on VoIP record
exten => 123,2,Dial(${ENUM})
; otherwise, go to priority 52 on TEL record
exten => 123,52,Dial(Zap4/${ENUM})
; otherwise, go to priority 102 because the lookup failed
exten => 123,102,Playback(im-sorry)
```

Eval( )

Evaluates any Asterisk variables located within a string

Eval(*newvar=string*)

Processes the given *string* and evaluates any variables contained in the string. The resulting value is assigned to the variable *newvar*.

This application is used in situations where a string is used in the dialplan, but any variables contained within it need to be evaluated first. This is often the case when the string is retrieved from a database or other external source.

```
; go through some convoluted steps to create a string that contains
; the unparsed variable ${UNIQUEID}
exten => 123,1,Set(ONE=\$)
exten => 123,2,Set(TWO=$[{UNIQUEID}])
; print the values to the console, to make sure it hasn't been parsed
exten => 123,3,NoOp(${ONE}${TWO})
; now evaluate the variables in the string
exten => 123,4,Eval(TEST=${ONE}${TWO})
; print the result to the console
exten => 123,5,NoOp(${TEST})
```

See Also

Exec( ), ExecIf( )

Exec( )

Executes an Asterisk application dynamically

Exec(*appname(arguments)*)

Allows an arbitrary application to be invoked even when not hard-coded into the dialplan. Returns whatever value the Asterisk *application* returns, or -2 when the called application cannot be found. The *arguments* are passed to the called *application*.

This application allows you to dynamically call applications by pulling them from a database or other external source.

```
exten => 123,1,Set(MYAPP=SayDigits(12345))
exten => 123,2,Exec(${MYAPP})
```

See Also

Eval( ), ExecIf( )

ExecIf( )

ExecIf(*expression*,*application*,*arguments*)

If *expression* is true, executes the given *application* with *arguments* as its arguments, and returns the result. For more information on standard Asterisk expressions, see Chapter 6 or the *README.variables* file in the *doc/* subdirectory of the Asterisk source.

If *expression* is false, execution continues at the next priority.

```
exten => 123,1,ExecIf($[ ${CALLERIDNUM} = 101],SayDigits,12345)
exten => 123,2,SayDigits(6789)
```

See Also

Exec( ), Eval( )

FastAGI( )

FastAGI(agi://*hostname*[:*port*][/*script*],*args*)

Executes an AGI-compliant program across the network. This application is very similar to AGI( ), except that it calls a specially written FastAGI script across a network connection. The main purposes for using FastAGI are to offload CPU-intensive AGI scripts to remote servers and to help reduce AGI script startup times (the FastAGI program is already running before Asterisk connects to it).

FastAGI( ) tries to connect directly to the running FastAGI program, which must already be listening for connections on the specified *port* on the server specified by *hostname*. If *port* is not specified, it defaults to port 4573. If *script* is specified, it is passed to the FastAGI program as the agi_network_script variable. The arguments specified by *args* will be passed to the program.

> See *agi/fastagi-test* in the Asterisk source directory for a sample Fast-AGI script. This should serve as a good roadmap for writing your own FastAGI programs.

Returns -1 if the application requested a hangup, or 0 on a non-hangup exit.

```
; connect to the sample fastagi-test program, which must already be running
; on the local machine
exten => 123,1,Answer( )
exten => 123,2,FastAGI(agi://localhost)

; connect to a FastAGI script on a host named "calvin" on port 8000, and pass along
; a script name of "testing", with the argument "12345"
exten => 124,1,Answer( )
exten => 124,2,FastAGI(agi://calvin:8000/testing,12345)
```

See Also

AGI( ), DeadAGI( )

Festival( )

Festival(*text*[,*intkeys*])

Connects to the locally running Festival server, sends it the text specified by *text*, and plays the resulting sound file back to the user. This application allows the caller to press a key (specified by *intkeys*) to immediately stop the playback and return the value of *intkeys*. If *intkeys* is set to any, Festival( ) will send control of the channel to the extension entered by the user.

See Chapter 10 for more in-depth information on using Festival with Asterisk.

You must start the Festival server before starting Asterisk, and you must use the Answer( ) application to answer the channel before calling Festival( ).

> For more information on using Festival from within Asterisk, see the *README.festival* file located in the *contrib/* subdirectory of the Asterisk source.

```
exten => 123,1,Answer( )
exten => 123,2,Festival('This is sample speech from Festival',#)
```

Flash( )

Flash( )

Sends a flash on a Zap channel. This is only a hack for people who want to perform transfers and other actions that require a flash via an AGI script. It is generally quite useless otherwise.

Returns 0 on success or -1 if this is not a Zap trunk.

```
exten => 123,1,Flash( )
```

ForkCDR( )

ForkCDR( )

Creates an additional Call Detail Record for the remainder of the current call.

This application is often used in calling-card applications to distinguish the inbound call (the original CDR) from the billable call time (the second CDR).

```
exten => 123,1,Answer( )
exten => 123,2,ForkCDR( )
exten => 123,3,Playback(tt-monkeys)
exten => 123,4,Hangup( )
```

See Also

AppendCDRUserField( ), NoCDR( ), ResetCDR( ), SetCDRUserField( )

GetCPEID( )

```
GetCPEID( )
```

Obtains the CPE ID and other information and displays it on the Asterisk console. This information is often needed in order to properly set up *zapata.conf* for on-hook operations with ADSI-capable telephones.

Returns -1 on hangup only.

```
; use this extension to get the necessary information to set up ADSI
; telephones
exten => 123,1,GetCPEID( )
```

See Also

ADSIProg( ), *adsi.conf*, *zapata.conf*

GetGroupCount( )

```
GetGroupCount([group][@category])
```

Counts the members in the given *group* (and optional *category*) and sets the ${GROUPCOUNT} variable to the corresponding value. If no *group* is specified, the current channel's group is used.

Use SetGroup( ) to assign a call as a member of a particular group.

Always returns 0.

```
; say the number of callers in the tech-support group
exten => 123,1,GetGroupCount(tech-support)
exten => 123,2,SayNumber(${GROUPCOUNT})
```

See Also

CheckGroup( ), GetGroupMatchCount( ), SetGroup( )

GetGroupMatchCount( )

```
GetGroupMatchCount(groupmatch[@category])
```

Counts the number of members in all groups matching the regular expression specified by *groupmatch*. The result is stored in the ${GROUPCOUNT} variable.

Note that standard regular expressions are used in the *groupmatch* argument.

Always returns 0.

```
; get the count of members in any group that starts with tech
exten => 123,1,GetGroupMatchCount(tech.*)
exten => 123,2,SayNumber($GROUPMATCH)
```

See Also

CheckGroup( ), GetGroupCount( ), SetGroup( )

Goto( )

Sends the call to the specified priority, extension, and context

Goto([[*context*,]*extension*,]*priority*)
Goto(*named_priority*)

Sends control of the current channel to the specified *priority*, optionally setting the destination *extension* and *context*.

Optionally, you can use the application to go to the named priority specified by the *named_priority* argument. Named priorities only work within the current extension.

Always returns 0, even if the given context, extension, or priority is invalid.

```
exten => 123,1,Answer( )
exten => 123,2,Set(COUNT=1)
exten => 123,3,SayNumber(${COUNT})
exten => 123,4,Set(COUNT=$[ ${COUNT} + 1 ])
exten => 123,5,Goto(3)

; same as above, but using a named priority
exten => 124,1,Answer( )
exten => 124,2,Set(COUNT=1)
exten => 124,3(repeat),SayNumber(${COUNT})
exten => 124,4,Set(COUNT=$[ ${COUNT} + 1 ])
exten => 124,5,Goto(repeat)
```

See Also

GotoIf( ), GotoIfTime( )

GotoIf( )

Conditionally goes to the specified priority

GotoIf(*condition*?*label1*:*label2*)

Sends the call to *label1* if *condition* is true or to *label2* if *condition* is false. Either *label1* or *label2* may be omitted (in that case, we just don't take the particular branch), but not both.

A label can be any one of the following:

- A priority, such as 10
- An extension and a priority, such as 123,10
- A context, extension, and priority, such as incoming,123,10
- A named priority within the same extension, such as passed

Each type of label is explained in the example below.

```
[globals]
; set TEST to something else besides 101 to see what GotoIf( )
; does when the condition is false
TEST=101
;
[incoming]
; set a variable
; go to priority 10 if ${TEST} is 101, otherwise go to priority 20
```

```
exten => 123,1,GotoIf($[ ${TEST} = 101 ]?10:20)
exten => 123,10,Playback(the-monkeys-twice)
exten => 123,20,Playback(tt-somethingwrong)
;
; same thing as above, but this time we'll specify an extension
; and a priority for each label
exten => 124,1,GotoIf($[ ${TEST} = 101 ]?123,10:123,20)
;
; same thing as above, but these labels have a context, extension, and
; priority
exten => 125,1,GotoIf($[ ${TEST} = 101 ]?incoming,123,10:incoming,123,20)
;
; same thing as above, but this time we'll go to named priorities
exten => 126,1,GotoIf($[ ${TEST} = 101 ]?passed:failed)
exten => 126,15(passed),Playback(the-monkeys-twice)
exten => 126,25(failed),Playback(the-monkeys-twice)
```

See Also

Goto( ), GotoIfTime( )

GotoIfTime( )   Conditionally branches, depending on the time and day

GotoIfTime(*times*,*days_of_week*,*days_of_month*,*months?label*)

Branches to the specified extension, if the current time matches the specified time. Each of the elements may be specified either as * (for always) or as a range.

The arguments to this application are:

times
: Time ranges, in 24-hour format

days_of_week
: Days of the week (mon, tue, wed, thu, fri, sat, sun)

days_of_month
: Days of the month (1-31)

months
: Months (jan, feb, mar, apr, etc.)

```
; If we're open, then go to the open context
; We're open from 9am to 6pm Monday through Friday
exten => s,1,GotoIfTime(09:00-17:59,mon-fri,*,*?open,s,1)
; We're also open from 9am to noon on Saturday
exten => s,2,GotoIfTime(09:00-11:59,sat,*,*?open,s,1)
; Otherwise, we're closed
exten => s,3,Goto(closed,s,1)
```

See Also

GotoIf( )

Hangup( )

Unconditionally hangs up the current channel

Hangup( )

Unconditionally hangs up the current channel.

Always returns -1.

```
exten => 123,1,Answer( )
exten => 123,2,Playback(im-sorry)
exten => 123,3,Hangup( )
```

See Also

Answer( )

HasNewVoicemail( )

Conditionally branches if there is new voicemail in the indicated voicemail box

HasNewVoicemail(*vmbox*[*@context*][*:folder*][*,varname*])

Similar to HasVoicemail( ). This application branches to priority n+101 (where n is the current priority) if there is new (unheard) voicemail in the voicemail box indicated by *vmbox*. The *context* argument corresponds to the voicemail context, and *folder* corresponds to a voicemail folder. If the voicemail folder is not specified, it defaults to the *INBOX* folder. If the *varname* argument is present, HasNewVoicemail( ) assigns the number of messages in the specified folder to that variable.

```
; check to see if there's unheard voicemail in INBOX of mailbox 123
; in the default voicemail context
exten => 123,1,Answer( )
exten => 123,2,HasNewVoicemail(123@default,COUNT)
exten => 123,3,Playback(vm-youhave)
exten => 123,4,Playback(vm-no)
exten => 123,5,Playback(vm-messages)
exten => 123,103,Playback(vm-youhave)
exten => 123,104,SayNumber($COUNT)
exten => 123,105,Playback(vm-messages)
```

See Also

HasVoicemail( ), MailboxExists( )

HasVoicemail( )

Conditionally branches if there is voicemail in the indicated voicemail box

HasVoicemail(*vmbox*[*@context*][*:folder*][*|varname*])

Branches to priority n+101 (where n is the current priority) if there is voicemail in the voicemail box indicated by *vmbox*. The *context* argument corresponds to the voicemail context, and *folder* corresponds to a voicemail folder. If the folder is not specified, it defaults to the *INBOX* folder. If the *varname* argument is passed, this application assigns the number of messages in the specified folder to that variable.

```
; check to see if there's any voicemail at all in INBOX of mailbox 123
; in the default voicemail context
```

256 | **Appendix B: Application Reference**

```
exten => 123,1,Answer( )
exten => 123,2,HasVoicemail(123@default,COUNT)
exten => 123,3,Playback(vm-youhave)
exten => 123,4,Playback(vm-no)
exten => 123,5,Playback(vm-messages)
exten => 123,103,Playback(vm-youhave)
exten => 123,104,SayNumber($COUNT)
exten => 123,105,Playback(vm-messages)
```

See Also

HasNewVoicemail( ), MailboxExists( )

IAX2Provision( ) Provisions a calling IAXy device

IAX2Provision([*template*])

Provisions a calling IAXy device (assuming that the calling entity is an IAXy) with the given *template*. If no template is specified, the default template is used. IAXy provisioning templates are defined in the *iaxprov.conf* configuration file.

Returns -1 on error or 0 on success.

```
; provision IAXy devices with the default template when they dial this extension
exten => 123,1,IAX2Provision(default)
```

ImportVar( ) Sets a variable based on a channel variable from a different channel

ImportVar(*newvar=channel,variable*)

Sets variable *newvar* to *variable* as evaluated on the specified *channel* (instead of the current channel). If *newvar* is prefixed with _, single inheritance is assumed. If prefixed with __, infinite inheritance is assumed.

```
; read the Caller ID information from channel Zap/1
exten => 123,1,Answer( )
exten => 123,1,ImportVar(cidinfo=Zap/1,CALLERID)
```

See Also

Set( )

LookupBlacklist( ) Performs a lookup of a Caller ID name/number from the blacklist database

LookupBlacklist( )

Looks up the Caller ID number on the active channel in the Asterisk database (family blacklist). If the number is found, and if there exists a priority n+101 (where n is the priority of the current instance), the channel will be set up to continue at that priority level. Otherwise, the application returns 0. If no Caller ID was received on the channel, it does nothing.

To add to the blacklist from the Asterisk CLI, type **database put blacklist** *name/number*.

```
; send blacklisted numbers to an endless loop
; otherwise, dial the number defined by the variable ${JOHN}
exten => s,1,Answer( )
exten => s,2,LookupBlacklist( )
exten => s,3,Dial(${JOHN})
exten => s,103,Playback(tt-allbusy)
exten => s,104,Wait(10)
exten => s,105,Goto(103)
```

LookupCIDName( )   Performs a lookup of a Caller ID name from the AstDB

LookupCIDName( )

Uses the Caller ID number on the active channel to retrieve the Caller ID name from the AstDB (family cidname). This application does nothing if no Caller"*ID was received on the channel. This is useful if you do not subscribe to Caller ID name delivery, or if you want to change the Caller ID names on some incoming calls.

Always returns 0.

```
; look up the Caller ID information from the AstDB, and pass it along
; to Jane's phone
exten => 123,1,Answer( )
exten => 123,2,LookupCIDName( )
exten => 123,3,Dial(SIP/Jane)
```

Macro( )   Calls a previously defined macro

Macro(*macroname*,*arg1*,*arg2*...)

Executes a macro defined in the context named macro-*macroname*, jumping to the s extension of that context and executing each step, then returning when the steps end.

The calling extension, context, and priority are stored in ${MACRO_EXTEN}, ${MACRO_CONTEXT}, and ${MACRO_PRIORITY}, respectively. Arguments *arg1*, *arg2*, etc. become ${ARG1}, ${ARG2}, etc. in the macro context.

Macro( ) returns -1 if any step in the macro returns -1, and 0 otherwise. If ${MACRO_OFFSET} is set at termination, this application will attempt to continue at priority MACRO_OFFSET+n+1 if such a step exists, and at n+1 otherwise. (In both cases, n stands for the current priority.)

If you call the Goto( ) application inside of the macro, the macro will terminate and control will go to the destination of the Goto( ).

```
; define a macro to count down from the specified value
[macro-countdown]
exten => s,1,Set(COUNT=${ARG1})
exten => s,2,While($[ ${COUNT} > 0])
exten => s,3,SayNumber(${COUNT})
exten => s,4,Set(COUNT=$[ ${COUNT} - 1 ])
exten => s,5,EndWhile( )
```

```
; call our macro with two different values
[example]
exten => 123,1,Macro(countdown,10)
exten => 124,1,Macro(countdown,5)
```

See Also

Goto( ), Chapter 6

MailboxExists( )

MailboxExists(*mailbox*[*@context*])

Conditionally branches to priority n+101 (where n is the current priority) if the voicemail box specified by the *mailbox* argument exists. You may pass a voicemail *context* if the mailbox is not in the default voicemail context.

```
exten => 123,1,Answer( )
exten => 123,2,MailboxExists(123@default)
exten => 123,3,Playback(im-sorry)
exten => 123,103,Voicemail(u123)
```

See Also

HasVoicemail( ), HasNewVoicemail( )

Math( )

Math(*returnvar,number1 operator number2*)

Performs a floating-point calculation on *number1* to *number2*, and assigns the result to the variable named *returnvar*. Valid operators are +, -, /, *, %, <, >, >=, <=, and ==, and they behave as their C equivalents.

Always returns 0.

```
; add two numbers, and say the result
exten => 123,1,Math(SUM,2+2)
exten => 123,2,SayNumber(${SUM})

; subtract two numbers, and say the difference
exten => 124,1,Math(DIFFERENCE,5-3)
exten => 124,2,SayNumber(${DIFFERENCE})
```

MeetMe( )

MeetMe([*confno*][,[*options*][,*pin*]])

Joins the caller on the current channel into the MeetMe conference specified by the *confno* argument. If the conference number is omitted, the user will be prompted to enter one.

The *options* string may contain zero or more of the characters in the following list.

m	Sets monitor-only mode (listen only, no talking).
t	Sets talk-only mode (talk only, no listening).
T	Sets talker detection (sent to Manager interface and MeetMe list).
i	Announces user join/leave.
p	Allows user to exit the conference by pressing #.
X	Allows user to exit the conference by entering a valid single-digit extension (set via the variable ${MEETME_EXIT_CONTEXT}), or the number of an extension in the current context if that variable is not defined.
d	Dynamically adds conference.
D	Dynamically adds conference, prompting for a PIN.
e	Selects an empty conference.
E	Selects an empty pinless conference.
v	Sets video mode.
r	Records conference (as ${MEETME_RECORDINGFILE} using format ${MEETME_ RECORDINGFORMAT}). The default filename is *meetme-conf-rec-*${CONFNO}-${UNIQUEID} and the default format is *.wav*.
q	Sets quiet mode (don't play enter/leave sounds).
M	Enables Music on Hold when the conference has a single caller.
x	Closes the conference when the last marked user exits.
w	Waits until the marked user enters the conference.
b	Runs the AGI script specified in ${MEETME_AGI_BACKGROUND}. Default: *conf-background.agi*. (Note: this does not work with non-Zap channels in the same conference.)
s	Presents the menu (user or admin) when * is received ("send" to menu).
a	Sets admin mode.
A	Sets marked mode.

If the *pin* argument is passed, the caller must enter that pin number to successfully enter the conference.

MeetMe( ) returns 0 if the caller presses # to exit (see option p); otherwise, it returns -1.

 A suitable Zaptel timing interface must be installed for MeetMe conferencing to work.

```
exten => 123,1,Answer( )
; add the caller to conference number 501 with pin 1234
exten => 123,2,MeetMe(501,DpM,1234)
```

See Also

MeetMeAdmin( ), MeetMeCount( )

MeetMeAdmin( )

MeetMeAdmin(*confno*,*command*[,*pin*])

Runs the specified MeetMe administration *command* on the specified conference. The *command* may be one of the following (note that the *pin* argument is used only for the k option):

K   Kicks all users out of the conference

k   Kicks one user (with the specified PIN as the third argument) out of the conference

e   Ejects the last user that joined

L   Locks the conference

l   Unlocks the conference

M   Mutes the conference

m   Unmutes the conference

N   Mutes the entire conference (except admin)

n   Unmutes the entire conference (except admin)

```
        ; mute conference 501
        exten => 123,1,MeetMeAdmin(501,M)

        ; kick user with PIN number 1234 from conference 501
        exten => 124,1,MeetMeAdmin(501,k,1234)
```

See Also

MeetMe( ), MeetMeCount( )

MeetMeCount( )

MeetMeCount(*confno*[,*variable*])

Plays back the number of users in the MeetMe conference identified by *confno*. If a variable is specified by the *variable* argument, playback will be skipped and the count will be assigned to *variable*.

Returns 0 on success or -1 on a hangup.

```
; count the number of users in conference 501, and assign that number to ${COUNT}
exten => 123,1,MeetMeCount(501,COUNT)
```

See Also

MeetMe( ), MeetMeAdmin( )

Milliwatt( )

Generates a 1,000-Hz tone

```
Milliwatt( )
```

Generates a constant 1,000-Hz tone at 0 dbm (mu-law). This application is often used for testing the audio properties of a particular channel.

```
; generate a milliwatt tone for testing
exten => 123,1,Milliwatt( )
```

See Also

Echo( )

Monitor( )

Monitors (records) the audio on the current channel

```
Monitor([file_format[:urlbase][,fname_base][,options]])
```

Starts monitoring a channel. The channel's input and output voice packets are logged to files until the channel hangs up or monitoring is stopped by the StopMonitor( ) application.

Monitor( ) takes the following arguments:

file_format
  Specifies the file format. If not set, defaults to wav.

fname_base
  If set, changes the filename used to the one specified.

options
  One of two options can be specified:

m

  When the recording ends, mix the two leg files into one and delete the original leg files. If the variable ${MONITOR_EXEC} is set, the application referenced in it will be executed instead of *soxmix*, and the raw leg files will *not* be deleted automatically. *soxmix* (or ${MONITOR_EXEC}) is handed three arguments: the two leg files and the filename for the target mixed file, which is the same as the leg filenames but without the in/out designator. If ${MONITOR_EXEC_ARGS} is set, the contents will be passed on as additional arguments to ${MONITOR_EXEC}. Both ${MONITOR_EXEC} and the m flag can be set from the administrator interface

b

  Don't begin recording unless a call is bridged to another channel.

Returns -1 if monitor files can't be opened or if the channel is already monitored; otherwise, returns 0.

```
exten => 123,1,Answer( )
; record the current channel, and mix the audio channels at the end of
; recording
exten => 123,2,Monitor(wav,monitor_test,mb)
exten => 123,3,SayDigits(12345678901234567890)
exten => 123,4,StopMonitor( )
```

ChangeMonitor( ), StopMonitor( )

MP3Player( )
Plays an MP3 file or stream

MP3Player(*location*)

Uses the *mpg123* program to play the given *location* to the caller. The specified *location* can be either a filename or a valid URL. The caller can exit by pressing any key.

 The correct version of *mpg123* must be installed for this application to work properly. Asterisk currently works best with *mpg123-0.59r*.

Returns -1 on hangup; otherwise, returns 0.

```
exten => 123,1,Answer( )
exten => 123,2,MP3Player(test.mp3)

exten => 123,1,Answer( )
exten => 123,2,MP3Player(http://server.tld/test.mp3)
```

MusicOnHold( )
Plays Music on Hold indefinitely

MusicOnHold(*class*)

Plays hold music specified by *class*, as configured in *musiconhold.conf*. If omitted, the default music class for the channel will be used. You can use the SetMusicOnHold( ) application to set the default music class for the channel.

Returns -1 on hangup; otherwise, does not return.

```
; transfer telemarketers to this extension to keep them busy
exten => 123,1,Answer( )
exten => 123,2,Playback(tt-allbusy)
exten => 123,3,MusicOnHold(default)
```

See Also

SetMusicOnHold( ), WaitMusicOnHold( )

NBScat( )
Plays an NBS local stream

NBScat( )

Uses the nbscat8k program to listen to the local Network Broadcast Sound (NBS) stream. (For more information, see the *nbs* module in Digium's CVS server.) The caller can exit by pressing any key.

Returns -1 on hangup; otherwise, does not return.

```
exten => 123,1,Answer( )
exten => 123,2,NBScat( )
```

NoCDR( )

Disables Call Detail Records for the current call

NoCDR( )

Disables CDRs for the current call.

```
; don't log calls to 555-1212
exten => 5551212,1,Answer( )
exten => 5551212,2,NoCDR( )
exten => 5551212,3,Dial(Zap/4/5551212)
```

See Also

AppendCDRUserField( ), ForkCDR( ), SetCDRUserField

NoOp( )

Does nothing

NoOp(*text*)

Does nothing—this application is simply a placeholder. As a side effect, the application evaluates *text* and prints the result to the Asterisk command-line interface, which can be useful for debugging.

 You don't have to place quotes around the text. If quotes are placed within the brackets, they will show up on the console.

```
exten => 123,1,NoOp(CallerID is ${CALLERID})
```

Park( )

Parks the current call

Park(*exten*)

Parks the current call (typically in combination with a supervised transfer to determine the parking space number). This application is always registered internally and does not need to be explicitly added into the dialplan, although you should include the parkedcalls context. Parking configuration is set in *features.conf*.

```
; park the caller in parking space 701
include => parkedcalls
exten => 123,1,Answer( )
exten => 123,2,Park(701)
```

See Also

ParkAndAnnounce, ParkedCall( )

ParkAndAnnounce( )

ParkAndAnnounce(*template*,*timeout*,*channel*[,*return_context*])

Parks the current call in the parking lot and announces the call over the specified *channel*. The *template* is a colon-separated list of files to announce; the word PARKED is replaced with the parking space number of the call. The *timeout* argument is the time in seconds before the call returns to the *return_context*. The *channel* argument is the channel to call to make the announcement. Console/dsp calls the console. The *return_context* argument is a GoTo( )-style label to jump the call back into after timeout, which defaults to n+1 (where n is the current priority) in the *return_context* context.

```
include => parkedcalls
exten => 123,1,Answer( )
exten => 123,2,ParkAndAnnounce(vm-youhave:a:pbx-transfer:at:vm-extension:
PARKED,120,Console/dsp)
exten => 123,3,Playback(vm-nobodyavail)
exten => 123,4,Playback(vm-goodbye)
exten => 123,5,Hangup( )
```

See Also

Park( ), ParkedCall( )

ParkedCall( )

ParkedCall(*exten*)

Connects the caller to the parked call in the parking space identified by *exten*. This application is always registered internally and does not need to be explicitly added into the dialplan, although you should include the parkedcalls context.

```
; pick up the call parked in parking space 701
exten => 123,1,Answer( )
exten => 123,2,ParkedCall(701)
```

See Also

Park( ), ParkAndAnnounce( )

PauseQueueMember( )

PauseQueueMember([*queuename*],*interface*)

Pauses (blocks calls for) a queue member. The specified interface will be paused in the given queue. This prevents any calls from being sent from the queue to the interface until it is unpaused by the UnpauseQueueMember( ) application or the Manager interface. If no *queuename* is given, the interface is paused in every queue it is a member of. If the *interface* is not in the named queue, or if no queue is given and the *interface* is not in any queue, it will jump to priority n+101 (where n is the current priority), if it exists.

Returns -1 if the interface is not found and no extension to jump to exists; otherwise, returns 0.

```
exten => 123,1,PauseQueueMember(,SIP/300)
exten => 124,1,UnpauseQueueMember(,SIP/300)
```

See Also

UnpauseQueueMember( )

Playback( )   Plays the specified audio file to the caller

Playback(*filename*[,*options*])

Plays back a given filename to the caller. The filename should not contain the file extension, as Asterisk will automatically choose the audio file with the lowest conversion cost. Zero or more *options* may also be included. The skip option causes the playback of the message to be skipped if the channel is not in the "up" state (i.e., it hasn't yet been answered). If skip is specified, the application will return immediately should the channel not be off-hook. Otherwise, unless noanswer is specified, the channel will be answered before the sound file is played. (Not all channels support playing messages while still on-hook.) Returns -1 if the channel was hung up. If the file does not exist, jumps to priority n+101 (where n is the current priority), if it exists.

```
exten => 123,1,Answer( )
exten => 123,2,Playback(tt-weasels)
```

See Also

Background( )

Playtones( )   Plays a tone list

Playtones(*tonelist*)

Plays a tone list. Execution immediately continues with the next step, while the tones continue to play. The *tonelist* is either the tone name defined in the *indications.conf* configuration file, or a specified list of frequencies and durations. See *indications.conf* for a description of the specification of a tone list.

Use the StopPlaytones( ) application to stop the tones playing.

```
; play a busy signal for two seconds, and then a congestion tone for two seconds
exten => 123,1,Playtones(busy)
exten => 123,2,Wait(2)
exten => 123,3,StopPlaytones( )
exten => 123,4,Playtones(congestion)
exten => 123,5,Wait(2)
exten => 123,6,StopPlaytones( )
exten => 123,7,Goto(1)
```

See Also

StopPlaytones( ), *indications.conf,* Busy( ), Congestion( ), Progress( ), Ringing( )

Prefix( )

Prefix(*digits*)

Prefixes the current extension with the digit string specified by *digits* and continues processing at the next priority for the new extension. So, for example, if priority 1 of extension 1212 is Prefix(555), 555 will be prepended to 1212 and the next step executed will be priority 2 of extension 5551212. If you switch into an extension that has no priority n+1 (where n is the current priority), Asterisk will treat it as though the user dialed an invalid extension.

Always returns 0.

```
exten => 1212,1,Prefix(555)
exten => 5551212,2,SayDigits(${EXTEN})
```

See Also

Suffix( )

PrivacyManager( )

PrivacyManager( )

If no Caller ID is received, answers the channel and asks the caller to enter his or her phone number. By default, the caller is given three attempts. If after three attempts the caller has not entered at least a 10-digit phone number, and if there exists a priority n+101 (where n is the current priority), the channel will be set up to continue at that priority level. Otherwise, it returns 0. If Caller ID was received on the channel, PrivacyManager( ) does nothing..

The *privacy.conf* configuration file changes the functionality of the PrivacyManger( ) application. It contains the following two lines:

maxretries

    Specifies the maximum number of attempts the caller is allowed to input a Caller ID number (default: 3)

minlength

    Specifies the minimum allowable digits in the input Caller ID number (default: 10)

```
exten => 123,1,Answer( )
exten => 123,2,PrivacyManager( )
exten => 123,3,Dial(Zap/1)
exten => 123,103,Playback(im-sorry)
exten => 123,104,Playback(vm-goodbye)
```

See Also

Zapateller( )

Progress( )

Progress( )

Requests that the channel indicate that in-band progress is available to the user.

Always returns 0.

```
; indicate progress to the calling channel
exten => 123,1,Progress( )
```

See Also

Busy( ), Congestion( ), Ringing( ), Playtones( )

Queue( )

Queue(*queuename*[,*options*[,*URL*[,*announceoverride*[,*timeout*]]]])

Places an incoming call into the call queue specified by *queuename*, as defined in *queues.conf*.

The *options* argument may contain zero or more of the following characters:

t   Allows the called user to transfer the call

T   Allows the calling user to transfer the call

d   Specifies a data-quality (modem) call (minimum delay)

h   Allows callee to hang up by hitting *

H   Allows caller to hang up by hitting *

n   Disallows retries on the timeout; exits this application and goes to the next step

r   Rings instead of playing MoH

In addition to being transferred, a call may be parked and then picked up by another user.

The *announceoverride* argument overrides the standard announcement played to queue agents before they answer the specified call.

The optional *URL* will be sent to the called party if the channel supports it.

The *timeout* will cause the queue to fail out after a specified number of seconds, checked between each *queues.conf* *timeout* and *retry* cycle.

Returns -1 if the originating channel hangs up, or if the call is bridged and either of the parties in the bridge terminates the call. If the queue is full, does not exist, or has no members, returns 0.

```
; place the caller in the techsupport queue
exten => 123,1,Answer( )
exten => 123,2,Queue(techsupport,t)
```

Random( )

Random([*probability*]:[[*context*,]*extension*,]*priority*)

Conditionally jumps to the specified *priority* (and optional *extension* and *context*), based on the specified *probability*. *probability* should be an integer between 1 and 100. The application will jump to the specified destination *priority* percent of the time.

```
; test your luck over and over again
exten => 123,1,Random(20:lucky,1)
exten => 123,2,Goto(unlucky,1)

exten => lucky,1,Playback(good)
exten => lucky,2,Goto(123,1)

exten => unlucky,1,Playback(bad)
exten => unlucky,2,Goto(123,1)
```

Read( )

Read(*variable*[,*filename*][,*maxdigits*][,*option*][,*attempts*][,*timeout*])

Reads a #-terminated string of digits a certain number of times from the user into the given *variable*.

Other arguments include:

filename
: Specifies the file to play before reading digits.

maxdigits
: Sets the maximum acceptable number of digits. If this argument is specified, the application stops reading after *maxdigits* have been entered (without requiring the user to press the # key). Defaults to 0- (no limit, wait for the user to press the # key). Any value below 0 means the same. The maximum accepted value is 255.

option
: Specify skip to return immediately if the line is not answered, or *noanswer* to read digits even if the line is not answered.

attempts
: If greater than 1, that many attempts will be made in the event that no data is entered.

timeout
: If greater than 0, that value will override the default timeout.

Returns -1 on hangup or error and 0 otherwise.

```
; read a two-digit number and repeat it back to the caller
exten => 123,1,Read(NUMBER,,2)
exten => 123,2,SayNumber(${NUMBER})
exten => 123,3,Goto(1)
```

See Also

SendDTMF( )

RealTime

RealTime(*family*,*colmatch*,*value*[,*prefix*])

Uses the RealTime configuration handler system to read data into channel variables. All unique column names (from the specified *family*) will be set as channel variables, with an optional *prefix* to the name (e.g., a prefix of var_ would make the column name become the variable ${var_name}).

```
; retrieve all columns from the sipfriends table where the name column
; matches "John", and prefix all the variables with "John_"
exten => 123,1,RealTime(sipfriends,name,John,John_)
; now, let's read the value of the column named "port"
exten => 123,2,SayNumber(${John_port})
```

See Also

RealTimeUpdate( )

RealTimeUpdate( )

RealTimeUpdate(*family*,*colmatch*,*value*,*newcol*,*newval*)

Uses the RealTime configuration handler system to update a value. The column *newcol* in *family* matching column *colmatch=value* will be updated to *newval*.

```
; this will update the port column in the sipfriends table to a new
; value of 5061, where the name column matches "John"
exten => 123,1,RealTimeUpdate(sipfriends,name,John,port,5061)
```

See Also

RealTime( )

Record( )

Record(*filename*:*format*,*silence*[,*maxduration*][,*options*]) (in Asterisk 1.0.x)
Record(*filename*.*format*,*silence*[,*maxduration*][,*options*]) (in Asterisk 1.2.x)

Records audio from the channel into the given *filename*. If the file already exists, it will be overwritten.

Optional arguments include:

format
: Specifies the format of the file type to be recorded. Valid formats include: g723, g729, gsm, h263, ulaw, alaw, vox, wav, and WAV.

silence
: Specifies the number of seconds of silence to allow before returning.

maxduration
: Specifies the maximum recording duration, in seconds. If missing or 0, there is no maximum.

options

May contain any of the following letters:

s

Skip recording if the line is not yet answered.

n

Do not answer, but record anyway if the line is not yet answered.

a

Append the recording to the existing recording rather than replacing it.

t

Use the alternate * terminator key instead of the default #.

If the filename contains %d, these characters will be replaced with a number incremented by one each time the file is recorded.

The user can press # to terminate the recording and continue to the next priority.

Returns -1 when the user hangs up.

```
; record the caller's name
exten => 123,1,Playback(pls-rcrd-name-at-tone)
exten => 123,2,Record(/tmp/name:gsm,3,30)
exten => 123,3,Playback(/tmp/name)
```

RemoveQueueMember( )   Dynamically removes queue members

RemoveQueueMember(*queuename*[,*interface*])

Dynamically removes the specified *interface* from the *queuename* call queue. If *interface* is not specified, this application removes the current interface from the queue.

If the interface is not in the queue and there exists a priority n+101 (where n is the current priority), the application will jump to that priority. Otherwise, it will return an error.

Returns -1 if there is an error.

```
; remove SIP/3000 from the techsupport queue
exten => 123,1,RemoveQueueMember(techsupport,SIP/3000)
```

See Also

AddQueueMember( )

ResetCDR( )   Resets the Call Detail Record

ResetCDR([*options*])

Causes the Call Detail Record to be reset. If the w option is specified, a copy of the current CDR will be stored before the current CDR is zeroed out.

Always returns 0.

```
; write a copy of the current CDR record, and then reset the CDR
exten => 123,1,Answer( )
exten => 123,2,Playback(tt-monkeys)
```

```
exten => 123,3,ResetCDR(w)
exten => 123,4,Playback(tt-monkeys)
```

See Also

ForkCDR( ), NoCDR( )

ResponseTimeout( )
Sets maximum timeout for awaiting response from caller

ResponseTimeout(*seconds*)

Sets the maximum amount of time permitted after falling through a series of priorities for a channel in which the caller may begin typing an extension. If the caller does not type an extension in this amount of time, control will pass to the t extension, if it exists; if not, the call will be terminated.

Always returns 0.

```
; allow callers three seconds to make a choice, before sending them
; to the 't' extension
exten => s,1,Answer( )
exten => s,2,ResponseTimeout(3)
exten => s,3,Background(enter-ext-of-person)

exten => t,1,Playback(im-sorry)
exten => t,1,Playback(goodbye)
```

See Also

AbsoluteTimeout( ), DigitTimeout( )

RetryDial( )
Attempts to place a call, and retries on failure

RetryDial(*announce,sleep,loops,technology/resource*[*&Technology2/resource2*...])
[,*timeout*][,*options*][,*URL*])

Attempts to place a call. If no channel can be reached, plays the file defined by *announce*, waiting *sleep* seconds to retry the call. If the specified number of attempts matches *loops*, the call will continue with the next priority in the dialplan. If *loops* is set to 0, the call will retry endlessly.

While waiting, a one-digit extension may be dialed. If that extension exists in either the context defined in ${EXITCONTEXT} (if defined) or the current one, the call will transfer to that extension immediately.

All arguments after *loops* are passed directly to the Dial( ) application.

```
; attempt to dial the number three times via IAX, retrying every five seconds
exten => 123,1,RetryDial(priv-trying,5,3,IAX2/VOIP/8885551212,30)
; if the caller presses 9 while waiting, dial the number on the Zap/4 channel
exten => 9,1,RetryDial(priv-trying,5,3,Zap/4/8885551212,30)
```

SeeAlso

Dial( )

Ringing( )

Ringing( )

Requests that the channel indicate ringing tone to the user. It is up to the channel driver to specify exactly how ringing is indicated.

Note that this application does not actually provide audio ringing to the caller. Use the Playtones( ) application to do this.

Always returns 0.

```
; indicate that the phone is ringing, even though it isn't
exten => 123,1,Ringing( )
exten => 123,2,Wait(5)
exten => 123,3,Playback(tt-somethingwrong)
```

See Also

Busy( ), Congestion( ), Progress( ), Ringing( ), Playtones( )

SayAlpha( )

SayAlpha(*string*)

Spells out the passed *string*, using the current language setting for the channel. See the SetLanguage( ) application to change the current language.

```
exten => 123,1,SayAlpha(ABC123XYZ)
```

See Also

SayDigits( ), SayNumber( ), SayPhonetic( ), SetLanguage( )

SayDigits( )

SayDigits(*digits*)

Says the passed digits, using the current language setting for the channel. See the SetLanguage( ) application to change the current language.

```
exten => 123,1,SayDigits(1234)
```

See Also

SayAlpha( ), SayNumber( ), SayPhonetic( ), SetLanguage( )

SayNumber( )

SayNumber(*digits*[,*gender*])

Says the specified number, using the current language setting for the channel. See the SetLanguage( ) application to change the current language.

Currently, syntax for the following languages is supported:

da   Danish

de   German

en   English

es   Spanish

fr   French

it   Italian

nl   Dutch

no   Norwegian

pl   Polish

pt   Portuguese

se   Swedish

tw   Taiwanese

If the current language supports different genders, you can pass the *gender* argument to change the gender of the spoken number. You can use the following *gender* arguments:

- Use the *gender* arguments f for female, m for male, and n for neuter in European languages such as Portuguese, French, Spanish, and German.
- Use the *gender* argument c for commune and n for neuter in Nordic languages such as Danish, Swedish, and Norwegian.
- Use the *gender* argument p for plural enumerations in German.

 For this application to work in languages other than English, you must have the appropriate sounds for the language you wish to use.

```
; say the number in English
exten => 123,1,SetLanguage(en)
exten => 123,2,SayNumber(1234)
```

See Also

SayAlpha( ), SayDigits( ), SayPhonetic( ), SetLanguage( )

SayPhonetic( )   Spells the specified string phonetically

SayPhonetic(*string*)

Spells the specified *string* using the NATO phonetic alphabet.

```
exten => 123,1,SayPhonetic(asterisk)
```

See Also

SayAlpha( ), SayDigits( ), SayNumber( )

SayUnixTime( )

SayUnixTime([*unixtime*][,[*timezone*][,*format*]])

Speaks the specified time according to the specified time zone and format. The arguments are:

unixtime
> The time, in seconds, since January 1, 1970. May be negative. Defaults to now.

timezone
> The time zone. See */usr/share/zoneinfo/* for a list. Defaults to the machine default.

format
> The format in which the time is to be spoken. See *voicemail.conf* for a list of formats. Defaults to "ABdY 'digits/at' IMp".

Returns 0, or -1 on hangup.

```
exten => 123,1,SayUnixTime(,,IMp)
```


SendDTMF( )

SendDTMF(*digits*[,*timeout_ms*])

Sends the specified DTMF digits on a channel. Valid DTMF digits include 0-9, *, #, and A-D. You may also use the letter w as a digit, which indicates a 500-millisecond wait. The *timeout_ms* argument is the amount of time between digits, in milliseconds. If not specified, *timeout_ms* defaults to 250 milliseconds.

Returns 0 on success or -1 on a hangup.

```
exten => 123,1,SendDTMF(3212333w222w366w3212333322321,250)
```

See Also

Read( )

SendImage( )

SendImage(*filename*)

Sends an image on a channel, if image transport is supported. If the channel does not support image transport, and there exists a priority n+101 (where n is the current priority), execution will continue at that step. Otherwise, execution will continue at the next priority level.

Returns 0 if the image was sent correctly or if the channel does not support image transport; otherwise, returns -1.

```
exten => 123,1,SendImage(logo.jpg)
```

See Also

SendText( ), SendURL( )

SendText( )

SendText(*text*)

Sends *text* on a channel, if text transport is supported. If the channel does not support text transport, and there exists a priority n+101 (where n is the current priority), execution will continue at that step. Otherwise, execution will continue at the next priority level.

Returns 0 if the text was sent correctly or if the channel does not support text transport; otherwise, returns -1.

```
exten => 123,1,SendText(Welcome to Asterisk)
```

See Also

SendImage( ), SendURL( )

SendURL( )

SendURL(*URL*[,*option*])

Requests that the client go to the specified URL. If the client does not support HTML transport, and there exists a priority n+101 (where n is the number of the current priority), execution will continue at that step. Otherwise, execution will continue at the next priority level.

Returns 0 if the URL was sent correctly or if the channel does not support HTML transport; otherwise, returns -1.

If the option wait is specified, execution will wait for an acknowledgment that the URL has been loaded before continuing and will return -1 if the peer is unable to load the URL.

```
exten => 123,1,SendURL(www.asterisk.org,wait)
```

See Also

SendImage( ), SendText( )

Set( )

Set(*n=value*)

Sets the variable *n* to the specified *value*. If the variable name is prefixed with _, single inheritance is assumed. If the variable name is prefixed with __, infinite inheritance is assumed. Inheritance is used when you want the outgoing channel to inherit the variable from the dialplan.

Variables set with this application are valid only in the current channel. Use the SetGlobalVar( ) application to set global variables.

```
; set a variable called DIALTIME, then use it
exten => 123,1,SetVar(DIALTIME=20)
exten => 123,1,Dial(Zap/4/5551212,,${DIALTIME})
```

See Also

SetGlobalVar( ), *README.variables*

SetAccount( )

SetAccount(*account*)

Sets the account code in the Call Detail Record, for billing purposes.

Always returns 0.

```
; set the account code to 4321 before dialing the boss
exten => 123,1,SetAccount(4321)
exten => 123,2,Dial(${BOSS})
```

See Also

SetAMAFlags( ), SetCDRUserField( ), AppendCDRUserField( )

SetAMAFlags( )

SetAMAFlags(*flag*)

Sets the AMA flags in the Call Detail Record for billing purposes, overriding any AMA settings in the channel configuration files. Valid choices are default, omit, billing, and documentation.

Always returns 0.

```
exten => 123,1,SetAMAFlags(billing)
```

See Also

SetAccount( ), SetCDRUserField( ), AppendCDRUserField( )

SetCallerID( )

SetCallerID(*clid*[,*a*])

Sets the Caller ID on the channel to a specified value. If the a argument is passed, ANI is also set to the specified value.

Always returns 0.

```
; set both the Caller ID and ANI
exten => 123,1,SetCallerID("John Q. Public <8885551212>",a)
```

See Also

SetCIDName( ), SetCIDNum( )

SetCallerPres( )

SetCallerPres(*presentation*)

Sets the Caller ID presentation flags on a Q931 PRI connection.

Valid presentations are:

allowed_not_screened
> Presentation Allowed, Not Screened

allowed_passed_screen
> Presentation Allowed, Passed Screen

allowed_failed_screen
> Presentation Allowed, Failed Screen

allowed
> Presentation Allowed, Network Number

prohib_not_screened
> Presentation Prohibited, Not Screened

prohib_passed_screen
> Presentation Prohibited, Passed Screen

prohib_failed_screen
> Presentation Prohibited, Failed Screen

prohib
> Presentation Prohibited, Network Number

unavailable
> Number Unavailable

Always returns 0.

```
exten => 123,1,SetCallerPres(allowed_not_screened)
exten => 123,2,Dial(Zap/g1/8885551212)
```

SeeAlso

SetCallerID( )

SetCDRUserField( )

SetCDRUserField(*value*)

Sets the CDR user field to the specified *value*. The CDR user field is an extra field that you can use for data not stored anywhere else in the record. CDR records can be used for billing purposes or for storing other arbitrary data about a particular call.

```
exten => 123,1,SetCDRUserField(testing)
exten => 123,2,Playback(tt-monkeys)
```

See Also

AppendCDRUserField( ), SetAccount( ), SetAMAFlags( )

SetCIDName( )

Sets the Caller ID name on the channel

SetCIDName(*cname*[,a])

Sets the Caller ID name on the current channel to *cname*, while preserving the original Caller ID number. This is useful for providing additional information to the called party. If the a option is used, ANI is also set.

Always returns 0.

```
exten => 123,1,SetCIDName("John Q. Public")
```

See Also

SetCallerID( ), SetCIDNum( )

SetCIDNum( )

Sets the Caller ID number for a channel

SetCIDNum(*cnum*[,a])

Sets the Caller ID number on the current channel to the number specified by *cnum*, while preserving the original Caller ID name. This is useful for providing additional information to the called party. The application sets ANI as well if the a flag is used.

Always returns 0.

```
exten => 123,1,SetCIDNum(8885551212)
```

See Also

SetCIDName( ), SetCallerID( )

SetGlobalVar( )

Sets a global variable to the specified value

SetGlobalVar(*n=value*)

Sets a global variable called *n* to the specified *value*. Global variables are available across channels.

```
; set the NUMRINGS global variable to 3
exten => 123,1,SetGlobalVar(NUMRINGS=3)
```

See Also

SetVar( )

SetGroup( )

Sets the channel group to the specified value

SetGroup(*groupname*[@*category*])

Sets the channel group to the specified *groupname* value. Equivalent to Set(GROUP=*group*). Used in conjunction with CheckGroup( ) to limit the number of calls accessing a particular resource. A group *category* may also be set.

Always returns 0.

```
; limit the number of concurrent receptionist calls to three
exten => s,1,SetGroup(receptionist)
exten => s,2,2,CheckGroup(3)
exten => s,3,Dial(${RECEPTIONIST})
exten => s,103,VoiceMail(u${RECEPTION_VM})
```

See Also

CheckGroup( ), GetGroupCount( ), GetGroupMatchCount( )

SetLanguage( )   Sets the channel's language

SetLanguage(*language*)

Sets the channel language to *language*. This information is used for the syntax in generation of numbers, and to choose a natural language file when available. For example, if
language is set to fr and the file *demo-congrats* is requested to be played, the file *fr/demo-
congrats* will be played if it exists; if not, the normal *demo-congrats* file will be played.

Always returns 0.

```
exten => s,1,SetLanguage(fr)
exten => s,2,SayNumber(1234)
exten => s,3,Playback(enter-ext-of-person)
```

SetMusicOnHold( )   Sets the default Music on Hold class for the current channel

SetMusicOnHold(*class*)

Sets the default *class* for Music on Hold for the current channel. When Music on Hold is
activated, this class will be used to select which music is played. Classes are defined in the
configuration file *musiconhold.conf*.

```
exten=s,1,Answer( )
exten=s,2,SetMusicOnHold(default)
exten=s,3,WaitMusicOnHold( )
```

See Also

WaitMusicOnHold( ), *musiconhold.conf*, MusicOnHold( )

SetRDNIS( )   Sets the RDNIS number on the current channel

SetRDNIS(*cnum*)

Sets the Redirected Dial Number ID Service (RDNIS) number on a call to the value specified by *cnum*. RDNIS is supported only on certain PRI lines.

Always returns 0.

```
exten => 123,1,SetRDNIS(8885551212)
exten => 123,2,Dial(Zap/4/5551234)
```

SetVar( )

SetVar(*n=value*)

Sets the variable *n* to the specified *value*. If the variable name is prefixed with _, single inheritance is assumed. If the variable name is prefixed with __, infinite inheritance is assumed. Inheritance is used when you want the outgoing channel to inherit the variable from the dialplan. Reprecated in favor of Set(), which has the same syntax.

Variables set with this application are valid only in the current channel. Use the SetGlobalVar( ) application to set global variables.

```
; set a variable called DIALTIME, then use it
exten => 123,1,SetVar(DIALTIME=20)
exten => 123,1,Dial(Zap/4/5551212,,${DIALTIME})
```

See Also

SetGlobalVar( ), *README.variables*

SIPAddHeader( )

SIPAddHeader(*Header: Content*)

Adds a header to a SIP call placed with the Dial( ) application. A nonstandard SIP header should begin with X-, such as X-Asterisk-Accountcode:. Use this application with care—adding the wrong headers may cause any number of problems.

Always returns 0.

```
exten => 123,1,SIPAddHeader(X-Asterisk-Testing: Just testing!)
exten => 123,2,Dial(SIP/123)
```

See Also

SIPGetHeader( )

SIPDtmfMode( )

SIPDtmfMode(*method*)

Changes the DTMF method for a SIP call. The *method* can be either inband, info, or rfc2833.

```
exten => 123,1,SIPDtmfMode(rfc2833)
exten => 123,2,Dial(SIP/123)
```

See Also

Appendix A

SIPGetHeader( )

Gets a SIP header from an incoming SIP call

SIPGetHeader(*var=headername*)

Sets a channel variable named *var* to the content of the *headername* SIP header. Skips to priority n+101 (where n is the current priority) if the specified header does not exist.

```
; get the "To" header and assign it to the variable called TESTING
exten => 123,1,SIPGetHeader(TESTING=To)
```

See Also

SIPAddHeader( )

SoftHangup( )

Performs a soft hangup of the requested channel

SoftHangup(*technology/resource,options*)

Hangs up the requested channel. Always returns 0. The *options* argument may contain the letter a, which causes all channels on the specified device to be hung up. Currently, the *options* argument may contain only one letter: a. Supplying the a argument causes all channels on the specified device to be hung up.

```
; hang up all calls using Zap/4 so we can use it
exten => 123,1,SoftHangup(Zap/4,a)
exten => 123,2,Wait(2)
exten => 123,3,Dial(Zap/4/5551212)
```

See Also

Hangup( )

StopMonitor( )

Stops monitoring a channel

StopMonitor( )

Stops monitoring (recording) a channel. This application has no effect if the channel is not currently being monitored.

```
exten => 123,1,Answer( )
exten => 123,2,Monitor(wav,monitor_test,mb)
exten => 123,3,SayDigits(12345678901234567890)
exten => 123,4,StopMonitor( )
```

See Also

Monitor( )

StopPlaytones( )

Stops playing a tone list

StopPlaytones( )

Stops playing the currently playing tone list.

```
exten => 123,1,Playtones(busy)
exten => 123,2,Wait(2)
exten => 123,3,StopPlaytones( )
exten => 123,4,Playtones(congestion)
exten => 123,5,Wait(2)
exten => 123,6,StopPlaytones( )
exten => 123,7,Goto(1)
```

See Also

Playtones( ), *indications.conf*

StripLSD( )   Strips the specified number of trailing (least significant) digits from the current extension

StripLSD(*count*)

Strips the trailing *count* digits from the channel's associated extension and continues processing at the next priority for the resulting extension. So, for example, if priority 1 of extension 5551212 is StripLSD(4), the last 4 digits will be stripped from 5551212 and the next step executed will be priority 2 of extension 555. If you switch into an extension that has no priority n+1 (where n is the current priority), the PBX will treat it as though the user dialed an invalid extension.

Always returns 0.

This application is deprecated and has been replaced with the substring expression ${EXTEN:*X*:*Y*}.

```
exten => 5551212,1,StripLSD(4)
exten => 555,2,SayDigits(${EXTEN})

; a better way of doing the same thing
exten => 5551234,1,SayDigits(${EXTEN::3})
```

See Also

StripMSD( ), *README.variables*, variable substring syntax

StripMSD( )   Strips the specified number of leading (most significant) digits from the current extension

StripMSD(*count*)

Strips the leading *count* digits from the channel's associated extension and continues processing at the next priority for the resulting extension. So, for example, if priority 1 of extension 5551212 is StripMSD(3), the first 3 digits will be stripped from 5551212 and the next step executed will be priority 2 of extension 1212. If you switch into an extension that has no priority n+1 (where n is the current priority), the PBX will treat it as though the user dialed an invalid extension.

Always returns 0.

This application is deprecated and has been replaced with the substring expression ${EXTEN:*X*:*Y*}.

```
exten => 5551212,1,StripMSD(3)
exten => 1212,2,SayDigits(${EXTEN})

; a better way of doing the same thing
exten => 5551234,1,SayDigits(${EXTEN:3})
```

See Also

StripLSD( ), *README.variables*, variable substring syntax

SubString( )   Saves substring digits in a given variable

SubString(*variable*=*string_of_digits*,*count1*,*count2*)

Assigns the substring of *string_of_digits* to a given variable. The parameter *count1* may be positive or negative. If it's positive, we skip the first *count1* digits from the left. If it's negative, we move *count1* digits from the end of the string to the left. The parameter *count2* indicates how many digits to take from the point that *count1* placed us. If *count2* is negative, that many digits are omitted from the end.

This application is deprecated. Instead, use ${EXTEN:*X*:*Y*}.

```
; here are some examples using SubString( ):
;assign the area code (3 first digits) to variable TEST
exten => 8885551212,1,SubString(TEST=8885551212,0,3)
; assign the last 7 digits to variable TEST
exten => 8885551212,1,SubString(TEST=8885551212,-7,7)
; assign all but the last 4 digits to variable TEST
exten => 8885551212,1,SubString(TEST=8885551212,0,-4)
;
; and here are the preferred alternatives:
;assign the area code (3 first digits) to variable TEST
exten => 8885551212,1,Set(TEST=${EXTEN::3})
; assign the last 7 digits to variable TEST
exten => 8885551212,1,Set(TEST=${EXTEN:-7:7})
; assign all but the last 4 digits to variable TEST
exten => 8885551212,1,Set(TEST=${EXTEN:6}
```

Suffix( )   Appends trailing digits to the current extension

Suffix(*digits*)

Appends the digit string specified by *digits* to the channel's associated extension and continues processing at the next priority for the new extension. So, for example, if priority 1 of extension 555 is Suffix(1212), 1212 will be appended to 555 and the next step executed will be priority 2 of extension 5551212. If you switch into an extension that has no priority n+1 (where n is the current priority), the PBX will treat it as though the user dialed an invalid extension.

Always returns 0.

```
exten => 555,1,Suffix(1212)
exten => 5551212,2,SayDigits(${EXTEN})
```

See Also

Prefix( )

System( ) Executes an operating system command

System(*command*)

Executes a *command* in the underlying operating system. If the command itself executes but is in error, and if there exists a priority n+101 (where n is the current priority), the execution of the dialplan will continue at that priority level.

This application is very similar to the TrySystem( ) application, except that it will return -1 if it is unable to execute the system command, whereas the TrySystem( ) application will always return 0.

```
exten => 123,1,System(echo hello > /tmp/hello.txt)
```

See Also

TrySystem( )

Transfer( ) Transfers the caller to a remote extension

Transfer(*exten*)

Requests that the remote caller be transferred to the given extension. If the transfer is *not* supported or successful and there exists a priority n+101 (where n is the current priority), that priority will be taken next.

```
; transfer calls from extension 123 to extension 130
exten => 123,1,Transfer(130)
```

TrySystem( ) Tries to execute an operating system command

TrySystem(*command*)

Attempts to execute a *command* in the underlying operating system. If the command itself executes but is in error, and if there exists a priority n+101 (where n is the current priority), the execution of the dialplan will continue at that priority level.

This application is very similar to the System( ) application, except that it always returns 0, whereas the System( ) application will return -1 if it is unable to execute the system command.

```
exten => 123,1,TrySystem(echo hello > /tmp/hello.txt)
```

See Also

System( )

TXTCIDName( )

Looks up a caller's name from a DNS TXT record

TXTCIDName(*CallerID*)

Looks up a caller's name via DNS and sets the variable ${TXTCIDNAME}. TXTCIDNAME will either be blank or return the value found in the TXT record in DNS. This application looks up the number via the ENUM sources listed in *enum.conf*.

```
exten => 123,1,TXTCIDName(8662331454)
exten => 123,2,SayAlpha(${TXTCIDNAME})
exten => 123,3,Playback(vm-goodbye)
```

UnpauseQueueMember( )

Unpauses a queue member

UnpauseQueueMember([*queuename*],*interface*)

Unpauses (resumes calls to) a queue member. This is the counterpart to PauseQueueMember( ), and it operates exactly the same way, except it unpauses instead of pausing the given interface.

```
exten => 123,1,PauseQueueMember(,SIP/300)
exten => 124,1,UnpauseQueueMember(,SIP/300)
```

See Also

PauseQueueMember( )

UserEvent( )

Sends an arbitrary event to the Manager interface

UserEvent(*eventname*[,*body*])

Sends an arbitrary event to the Manager interface, with an optional body representing additional arguments. The format of the event is:

```
Event: UserEvent<specified event name>
Channel: <channel name>
Uniqueid: <call uniqueid>
[body]
```

If the body is not specified, only the Event, Channel, and Uniqueid fields will be present.

Always returns 0.

```
exten => 123,1,UserEvent(BossCalled,${CALLERIDNAME} has called the boss!)
exten => 123,2,Dial(${BOSS})
```

See Also

manager.conf, Asterisk Manager interface

Verbose( )

Verbose([*level*,]*message*)

Sends the specified *message* to verbose output. The *level* must be an integer value. If not specified, *level* defaults to 0.

Always returns 0.

```
exten => 123,1,Verbose(Somebody called extension 123)
exten => 123,2,Playback(extension)
exten => 123,3,SayDigits(${EXTEN})
```

VMAuthenticate( )

VMAuthenticate([*mailbox*][@*context*])

Behaves identically to the Authenticate( ) application, with the exception that the passwords are taken from *voicemail.conf*.

If *mailbox* is specified, only that mailbox's password will be considered valid. If *mailbox* is not specified, the channel variable ${AUTH_MAILBOX} will be set with the authenticated mailbox.

```
; authenticate off of any mailbox password, and tell us the matching
; mailbox number
exten => 123,1,VMAuthenticate( )
exten => 123,2,SayDigits(${AUTH_MAILBOX})
```

See Also

Authenticate( ), *voicemail.conf*

VoiceMail( )

VoiceMail([s|u|b]*mailbox*[@*context*][&*mailbox*[@*context*]][...])

Leaves voicemail for a given *mailbox* (must be configured in *voicemail.conf*).

If the mailbox is preceded by s, instructions for leaving the message will be skipped. If it is preceded by u, the "unavailable" message (*/var/lib/asterisk/sounds/vm/exten/unavail*) will be played, if it exists. If the mailbox is preceded by b, the busy message will be played (that is, *busy* instead of *unavail*).

If the caller presses 0 (zero) during the prompt, the call jumps to the o (lower-case letter o) extension in the current context.

If the caller presses * during the prompt, the call jumps to extension a in the current context. This is often used to send the caller to a personal assistant.

If the requested mailbox does not exist, and there exists a priority n+101 (where n is the current priority), that priority will be taken next.

When multiple mailboxes are specified, the unavailable or busy message will be taken from the first mailbox specified.

Returns -1 on error or mailbox not found, or if the user hangs up; otherwise, returns 0.

```
; send caller to unavailable voicemail for mailbox 123
exten => 123,1,VoiceMail(u123)
```

See Also

VoiceMailMain( ), *voicemail.conf*

VoiceMailMain( ) Enters the voicemail system

VoiceMailMain([[s|p]*mailbox*][@*context*])

Enters the main voicemail system for the checking of voicemail. Passing the *mailbox* argument will stop the voicemail system from prompting the user for the mailbox number.

If the mailbox is preceded by the letter s, the password check will be skipped. If the mailbox is preceded by the letter p, the supplied mailbox will be prepended to the user's entry and the resulting string will be used as the mailbox number. This is useful for virtual hosting of voicemail boxes. If a *context* is specified, logins are considered in that voicemail context only.

Returns -1 if the user hangs up; otherwise, returns 0.

```
; go to voicemail menu for mailbox 123 in the default voicemail context
exten => 123,1,VoiceMailMain(123@default)
```

See Also

VoiceMail( ), *voicemail.conf*

Wait( ) Waits for a specified number of seconds

Wait(*seconds*)

Waits for the specified number of *seconds*, then returns 0. You can pass fractions of a second (e.g., 1.5 = 1.5 seconds).

```
; wait 1.5 seconds before playing the prompt
exten => s,1,Answer( )
exten => s,2,Wait(1.5)
exten => s,3,Background(enter-ext-of-person)
```

WaitExten( ) Waits for an extension to be entered

WaitExten([*seconds*])

Waits for the user to enter a new extension for the specified number of seconds, then returns 0. You can pass fractions of a second (e.g., 1.5 = 1.5 seconds). If unspecified, the default extension timeout will be used.

```
; wait 15 seconds for the user to dial an extension
exten => s,1,Answer( )
exten => s,2,Playback(enter-ext-of-person)
exten => s,3,WaitExten(15)
```

WaitForRing( ) Waits the specified number of seconds for a ring

```
WaitForRing(timeout)
```

Waits at least *timeout* seconds after the next ring has completed.

Returns 0 on success or -1 on hangup.

```
; wait five seconds for a ring, and then send some DTMF digits
exten => 123,1,Answer( )
exten => 123,2,WaitForRing(5)
exten => 123,3,SendDTMF(1234)
```

WaitForSilence( ) Waits for a specified amount of silence

```
WaitForSilence(wait[,repeat])
```

Waits for *repeat* instances of *wait* milliseconds of silence. If *repeat* is omitted, the application waits for a single instance of *wait* milliseconds of silence.

```
; wait for three instances of 300 ms of silence
exten => 123,WaitForSilence(300,3)
```

WaitMusicOnHold( ) Waits the specified number of seconds, playing Music on Hold

```
WaitMusicOnHold(delay)
```

Plays hold music for the specified number of seconds. If no hold music is available the delay will still occur, but with no sound.

Returns 0 when done, or -1 on hangup.

```
; allow caller to hear Music on Hold for five minutes
exten => 123,1,Answer( )
exten => 123,2,WaitMusicOnHold(300)
exten => 123,3,Hangup( )
```

See Also

SetMusicOnHold( ), *musiconhold.conf*

While( ) Starts a while loop

```
While(expr)
```

Starts a while loop. Execution will return to this point when EndWhile( ) is called, until *expr* is no longer true. If a condition is met causing the loop to exit, it continues on past the EndWhile( ).

```
exten => 123,1,Set(COUNT=1)
exten => 123,2,While($[ ${COUNT} < 5 ])
exten => 123,3,SayNumber(${COUNT})
exten => 123,4,Set(COUNT=$[ ${COUNT} + 1 ])
exten => 123,5,EndWhile( )
```

See Also

EndWhile( ), GotoIf( )

Zapateller( )

Uses a special information tone to block telemarketers

Zapateller(*options*)

Generates a special information tone to block telemarketers and other computer-dialed calls from bothering you.

The *options* argument is a pipe-delimited list of options. The following options are available:

answer

Causes the line to be answered before playing the tone

nocallerid

Causes Zapateller to play the tone only if no Caller ID information is available

```
; answer the line, and play the SIT tone if there is no Caller ID information
exten => 123,1,Zapateller(answer|nocallerid)
```

See Also

PrivacyManager( )

ZapBarge( )

Barges in on (monitors) a Zap channel

ZapBarge([*channel*])

Barges in on a specified Zap *channel*, or prompts if one is not specified. The people on the channel won't be able to hear you and will have no indication that their call is being monitored.

If *channel* is not specified, you will be prompted for the channel number. Enter **4#** for Zap/4, for example.

Returns -1 when the caller hangs up and is independent of the state of the channel being monitored.

```
exten => 123,1,ZapBarge(Zap/2)
exten => 123,2,Hangup( )
```

See Also

ZapScan( )

ZapRAS( )

Executes the Zaptel ISDN Remote Access Server

ZapRAS(*args*)

Executes an ISDN RAS server using *pppd* on the current channel. The channel must be a clear channel (i.e., PRI source) and a Zaptel channel to be able to use this function. (No modem emulation is included.)

Your *pppd* must be patched to be Zaptel-aware. *args* is a pipe-delimited list of arguments. Always returns -1.

This application is only for use on ISDN lines, and your kernel must be patched to support ZapRAS( ). You must also have *ppp* support in your kernel.

```
exten => 123,1,Answer( )
exten => 123,1,ZapRas(debug|64000|noauth|netmask|255.255.255.0|10.0.0.1:10.0.0.2)
```

ZapScan( )

ZapScan([*group*])

Allows a call center manager to monitor Zap channels in a convenient way. Use # to select the next channel, and use * to exit. You may limit scanning to a particular channel group by setting the *group* argument.

```
exten => 123,1,ZapScan( )
```

See Also

ZapBarge( )

APPENDIX C
AGI Reference

ANSWER

ANSWER

Answers the channel (if it is not already in an answered state).

Return values:

-1   Failure

0   Success

CHANNEL STATUS

CHANNEL STATUS [*channelname*]

Queries the status of the channel indicated by *channelname* or, if no channel is specified, the current channel.

Return values:

0   Channel is down and available

1   Channel is down, but reserved

2   Channel is off-hook

3   Digits have been dialed

4   Line is ringing

5   Line is up

6   Line is busy

DATABASE DEL

DATABASE DEL *family key*

Deletes an entry from the Asterisk database for the specified family and key.

Return values:

0   Failure

1   Success

DATABASE DELTREE

DATABASE DELTREE *family* [*keytree*]

Deletes a family and/or keytree from the Asterisk database.

Return values:

0   Failure

1   Success

DATABASE GET

DATABASE GET *family key*

Retrieves a value from the Asterisk database for the specified family and key.

Return values:

0   Not set

1 (*value*)
  Value is set (and is included in parentheses)

DATABASE PUT

DATABASE PUT *family key value*

Adds or updates an entry in the Asterisk database for the specified family and key, with the specified value.

Return values:

0   Failure

1   Success

EXEC

EXEC *application options*

Executes the specified dialplan application, including options.

Return values:

-2   Failure to find the application

value
  Return value of the application

GET DATA

GET DATA *filename* [*timeout*] [*max_digits*]

Plays the audio file specified by *filename* and accepts DTMF digits, up to the limit set by *max_digits*. Similar to the Background( ) dialplan application.

Return value:

value
> Digits received from the caller

GET FULL VARIABLE

GET FULL VARIABLE *variablename* [*channelname*]

If the variable indicated by *variablename* is set, returns its value in parentheses. This command understands complex variable names and built-in variable names, unlike GET VARIABLE.

Return values:

0   No channel, or variable not set

1 (*value*)
> Value is retrieved (and is included in parentheses)

GET OPTION

GET OPTION *filename* *escape_digits* [*timeout*]

Behaves the same as STREAM FILE, but has a *timeout* option (in seconds).

Return value:

value
> ASCII value of digits received, in decimal

GET VARIABLE

GET VARIABLE *variablename*

If the variable is set, returns its value in parentheses. This command does not understand complex variables or built-in variables; use the GET FULL VARIABLE command if your application requires these types of variables.

Return values:

0   No channel, or variable not set

1 (*value*)
> Value is retrieved (and is included in parentheses)

HANGUP

HANGUP [*channelname*]

Hangs up the specified channel or, if no channel is given, the current channel.

Return values:

-1   Specified channel does not exist

1   Hangup was successful

NOOP

NoOp [*text*]

Performs no operation. As a side effect, this command prints *text* to the Asterisk console. Usually used for debugging purposes.

Return value:

0   No channel, or variable not set

RECEIVE CHAR

RECEIVE CHAR *timeout*

Receives a character of text on a channel. Specify a *timeout* in milliseconds as the maximum amount of time to wait for input, or set to 0 to wait infinitely. Note that most channels do not support the reception of text.

Return values:

-1 (hangup)
    Failure or hangup

char (timeout)
    Timeout

value
    ASCII value of character, in decimal

RECORD FILE

RECORD FILE *filename format escape_digits timeout* [*offset_samples*] [BEEP] [s=*silence*]

Records the channel audio to the specified file until the reception of a defined escape (DTMF) digit. The *format* argument defines the type of file to be recorded (wav, gsm, etc.). The *timeout* argument is the maximum number of milliseconds the recording can last, and can be set to -1 for no timeout. The *offset_samples* argument is optional; if provided, it will seek to the offset without exceeding the end of the file. The *silence* argument is the number of seconds of silence allowed before the function returns despite the lack of DTMF digits or reaching the timeout. The silence value must be preceded by s= and is also optional.

Return values:

-1 Failure

0 Successful recording

SAY ALPHA

SAY ALPHA *number escape_digits*

Says a given character string, returning early if any of the given DTMF digits are received on the channel.

Return values:

-1 Error or hangup

0 Playback completed without being interrupted by an escape digit

value
  ASCII value of digit (if pressed), in decimal

SAY DATE

SAY DATE *date escape_digits*

Says a given *date*, returning early if any of the given DTMF digits are received on the channel. The *date* is the number of seconds elapsed since 00:00:00 on January 1, 1970, Coordinated Universal Time (UTC).

Return values:

-1 Error or hangup

0 Playback completed without being interrupted by an escape digit

value
  ASCII value of digit (if pressed), in decimal

SAY DATETIME

SAY DATETIME *datetime escape_digits* [*format*] [*timezone*]

Says the given *datetime*, returning early if any of the given DTMF digits are received on the channel. The *datetime* is the number of seconds elapsed since 00:00:00 on January 1, 1970, Coordinated Universal Time (UTC). The optional *format* argument is the format in which the time should be spoken. (See *voicemail.conf* for a complete description of the format options.) *format* defaults to "ABdY 'digits/at' IMp". Acceptable values for *timezone* can be found in */usr/share/zoneinfo/*. *timezone* defaults to the default time zone of the Asterisk server.

Return values:

-1 Error or hangup

0 Playback completed without being interrupted by an escape digit

value
  ASCII value of digit (if pressed), in decimal

SAY DIGITS

SAY DIGITS *number escape_digits*

Says a given digit string, returning early if any of the given DTMF digits are received on the channel.

Return values:

-1 Error or hangup

0   Playback completed without being interrupted by an escape digit

value
   ASCII value of digit (if pressed), in decimal

SAY NUMBER

SAY NUMBER *number escape_digits*

Says a given number, returning early if any of the given DTMF digits are received on the channel.

Return values:

-1 Error or hangup

0   Playback completed without being interrupted by an escape digit

value
   ASCII value of digit (if pressed), in decimal

SAY PHONETIC

SAY PHONETIC *string escape_digits*

Says a given character string with phonetics, returning early if any of the given DTMF digits are received on the channel.

Return values:

-1 Error or hangup

0   Playback completed without being interrupted by an escape digit

value
   ASCII value of digit (if pressed), in decimal

SAY TIME

SAY TIME *time escape_digits*

Says the indicated *time*, returning early if any of the given DTMF digits are received on the channel. The *time* is the number of seconds elapsed since 00:00:00 on January 1, 1970, Coordinated Universal Time (UTC).

Return values:

-1   Error or hangup

0   Playback completed without being interrupted by an escape digit

value
   ASCII value of digit (if pressed), in decimal

SEND IMAGE

SEND IMAGE *image*

Sends the given image on the current channel. Most channels do not support the transmission of images. Image names should not include extensions.

Return values:

-1   Error or hangup

0I   Image sent, or channel does not support sending an image

SEND TEXT

SEND TEXT *"text_to_send"*

Sends the specified text on the current channel. Most channels do not support the transmission of text. Text consisting of more than one word should be placed in quotes, since the command accepts only a single argument.

Return values:

-1   Error or hangup

0I   Text sent, or channel does not support sending text

SET AUTOHANGUP

SET AUTOHANGUP *time*

Causes the channel to automatically be hung up once *time* seconds have elapsed. Of course, it can be hung up before then as well. Setting *time* to 0 will cause the autohangup feature to be disabled on this channel.

Return value:

0   Autohangup has been set

SET CALLERID

SET CALLERID *number*

Changes the Caller ID of the current channel.

Return value:

1   Caller ID has been set

SET CONTEXT

SET CONTEXT *context*

Sets the *context* for continuation upon exiting the AGI application.

Return value:

0   Context has been set

SET EXTENSION

SET EXTENSION *extension*

Changes the *extension* for continuation upon exiting the AGI application.

Return value:

0   Extension has been set

SET MUSIC ON

SET MUSIC ON [on|off] [*class*]

Enables/disables the Music on Hold generator. If *class* is not specified, the default Music on Hold class will be used.

Return value:

0   Always returns 0

SET PRIORITY

SET PRIORITY *priority*

Changes the priority for continuation upon exiting the AGI application. *priority* must be a valid priority or label.

Return value:

0   Extension has been set

SET VARIABLE

SET VARIABLE *variablename value*

Sets or updates the *value* for the variable name specified by *variablename*. If the variable does not exist, it is created.

Return value:

1   Variable has been set

STREAM FILE

STREAM FILE *filename escape_digits* [*sample_offset*]

Play the audio file indicated by *filename*, allowing playback to be interrupted by the digits specified by *escape_digits*, if any. Use double quotes for the digits if you wish none to be permitted. If *sample_offset* is provided, the audio will seek to *sample_offset* before playback starts.

Remember, the file extension must not be included in the filename.

Return values:

0   Playback completed with no digit pressed

-1   Error or hangup

value
     ASCII value of digit (if pressed), in decimal

TDD MODE

TDD MODE [on|off]

Enable/disable Telecommunications Devices for the Deaf (TDD) transmission/reception on this channel.

Return values:

0   Channel not TDD-capable

1   Success

VERBOSE

VERBOSE *message level*

Sends *message* to the console via the verbose message system. The *level* argument is the minimum verbosity level at which the message will appear on the Asterisk command-line interface.

Return value:

0   Always returns 0

WAIT FOR DIGIT

WAIT FOR DIGIT *timeout*

Waits up to *timeout* milliseconds for the channel to receive a DTMF digit. Use -1 for the *timeout* value if you want the call to block indefinitely.

Return values:

-1   Error or channel failure

0   Timeout

value
     ASCII value of digit (if pressed), in decimal

Configuration Files

This appendix contains a reference to the configuration files not covered in the previous appendixes. If you are looking for VoIP channel configurations, refer to Appendix A. For a dialplan reference, you'll want to use Appendix B.

A configuration file is required for each Asterisk module you wish to use. These *.conf* files contain channel definitions, describe internal services, define the locations of other modules, or relate to the dialplan. You do not need to configure all of them to have a functioning system, only the ones required for your configuration. Although Asterisk ships with samples of all of the configuration files, it is possible to start Asterisk without any of them. This will not provide you with a working system, but it clearly demonstrates the modularity of the platform.

If no *.conf* files are found, Asterisk will make some decisions with respect to modules. For example, the following steps are always taken:

- The Asterisk Event Logger is loaded, and events are logged to */var/log/asterisk/event_log*.
- Manager actions are registered.
- The PBX core is initialized.
- The RTP port range is allocated from 5,000 through 31,000.
- Several built-in applications are loaded, such as Answer( ), Background( ), GotoIf( ), NoOp( ), and Set( ).
- The dynamic loader is started—this is the engine responsible for loading modules defined in *modules.conf*.

This appendix starts with an in-depth look at the *modules.conf* configuration file. We'll then briefly examine all the other files that you may need to configure for your Asterisk system.

modules.conf

The *modules.conf* file controls which modules are loaded or not loaded at Asterisk startup. This is done through the use of the load => or noload => constructs.

 This file is a key component to building a secure Asterisk installation: best practice suggests that only required modules be loaded.

The *modules.conf* file always starts with the [modules] header. The autoload statement tells Asterisk whether to automatically load all modules contained within the modules directory or to load only those modules specifically defined by load => statements. We recommend you manually load only those modules you need, but many people find it easier to let Asterisk attempt to autoload whatever it finds in */usr/lib/asterisk/modules*. You can then exclude certain modules with noload => statements.

Here's a sample *modules.conf* file:

```
[modules]
autoload=no                   ; set this to yes and Asterisk will load any
                              ; modules it finds in /usr/lib/asterisk/modules

load => res_adsi.so
load => pbx_config.so         ; Requires: N/A
load => chan_iax2.so          ; Requires: res_crypto.so, res_features.so
load => chan_sip.so           ; Requires: res_features.so
load => codec_alaw.so         ; Requires: N/A
load => codec_gsm.so          ; Requires: N/A
load => codec_ulaw.so         ; Requires: N/A
load => format_gsm.so         ; Requires: N/A
load => app_dial.so           ; Requires: res_features.so, res_musiconhold.so
```

Since we assume Asterisk is built on Linux, all the module names we use end in a *.so* extension. However, this may not be the case if you have built Asterisk on a different operating system.

As of this writing, there are eight module types: *resources*, *applications*, *Call Detail Record database connectors*, *channels*, *codecs*, *formats*, *pbx modules*, and *standalone functions*. Let's take a look at each of them.

Resources

A *resource* provides a connection to a static repository of a particular type of information, such as a unique regional requirement or a library of constant elements. This information must be configurable for each system, but once loaded it doesn't need to change in the course of normal operations.

For each resource below, we have outlined the applications and features it provides to other Asterisk modules We've indicated the *.conf* file used to define the resource, where needed; if no file is listed, then a configuration file isn't required. The resource modules are:

res_adsi.so
> Configuration file: *adsi.conf*
>
> Provides: ADSI functions to `ADSIProg( )` and `Voicemail( )`

res_agi.so
> Provides: `DeadAGI( )`, `EAGI( )`, `AGI( )`

res_crypto.so
> Provides: Loads public and private keys located in */var/lib/asterisk/keys/*

res_features.so
> Configuration file: *features.conf*
>
> Provides: `ParkedCall( )`, `Park( )`

res_indications.so
> Configuration file: *indications.conf*
>
> Provides: `Playtones( )`, `StopPlaytones( )`

res_monitor.so
> Provides: `Monitor( )`, `StopMonitor( )`, `ChangeMonitor( )`, action `Monitor`, action `StopMonitor`, action `ChangeMonitor`

res_musiconhold.so
> Configuration file: *musiconhold.conf*
>
> Provides: `MusicOnHold( )`, `WaitMusicOnHold( )`, `SetMusicOnHold( )`, `StartMusicOnHold( )`, `StopMusicOnHold( )`

res_odbc.so
> Configuration file: *res_odbc.conf*
>
> Provides: Connectivity information to the ODBC* driver—the purpose is to store configuration file information in a database and retrieve that information from the database; however, a reload is required to make changes take effect

Applications

If you build an Asterisk dialplan of any size, you are going to use at least one—and more likely dozens—of applications.† If an application is never going to be used, it is not strictly required that it be loaded. For performance-challenged systems (or if you

---

* Open DataBase Connectivity (ODBC) is a standard by which access to a database can be provided.

† To be of any use, a self-contained dialplan will always require several applications. Some folks, however, use the dialplan for no other purpose than to pass control to an external application. In this case, it would be possible to have the dialplan use no application other than `AGI( )`. We're not recommending that you do this, but again, it demonstrates Asterisk's enormous flexibility.

just like to keep it lean), you may elect to load only those applications that are referenced in your dialplan.

For each application module, we will define any resource requirements and name the applications that the module provides. Unless we have stated otherwise, the application does not require a configuration file or any other modules. The available application modules are:

app_adsiprog.so
> Requires: *res_adsi.so*
>
> Provides: ADSIProg( )

app_alarmreceiver.so
> Provides: AlarmReceiver( )

app_authenticate.so
> Provides: Authenticate( )

app_cdr.so
> Provides: NoCDR( )

app_chanisavail.so
> Provides: ChanIsAvail( )

app_chanspy.so
> Provides: ChanSpy( )

app_controlplayback.so
> Provides: ControlPlayback( )

app_curl.so
> Provides: Curl( )

app_cut.so
> Provides: Cut( )

app_db.so
> Provides: DBget( ), DBput( ), DBdel( ), DBdeltree( )

app_dial.so
> Requires: *res_features.so*, *res_musiconhold.so*
>
> Provides: Dial( ), RetryDial( )

app_dictate.so
> Provides: Dictate( )

app_directory.so
> Provides: Directory( )

app_disa.so
> Provides: DISA( )

app_dumpchan.so
> Provides: DumpChan( )

app_echo.so
> Provides: Echo( )

app_enumlookup.so
> Configuration file: *enum.conf*

> Provides: EnumLookup( )

app_eval.so
> Provides: Eval( )

app_exec.so
> Provides: Exec( )

app_festival.so
> Provides: Festival( )

app_forkcdr.so
> Provides: ForkCDR( )

app_getcpeid.so
> Requires: *res_adsi.so*

> Provides: GetCPEID( )

app_groupcount.so
> Provides: GetGroupCount( ), SetGroup( ), CheckGroup( ), GetGroupMatchCount( )

app_hasnewvoicemail.so
> Provides: HasVoicemail( ), HasNewVoicemail( )

app_ices.so
> Provides: ICES( )

app_image.so
> Provides: SendImage( )

app_lookupblacklist.so
> Provides: LookupBlacklist( )

app_lookupcidname.so
> Provides: LookupCIDName( )

app_macro.so
> Provides: Macro( ), MacroExit( ), MacroIf( )

app_math.so
> Provides: Math( )

app_md5.so
> Provides: MD5( ), MD5Check( )

app_milliwatt.so
> Provides: Milliwatt( )

app_mp3.so
> Provides: MP3Player( )

app_nbscat.so
    Provides: NBScat( )

app_parkandannounce.so
    Requires: *res_features.so*

    Provides: ParkAndAnnounce( )

app_playback.so
    Provides: Playback( )

app_privacy.so
    Provides: PrivacyManager( )

app_queue.so
    Requires: *res_features.so*, *res_monitor.so*, *res_musiconhold.so*

    Provides: Queue( ), AddQueueMember( ), RemoveQueueMember( ), PauseQueueMember( ), UnpauseQueueMember( ), action Queues, action QueueStatus, action QueueAdd, action QueueRemove, action QueuePause

app_random.so
    Provides: Random( )

app_read.so
    Provides: Read( )

app_readfile.so
    Provides: ReadFile( )

app_realtime.so
    Provides: RealTime( ), RealTimeUpdate( )

app_record.so
    Provides: Record( )

app_sayunixtime.so
    Provides: SayUnixTime( ), DateTime( )

app_senddtmf.so
    Provides: SendDTMF( )

app_sendtext.so
    Provides: SendText( )

app_setcallerid.so
    Provides: SetCallerPres( ), SetCallerID( )

app_setcdruserfield.so
    Provides: SetCDRUserField( ), AppendCDRUserField( ), action SetCDRUserField

app_setcidname.so
    Provides: SetCIDName( )

app_setcidnum.so
    Provides: SetCIDNum( )

app_setrdnis.so
    Provides: SetRDNIS( )

app_settransfercapability.so
    Provides: SetTransferCapability( )

app_sms.so
    Provides: SMS( )

app_softhangup.so
    Provides: SoftHangup( )

app_striplsd.so
    Provides: StripLSD( )

app_substring.so (deprecated)
    Provides: SubString( )

app_system.so
    Provides: System( ), TrySystem( )

app_talkdetect.so
    Provides: BackgroundDetect( )

app_test.so
    Provides: TestClient( ), TestServer( )

app_transfer.so
    Provides: Transfer( )

app_txtcidname.so
    Configuration file: *enum.conf*

    Provides: TXTCIDName( )

app_url.so
    Provides: SendURL( )

app_userevent.so
    Provides: UserEvent( )

app_verbose.so
    Provides: Verbose( )

app_voicemail.so
    Configuration file: *voicemail.conf*

    Requires: *res_adsi.so*

    Provides: VoiceMail( ), VoiceMailMain( ), MailboxExists( ), VMAuthenticate( )

app_waitforring.so
    Provides: WaitForRing( )

app_waitforsilence.so
    Provides: WaitForSilence( )

app_while.so
> Provides: While( ), ExecIf( ), EndWhile( )

app_zapateller.so
> Provides: Zapateller( )

Database-Stored Call Detail Records

Asterisk normally stores Call Detail Records (CDRs) in a Comma-Separated Values (CSV) file.[*] If you want CDRs to be stored in a database, you'll need to load the appropriate module and define the relevant *.conf* file.

For each module below, we state the database type it supports, and specify the configuration file, if required. The CDR database connector modules are:

cdr_csv.so
> Provides: CSV CDR backend

cdr_custom.so
> Configuration file: *cdr_custom.conf*
>
> Provides: Customizable CSV CDR backend

cdr_manager.so
> Configuration file: *cdr_manager.conf*
>
> Provides: Asterisk Call Manager CDR backend

cdr_odbc.so[†]
> Configuration file: *cdr_odbc.conf*
>
> Provides: ODBC CDR backend

cdr_pgsql.so
> Configuration file: *cdr_pgsql.conf*
>
> Provides: PostgreSQL CDR backend

Channels

Next, let's take a look at the channel modules. For each channel module, we identify dependencies and list the capabilities the module provides. We show the configuratin file, if one is required. The available modules are:

---

[*] Information stored in a text file as Comma-Separated Values can be imported into pretty much any spreadsheet or database (yes, even stuff from Microsoft). This makes the CSV format extremely portable.

[†] The *cdr_odbc* connector could theoretically replace all of the other database-specific connectors—however, people may prefer to use specific connectors due to performance differences, stability issues, personal preference, backward-compatibility, and so forth. Many options are available. If you are familiar with databases, Asterisk gives you lots of choices.

chan_agent.so

    Configuration file: *agents.conf*

    Requires: *res_features.so*, *res_monitor.so*, *res_musiconhold.so*

    Provides: channel Agent, `AgentLogin( )`, `AgentCallbackLogin( )`, `AgentMonitorOutgoing( )`, action `Agents`

chan_features.so

    Provides: channel Feature

chan_iax2.so

    Configuration file: *iax.conf*, *iaxprov.conf*

    Requires: *res_crypto.so*, *res_features.so*, *res_musiconhold.so*

    Provides: channel IAX2, `IAX2Provision( )`, function `IAXPEER`, action `IAXPEERS`, action `IAXnetstats`

chan_local.so

    Provides: channel Local

chan_mgcp.so

    Configuration file: *mgcp.conf*

    Requires: *res_features.so*

    Provides: channel MGCP

chan_modem.so

    Configuration file: *modem.conf*

    Provides: channel Modem

chan_modem_aopen.so

    Requires: *chan_modem.so*

    Provides: A/Open (Rockwell Chipset) ITU-2 VoiceModem Driver

chan_modem_bestdata.so

    Requires: *chan_modem.so*

    Provides: BestData (Conexant V.90 Chipset) VoiceModem Driver

chan_modem_i4l.so

    Requires: *chan_modem.so*

    Provides: ISDN4Linux Emulated Modem Driver

chan_oss.so

    Provides: channel Console (soundcard required)

chan_phone.so

    Configuration file: *phone.conf*

    Provides: channel Phone

chan_sip.so

    Configuration file: *sip.conf*, *sip_notify.conf*

    Requires: *res_features.so*

Provides: channel SIP, `SIPDtmfMode()`, `SIPAddHeader()`, `SIPGetHeader()`, action `SIPpeers`, action `SIPshowpeer`, function `SIP_HEADER`

chan_skinny.so

Configuration file: *skinny.conf*

Requires: *res_features.so*

Provides: channel Skinny

chan_zap.so

Configuration file: *zapata.conf*

Requires: *res_features.so*

Provides: channel Zap, action `ZapDialOffHook`, action `ZapDNDoff`, action `ZapDNDon`, action `ZapHangup`, action `ZapShowChannels`, action `ZapTransfer`

Codecs

There are several acceptable ways to pass audio information in digital form. The formulas used to encode and decode (or compress and decompress) this information are collectively referred to as *codecs*. Most of Asterisk's codecs are provided free of license requirements; however, some (such as G.729) are encumbered by patents and thus must be licensed before they can be used.

Asterisk will load these codecs without complaint, but if you attempt to transcode a channel using an unlicensed codec, your calls will be dropped as soon as they connect.

Here, then, are the codec modules—if there are parameters that can be defined, they will be configurable in the *codecs.conf* file:

codec_a_mu.so

Provides: translator `alawtoulaw`, translator `ulawtoalaw`

codec_adpcm.so

Configuration file: *codecs.conf*

Provides: translator `adpcmtolin`, translator `lintoadpcm`

codec_alaw.so

Configuration file: *codecs.conf*

Provides: translator `alawtolin`, translator `lintoalaw`

codec_g726.so

Configuration file: *codecs.conf*

Provides: translator `g726tolin`, translator `lintog726`

codec_gsm.so

Configuration file: *codecs.conf*

Provides: translator `gsmtolin`, translator `lintogsm`

codec_ilbc.so

> Configuration file: not required

> Provides: translator `ilbctolin`, translator `lintoilbc`

codec_lpc10.so

> Configuration file: *codecs.conf*

> Provides: translator `lpc10tolin`, translator `lintolpc10`

codec_ulaw.so

> Configuration file: *codecs.conf*

> Provides: translator `ulawtolin`, translator `lintoulaw`

Formats

Formats are essentially the same as codecs, except that they relate to handling files instead of live media streams. If you are talking to someone, a codec (or two) will be employed. If you are leaving a voicemail or listening to Music on Hold, a format will be involved.

Here are the current Asterisk formats. Formats do not have associated configuration files:

format_g723.so

> Provides: format `g723sf`

format_g726.so

> Provides: format `g726-40`, format `g726-32`, format `g726-24`, format `g726-16`

format_g729.so

> Provides: format `g729`

format_gsm.so

> Provides: format `gsm`

format_h263.so

> Provides: format `h263`

format_ilbc.so

> Provides: format `ilbc`

format_jpeg.so

> Provides: format `jpg`

format_pcm.so

> Provides: format `pcm`

format_pcm_alaw.so

> Provides: format `alaw`

format_sln.so

> Provides: format `sln`

format_vox.so

> Provides: format `vox`

format_wav.so
> Provides: format wav

format_wav_gsm.so
> Provides: format wav49

PBX Core Modules

The PBX modules deliver the core functionality of the system. For each module, we show the services it provides, and list the configuration file, if one is required. At minimum, *config*, *functions*, and *spool* are required. *dundi*, *loopback*, and *realtime* are needed only if you are going to make use of their capabilities. The PBX core modules are:

pbx_config.so
> Configuration file: *extensions.conf*
> Provides: Loads dialplan into memory

pbx_dundi.so
> Configuration file: *dundi.conf*
> Requires: *res_crypto.so*
> Provides: DUNDiLookup( )

pbx_functions.so
> Configuration file: not required
> Provides: function CDR, function CHECK_MD5, function DB, function DB_EXISTS, function ENV, function EVAL, function EXISTS, function FIELDQTY, function GROUP_COUNT, function GROUP_MATCH_COUNT, function GROUP, function GROUP_LIST, function IF, function ISNULL, function LANGUAGE, function LEN, function MD5, function REGEX, function STRFTIME, function SET, function TIMEOUT

pbx_loopback.so
> Provides: Loopback switch

pbx_realtime.so
> Provides: Realtime switch

pbx_spool.so
> Provides: Outgoing spool support

Standalone Functions

There is currently only one standalone function available. This function operates identically to those in *pbx_functions.so*, but because it is standalone, it can be loaded (or not) completely independently of the *pbx* functions. The function is:

func_callerid.so
> Configuration file: not required
> Provides: function CALLERID

adsi.conf

The Analog Display Services Interface (ADSI) was designed to allow telephone companies to deliver enhanced services across analog telephone circuits. In Asterisk, you can use this file to send ADSI commands to compatible telephones. Please note that the phone must be directly connected to a Zapata channel. ADSI messages cannot be sent across a VoIP connection to a remote analog phone.

The *res_adsi.so* module is required for the Voicemail( ) application; however, the *adsi.conf* file is not necessarily used. Detailed information about ADSI is not publicly available, and documentation needs to be purchased from Telcordia.

adtranvofR.conf

Prior to Voice over IP, Voice over Frame Relay (VoFR) enjoyed brief fame as a means of carrying packetized voice. Supporting VoFR through Adtran equipment is part of the history of Asterisk.

This feature is no longer popular in the community, though, so it may be difficult to find support for it.

agents.conf

This file allows you to create and manage agents for your call center. If you are using the Queue( ) application, you may want to configure agents for the queue. The *agents. conf* file is used to configure the AGENT channel driver.

The [general] section in *agents.conf* currently contains only one parameter. The persistentagents=yes parameter tells Asterisk to save the status of agents who use the callback feature of queues in the local Asterisk database. A logged-in remote agent will then remain logged in across a reboot (unless removed from the database through some other means).

The following parameters, which are specified in the [agents] section, are used to define agents and the way the system interacts with them. The settings apply to all agents, unless otherwise specified in the individual agent definitions:

ackcall
> Accepts the arguments yes and no. If set to yes, requires a callback agent to acknowledge login by pressing the # key after logging in. This works in conjunction with the AgentCallbackLogin( ) application.

autologoff
> Accepts an argument (in seconds) defining how long an agent channel should ring for before the agent is deemed unavailable and logged off.

group

Defines the groups to which an agent belongs, specified with integers. Specify that an agent belongs to multiple groups by separating the integers with commas.

musiconhold => *class*

Accepts a Music on Hold class as its argument. This setting applies to all agents.

updatecdr

Accepts the arguments yes and no. Used to define whether the source channel in the CDRs should be set to agent/agent_id to determine which agent generated the calls.

wrapuptime

Accepts an argument (in milliseconds) specifying the amount of time to wait after an agent has finished a call before that agent can be considered available to answer another call.

The remaining parameters are also specified in the [agents] section, but they are global to the *chan_agent* channel driver and thus cannot be defined on a per-agent basis:

createlink

Accepts the arguments yes and no. Inserts the name of the created recording in the CDR user field.

custom_beep

Accepts a filename as its argument. Can be used to define a custom notification tone to signal to an always-connected agent that there is an incoming call.

recordagentcalls

Accepts the arguments yes and no. Defines whether or not agent calls should be recorded.

recordformat

Defines the format to record files in. The argument specified should be wav, gsm, or wav49. The default recording format is wav.

savecallsin

Accepts a filesystem path as its argument. Allows you to override the default path of */var/spool/asterisk/monitor/* with one of your choosing.

Since the storage of calls will require a large amount of hard drive space, you will want to define a strategy to handle storing and managing these recordings.

This location should probably reside on a separate volume; one with very high performance characteristics.

urlprefix

Accepts a string as its argument. The string can be formed as a URL and is appended to the start of the text to be added to the name of the recording.

The final parameter is used to define agents. As in the *zapata.conf* file, configuration parameters are inherited from above the agent => definition. Agents are defined with the following format:

```
agent => agent_id,agent_password,name
```

For example, we can define agent Happy Tempura with the agent ID 1000 and password 1234, as follows.

```
agent => 1000,1234,Happy Tempura
```

Be aware that an *agents.conf* file is a complement to the queue configuration process. The most critical configuration file for your queues is *queues.conf*. You can configure a very basic queue without *agents.conf*.

alarmreceiver.conf

 The AlarmReceiver( ) application is not approved by Underwriter's Laboratory (UL) and should not be used as the primary or sole means of receiving alarm messages or events. This application is not guaranteed to be reliable, so don't depend on it unless you have extensively tested it. Use of this application without extensive testing may place your life and/or property at risk.

The *alarmreceiver.conf* file is used by the AlarmReceiver( ) application, which allows Asterisk to accept alarms using the SIA (Ademco) Contact ID protocol. When a call is received from an alarm panel, it should be directed to a context that calls the AlarmReceiver( ) application. In turn, AlarmReceiver( ) will read the *alarmreceiver.conf* configuration file and perform the configured actions as required. All parameters are specified under the [general] heading.

The sample configuration file will contain the current settings for this application and is very well documented.

alsa.conf

The *alsa.conf* file is used to configure Asterisk to use the Advanced Linux Sound Architecture (ALSA) to provide access to a sound card, if desired. You can use this file to configure the CONSOLE channel, which is most commonly used to create an overhead paging system (although, as with any other channel, there are all kinds of creative ways this can be used). Keep in mind that the usefulness of the ALSA channel by itself is limited due to its lack of a user interface.*

---

* Yes, we are aware that the user interface to the channel interface is the Asterisk CLI; however, this is not usable as a telephone and therefore does not meet the criteria of an interface from the perspective of a telephone user.

asterisk.conf

The *asterisk.conf* file defines the locations for the configuration files, the spool directory, and the modules, as well as a location to write log files to. The default settings are recommended unless you understand the implications of changing them. The *asterisk.conf* file is generated automatically when you run the make samples command, based on information it collects about your system. It will contain a [directories] section such as the following:

```
[directories]
astetcdir => /etc/asterisk
astmoddir => /usr/lib/asterisk/modules
astvarlibdir => /var/lib/asterisk
astagidir => /var/lib/asterisk/agi-bin
astspooldir => /var/spool/asterisk
astrundir => /var/run
astlogdir => /var/log/asterisk
```

Additionally, you can specify an [options] section, which will allow you to define startup options (command-line switches) in the configuration file. The following example shows the available options and the command-line switches that they effectively enforce:

```
[options]
verbose=<value>           ; starting verbosity level (-v)
debug=yes|no|<val>        ; turn debugging on or off (or value in 1.2) (-d)
nofork=yes|no             ; don't fork a background process (-f)
console=yes|no            ; load the Asterisk console (-c)
highpriority=yes|no       ; run with high priority (-p)
initcrypto=yes|no         ; initialize crypto at start (-i)
nocolor=yes|no            ; disable ANSI colors on the console (-n)
dumpcore=yes|no           ; dump a core file on failure (-g)
quiet=yes|no              ; run quietly (-q)
cache_record_files=yes|no ; cache files recorded with Record() in an alternative
                          ; directory in conjunction with record_cache_dir
record_cache_dir=<dir>    ; directory in which to cache files recorded with
                          ; Record () until completion
execincludes=yes|no       ; enable support of #exec includes in configuration
                          ; files (off by default)
```

cdr.conf

The *cdr.conf* file is used to enable call detail record logging to a database. Storing call records is useful for all sorts of purposes, including billing, fraud prevention, QoS evaluations, and more. *cdr.conf* contains some general parameters that are not specific to any particular database, but rather indicate how Asterisk should handle the passing of information to the database. All options are under the [general] heading of the *cdr.conf* file:

batch
> Accepts the arguments yes and no. Allows Asterisk to write data to a buffer instead of writing to the database at the end of every call, to reduce load on the system.

 Note that if the system dies unexpectedly when this option is set to yes, data loss may occur.

enable
> Accepts the arguments yes and no. Specifies whether or not to use CDR logging. If set to no, this will override any CDR module explicitly loaded. The default is yes.

safeshutdown
> Accepts the arguments yes and no. Setting safeshutdown to yes will prevent Asterisk from shutting down completely until the buffer is flushed and all information is written to the database. If this parameter is set to no and you shut down Asterisk with information still residing in the buffers, that information will likely be lost.

scheduleronly
> Accepts the arguments yes and no. If you are generating a massive volume of CDRs on a system that is pushing them to a remote database, setting scheduleronly to yes may be of benefit. Since the scheduler cannot start a new task until the current one is finished, slow CDR writes may adversely affect other processes needing the scheduler. This setting will instruct Asterisk to handle CDR writes in a new thread, essentially assigning a dedicated scheduler to this function. In normal operation, this would yield very little benefit.

size
> Accepts an integer as its argument. Defines the number of CDRs to accumulate in the buffer before writing to the database. The default is 100.

time
> Accepts an integer (in seconds) as its argument. Sets the number of seconds before Asterisk flushes the buffer and writes the CDRs to the database, regardless of the number of records in the buffer (as defined by size). The default is 300 seconds (5 minutes).

cdr_manager.conf

The *cdr_manager.conf* file simply contains a [general] heading and a single option, enabled, which you can use to specify whether or not the Asterisk Manager API generates CDR events. If you want CDR events to be generated, you will need the following lines in your *cdr_manager.conf* file:

```
[general]
enabled=yes
```

The Manager API will then output CDR events containing the following fields:

```
Event: Cdr
AccountCode:
Source:
Destination:
DestinationContext:
CallerID:
Channel:
DestinationChannel:
LastApplication:
LastData:
StartTime:
AnswerTime:
EndTime:
Duration:
BillableSeconds:
Disposition:
AMAFlags:
UniqueID:
UserField:
```

cdr_odbc.conf

Asterisk can store CDR data in a local or remote database via the ODBC interface. The *cdr_odbc.conf* file contains the information Asterisk needs to connect to the database. The *cdr_odbc.so* module will attempt to load the *cdr_odbc.conf* file, and if information is found for connecting to a database, the CDR data will be recorded there.

 If you are going to use a database for storing CDR data, you will have to select *one* of the many that are available. Asterisk does not like having multiple CDR databases to connect to, so do not have extra *cdr_.conf* files hanging about your Asterisk configuration directory.

cdr_pgsql.conf

Asterisk can store CDR data in a PostgreSQL database via the *cdr_pgsql.so* module. When the module is loaded the necessary information will be read from the *cdr_pgsql.conf* file, and Asterisk will connect to the PostgreSQL database to write and store CDR data.

cdr_tds.conf

Asterisk can also store CDR data to a FreeTDS database (including MS SQL) with the use of the *cdr_tds.so* module. The configuration file *cdr_tds.conf* is read once the module is loaded. Upon a successful connection, CDR data will be written to the database.

---

codecs.conf

Most codecs do not have any configurable parameters—they are what they are, and that's all they are.

Some codecs, however, are capable of behaving in different ways. This primarily means that they can be optimized for a particular goal, such as cutting down on latency, making best use of a network, or perhaps delivering high quality.

The *codecs.conf* file is fairly new in Asterisk, and as of this writing it allows configuration of Speex parameters only. The settings are self-explanatory, as long as you are familiar with the Speex protocol (see *http://www.speex.org*).

codecs.conf also allows you to configure Packet Loss Concealment (PLC). You need to define a [plc] section and indicate genericplc => true. This will cause Asterisk to attempt to interpolate any packets that are missed. (Enabling this functionality will incur a small performance penalty.)

dnsmgr.conf

This file is used to configure whether Asterisk should perform DNS lookups on a regular basis, and how often those lookups should be performed.

dundi.conf

The DUNDi protocol is used to dynamically look up the VoIP address of a phone number on a network, and to connect to that number. Unlike the ENUM standard, DUNDi has no central authority. The *dundi.conf* file contains DUNDi extensions used to control what is advertised; it also contains the peers to whom you will submit lookup requests and from whom you will accept lookup requests. The DUNDi protocol was explored in Chapter 10.

enum.conf

The Electronic Numbering (ENUM) system is used in conjunction with the Internet's DNS system to map E.164 ITU standard (ordinary telephone) numbers to email addresses, web sites, VoIP addresses, and the like. An ENUM number is created in DNS by reversing the phone number, separating each digit with a period, and appending *e164.arpa* (the primary DNS zone). If you want Asterisk to perform ENUM lookups, configure the domain(s) in which to perform the lookups within the *enum.conf* file. In addition to the official *e164.arpa* domain, you can have Asterisk perform lookups in the publicly accessible *e164.org* domain.

extconfig.conf

Asterisk can write configuration data to and load configuration data from a database using the external configuration engine (also known as *realtime*). This enables you to map external configuration files (static mappings) to a database, allowing the information to be retrieved from the database. It also allows you to map special runtime entries that permit the dynamic creation and loading of objects, entities, peers, and so on without a reload. These mappings are assigned and configured in the *extconfig.conf* file, which is used by both *res_odbc* and *realtime*.

extensions.conf

At the center of every good universe is a dialplan. The *extensions.conf* file is the means by which you tell Asterisk how you want calls to be handled. The dialplan contains a list of instructions that, unlike traditional telephony systems, is entirely customizable. The dialplan is so important that rather than defining it in this appendix, we have dedicated all of Chapters 5 and 6, as well as Appendix B, to this topic. Go forth, read, and enjoy!

features.conf

features.conf, the file formally known as *parking.conf*, contains configuration information related to call parking and call transfers. Call parking configuration options include:

- The extension to dial to park calls (`parkext =>`)
- The extension range to park calls in (`parkpos =>`)
- Which context to park calls in (`context =>`)
- How long a call can remain parked for before ringing the extension that parked it (`parkingtime =>`)
- The sound file played to the parked caller when the call is removed from parking (`courtesytone =>`)
- ADSI parking announcements (`asdipark=yes|no`)

In addition to the call parking options, in this file you can configure the button mappings for blind transfers, attended transfers, one-touch recording, disconnections, and the pickup extension (which allows you to answer a remotely ringing extension).

festival.conf

The Festival text-to-speech engine allows Asterisk to read text files to the end user with a computer-generated voice. Festival is covered in Chapter 10.

iax.conf

Similar to *sip.conf*, the *iax.conf* file is where you configure options related to the IAX protocol. Your end devices and service providers are also configured here. *iax.conf* is covered in detail in Appendix A.

iaxprov.conf

This file is used by Asterisk to allow the system to upgrade the firmware on an IAXy device.

indications.conf

The *indications.conf* file is used to tell Asterisk how to generate the various telephone sounds common in different parts of the world—a dial tone in England sounds very different from a dial tone in Canada, but your Asterisk system will be pleased to make the sounds you want to hear. This file consists of a list of sounds a telephone system might need to produce (dial tone, busy signals, and so forth), followed by the frequencies used to generate those sounds.

By default (and without an *indications.conf* file), Asterisk will use the tones common in North America. You can change the default country for your system by specifying the two-letter country code in the [general] section. Supported country codes are listed in the *indications.conf.sample* file located in */usr/src/asterisk/configs*. If you have the required information, your country can easily be added. Here's what the configuration for North America looks like:

```
[general]
country=us
;
[us]
description = United States / North America
ringcadance = 2000,4000
dial = 350+440
busy = 480+620/500,0/500
ring = 440+480/2000,0/4000
congestion = 480+620/250,0/250
callwaiting = 440/300,0/10000
dialrecall = !350+440/100,!0/100,!350+440/100,!0/100,!350+440/100,!0/100,350+440
record = 1400/500,0/15000
info = !950/330,!1400/330,!1800/330,0
```

logger.conf

The *logger.conf* file specifies the type and verbosity of messages logged to the various log files in the */var/log/asterisk/* directory. It has two sections, [general] and [logfile].

[general]

Settings under the [general] section are used to customize the output of the logs (and can safely be left blank, as the defaults serve most people very well). However, if you love to customize such things, read on.

You can define exactly how you want your timestamps to look through the use of the dateformat parameter:

```
dateformat=%F %T
```

The Linux man page for strftime(3) lists all of the ways you can do this.

If you want to append your system's hostname to the names of the log files, set appendhostname=yes. This can be useful if you have a lot of systems delivering log files to you.

If for some reason you do not want to log events from your queues, you can set queue_log=no.

If generic events do not interest you, instruct Asterisk to omit them from the by setting event_log=no.

[logfiles]

The [logfiles] section defines the types of information you wish to log. There are multiple ranks for the various bits of information that will be logged, and it can be desirable to separate log entries into different files. The general format for lines in the [logfiles] section is *filename* => *levels*, where *filename* is the name of the file to save the logged information to and *levels* are the types of information you wish to save.

> Using console for the *filename* is a special exception that allows you to control the type of information sent to the Asterisk console.

A sample [logfiles] section might look like this:

```
[logfiles]
console => notice,warning,error
messages => notice,warning,error
```

You can specify logging of the following types of information:

debug

> Enabling debugging gives far more detailed output about what is happening in the system. For example, with debugging enabled, you can see what DTMF tones the users entered while accessing their voicemail boxes. Debugging information should be logged only when you are actually debugging something, as it will create massive log files very rapidly.

verbose

> When you connect to the Asterisk console and set a verbosity of 3 or higher, you'll see output on the console showing what Asterisk is doing. You can save this output to a log file by adding a line such as verbose_log => verbose to your *logger.conf* file. Note that a high amount of verbosity can quickly eat up hard drive space.

notice

> A *notice* is used to inform you of minor changes to the system, such as when a peer changes state. It is normal to see these types of messages, and the events they indicate generally have no adverse effects on the server.

warning

> A *warning* happens when Asterisk attempts to do something and is unsuccessful. These types of errors are usually not fatal, but they should be investigated, especially if a lot of them are seen.

error

> *Errors* are often related to Out of Memory errors. They generally indicate serious problems that may lead to Asterisk to crashing or freezing.

manager.conf

The Asterisk Manager interface is an API that external programs can use to communicate with and control Asterisk, much as you would do from the Asterisk console.

 The Manager gives programs the ability to run commands and request information from the Asterisk server. However, it is not very secure—its authentication mechanism uses plain-text passwords, and all connected terminals receive all events. The Asterisk Manager should be used only on a trusted local area network, or locally on the box. The permit and deny constructs allow you to restrict access to certain extensions or subnets.

Many of the available graphical interfaces to Asterisk—such as the Flash Operator Panel—use the Manager to pull data and determine the status of applications. The *manager.conf* file defines the way programs authenticate with the Manager.

The Manager commands (which you can list by typing **show manager commands** at the Asterisk console) have varying degrees of privilege. You can control the read and write permissions for these commands with the use of the read and write options in the *manager.conf* file.

Here's a sample *manager.conf* file:

```
[general]
enabled = no
port = 5038
bindaddr = 0.0.0.0
```

```
[magma]
secret = welcome
deny=0.0.0.0/0.0.0.0
permit= 192.168.1.0/255.255.255.0
read = system,call,log,verbose,command,agent,user
write = system,call,log,verbose,command,agent,user
```

meetme.conf

MeetMe is one of the more remarkable applications in Asterisk. This rather simple concept has proven to be extremely expensive to implement in every other PBX, but what seems like a big deal to them is simple to Asterisk. Whether by using a dedicated server, or through the use of a service, Asterisk now delivers this functionality as a standard application.

MeetMe conferences can be created either dynamically, with the d flag in the Dial( ) application, or statically in the *meetme.conf* file. The format for creating conference rooms is as follows:

```
conf => conference_number[,pin][,administrator_pin]
```

All conferences must be defined under the [rooms] section header.

```
[rooms]
conf => 4569
conf => 5060,54377017
conf => 3389,4242,1337
conf => 333,,2424
```

mgcp.conf

The Media Gateway Control Protocol (MGCP) has only primitive support in Asterisk. This is likely due to the fact that SIP has stolen the limelight from every other VoIP protocol (except IAX, of course). Because of this, you should attempt to use Asterisk's MCGP channel in a production environment only if you are prepared to perform extensive testing, are willing to pay to have features and patches implemented within your time frames, and have in-house expertise with the protocol.

Having said that, we are not prepared to pronounce MGCP dead. SIP is not yet the panacea it has been touted as, and MGCP has proven itself to be very useful in carrier backbone environments. Many believe MGCP will fill a niche or void that has not yet been discovered, and we remain interested in it.

modem.conf

The *modem.conf* file is used by Asterisk to communicate with ISDN-BRI interfaces through the ISDN4Linux driver. Since ISDN4Linux lacks many core ISDN features,

---

it is not generally used. For BRI, the most popular add-on seems to be *chan_capi*, available from *http://www.junghanns.net*.

musiconhold.conf

The *musiconhold.conf* file is used to configure different classes of music and their locations for use in Music on Hold applications. Asterisk makes use of the *mpg123* application to play music to channels. You can specify arguments for a class, allowing you to use an external application to stream music either locally or over a network. Recently, native Music on Hold has been implemented, allowing Asterisk to play music without any external processes. If the file is available in the same format as the codec of the active channel, no transcoding will occur.

osp.conf

The Open Settlement Protocol (OSP) is officially documented in ETSI TS 101 321, a European Telecommunication Standards Institute (ETSI) document that came out of the work of the TIPHON working group. As far as we can tell, OSP is another attempt to apply old-style telecom thinking to disruptive technologies.

oss.conf

The *oss.conf* file is used to configure Asterisk to use the Open Sound System (OSS) driver to allow communications with the sound card via the CONSOLE channel. Note that ALSA is now the preferred interface for the CONSOLE channel.

phone.conf

The *phone.conf* file is used to configure a Quicknet PhoneJACK card. The PhoneJACK card seems to provide something like an FXS interface, in that you can plug an analog telephone into it and pass calls through Asterisk.

privacy.conf

The *privacy.conf* file is used to control the maximum number of tries a user has to enter his 10-digit telephone number in the PrivacyManager() application. The PrivacyManager() application determines if a Caller ID is set for the incoming call. If the user fails to enter his 10-digit number within the number of tries configured in *privacy.conf*, the call is sent to priority n + 101 (if it exists). If the Caller ID is set, the application does nothing.

queues.conf

Asterisk provides basic call center functionality via its queueing system, but those who are using it in more mission-critical environments often report that their solutions required customization. You can do this customization in the *queues.conf* file.

The [general] section of *queues.conf* contains settings that will apply to all queues. Currently, the only parameter that is supported is persistentmembers. If this parameter is set to yes, a member that is added to the system via the AddQueueMember() application will be stored in the AstDB, and therefore retained across a restart.

You can define a queue by placing its name inside of square brackets ([]). Within each queue, the following parameters are available:

musiconhold

> This parameter allows you to configure which Music on Hold class (configured in *musiconhold.conf*) to use for the queue.

announce

> When a call is presented to a member of the queue, the prompt specified by announce will be played to that agent before the caller is connected. This can be useful for agents who are logged into more than one queue. You can specify either the full path to the file, or a path relative to */var/lib/asterisk/sounds/*.

strategy

> Asterisk can use six strategies to distribute calls to agents:
>
> ringall
>
> > The queue rings every available agent and connects the call to whichever agent answers first (this is the default).
>
> roundrobin
>
> > The queue cycles through the agents until it finds one who is available to take the call. roundrobin does not take into account the workload of the agents. Also, because roundrobin always starts with the first agent in the queue, this strategy is suitable only in an environment where you want your higher-ranked agents to handle all calls unless they are busy, in which case the lower-ranked agents may get a call.
>
> leastrecent
>
> > The call is presented to the agent who has not been presented a call for the longest period of time.
>
> fewestcalls
>
> > The call is presented to the agent who has received the least amount of calls. This strategy does not take into account the actual agent workloads; it only considers the number of calls they have taken (for example, an agent who has had 3 calls that each lasted for 10 minutes will be preferred over an agent who has had 5 calls each lasting 2 minutes).

random

> As its name suggests, the random strategy chooses an agent at random. In a small call center, this strategy may prove to be the most fair.

rrmemory

> The queue cycles through each agent, keeping track of which agent last received a call (this strategy is known as *round-robin memory*). This ensures that call presentation cycles through the agents as fairly as possible.

servicelevel

> In a call center, the service level represents the maximum amount of time a caller should ideally have to wait before being presented to an agent. For example, if servicelevel is set to 60 and the service level percentage is 80%, that means 80% of the calls that came into the queue were presented to an agent in less than 60 seconds.

context

> If a context is assigned to a queue, the caller will be able to press a single digit to exit to the corresponding extension within the configured context, if it exists. This action takes the caller out of the queue, which means that she will lose her place in the queue—be aware of this when you use this feature.

timeout

> The timeout value defines the maximum amount of time (in seconds) to let an agent's phone ring before deeming the agent unavailable and placing the call back into the queue.

retry

> When a timeout occurs, the retry value specifies how many seconds to wait before presenting the call again to an available agent.

weight

> The weight parameter assigns a rank to the queue. If calls are waiting in multiple queues, those queues with the highest weight values will be presented to agents first. When you are designing your queues, be aware that this strategy can prevent a call in a lower-weighted queue from ever being answered. Always ensure that calls in lower-weighted queues eventually get promoted to higher-weighted queues to ensure that they don't have to hold forever.

wrapuptime

> You can configure this parameter to allow agents a few seconds of downtime after completing a call before the queue presents them with another call.

maxlen

> maxlen is the maximum number of calls that can be added to the queue before the call goes to the next priority of the current extension.

announce-frequency

> The announce-frequency value (defined in seconds) determines how often to announce to the caller his place in the queue and estimated hold time.

announce-holdtime

There are three possible values for this parameter: yes, no, and once. The announce-holdtime parameter determines whether or not to include the estimated hold time within the position announcement. If set to once, it will be played to the caller only once.

monitor-format

This parameter accepts three possible values: wav, gsm, and wav49. By enabling this option, you are telling Asterisk that you wish to record all completed calls in the queue in the format specified. If this option is not specified, no calls will be recorded.

monitor-join

The Monitor( ) application in Asterisk normally records either end of the conversation in a separate file. Setting monitor-join to yes instructs Asterisk to merge the files at the end of the call.

joinempty

This parameter accepts three values: yes, no, and strict. It allows you to determine whether callers can be added to a queue based on the status of the members of the queue. The strict option will not allow callers to join the queue if all members are unavailable.

leavewhenempty

This parameter determines whether you want your holding callers to be removed from the queue when the conditions preventing a caller from joining exist (i.e., when all of your agents log out and go home).

eventwhencalled

Set eventwhencalled to yes if you wish to have queue events presented on the Manager interface.

eventmemberstatusoff

Setting this parameter to no will generate extra information pertaining to each queue member.

reportholdtime

If you set this parameter to yes, the amount of time the caller held before being connected will be announced to the answering agent.

memberdelay

This parameter defines whether a delay will be inserted between the time when the queue identifies a free agent and the time when the call is connected to that agent.

member => member_name

Members of a queue can be either channel types or agents. Any agents you list here must be defined in the *agents.conf* file.

res_odbc.conf

The purpose of the *res_odbc.so* module is to store configuration file information in a database and retrieve that information from the database; however a reload is required to make changes take effect. The *res_odbc.conf* file specifies how to access the table within the database. The *extconfig.conf* file is used to determine how to connect to the database.

rpt.conf

The *rpt.conf* file is used to configure Jim Dixon's newest science project. Jim's Radio Repeater Application (*app_rpt*) allows Asterisk to communicate using VoIP via radio repeater technology. This allows people to efficiently provide large-area coverage of wireless networking and routing information to the Amateur Radio public through their local high-speed Internet connections.

rtp.conf

The *rtp.conf* file controls the Real-time Transport Protocol (RTP) ports that Asterisk uses to generate and receive RTP traffic. The RTP protocol is used by SIP, H.323, MGCP, and possibly other protocols to carry media between endpoints.

The default *rtp.conf* file uses the RTP port range of 10,000 through 20,000. However, this is far more ports than you're likely to need, and many network administrators may not be comfortable opening up such a large range in their firewalls. You can limit the RTP port range by changing the upper and lower bound limits within the *rtp.conf* file.

For every bidirectional SIP call between two endpoints, five ports are generally used: port 5060 for SIP signaling, one port for the data stream and one port for the Real-Time Control Protocol (RTCP) in one direction, and an additional two ports for the data stream and RTCP in the opposite direction.

UDP datagrams contain a 16-bit field for a Cyclic Redundancy Check (CRC), which is used to verify the integrity of the datagram header and its data. It uses polynomial division to create the 16-bit checksum from the 64-bit header. This value is then placed into the 16-bit CRC field of the datagram, which the remote end can then use to verify the integrity of the received datagram.

Setting rtpchecksums=no requests that the OS not do UDP checksum creating/checking for the sockets used by RTP. If you add this option to the sample *rtp.conf* file, it will look like this:

```
[general]
rtpstart=10000
rtpend=20000
rtpchecksums=no
```

sip.conf

The *sip.conf* file defines all the SIP protocol options for Asterisk. The authentication for endpoints, such as SIP phones and service providers, is also configured in this file. Asterisk uses the *sip.conf* file to determine which calls you are willing to accept and where those calls should go in relation to your dialplan. Many SIP-related options are configured in *sip.conf*, which was covered in depth in Appendix A.

sip_notify.conf

Asterisk has the ability to reboot a SIP phone remotely by sending it a specially formatted, manufacturer-specific NOTIFY message (defined in *sip_notify.conf*) consisting of an event. The phone receives this event, which it interprets as a reboot request. Other phones are supported, but as of this writing only phones by Polycom have been verified to work with this method.

skinny.conf

If you wish to connect to phones using Cisco's proprietary Skinny Client Control Protocol (SCCP), you can use the *skinny.conf* file to define the parameters and channels that will use it. However, since the Asterisk community uses the SIP image on their Cisco phones, you may find it difficult to find community support for this channel type.

voicemail.conf

The *voicemail.conf* file controls the Asterisk voicemail system (called Comedian Mail). It consists of three main sections. The first, called [general], sets the general system-wide settings for the voicemail system. The second, called [zonemessages], allows you to configure different voicemail zones, which are a collection of time and time zone settings. The third and final section is where you create one or more groups of voicemail boxes, each containing the mailbox definitions.

(For more information on adding voicemail capabilities to your dialplan, see Chapter 6.)

General Voicemail Settings

The [general] section of *voicemail.conf* contains a plethora of options that affect the entire voicemail system:

format
> Lists the codecs that should be used to save voicemail messages. Codecs should be separated with the pipe character (|). The first format specified is the format used when attaching a voicemail message to an email. Defaults to wav49|gsm|wav.

serveremail

Provides the email address from which voicemail notifications should be sent.

attach

Specifies whether or not Asterisk should attach the voicemail sound file to the voicemail notification email.

maxmessage

Sets the maximum length of a voicemail message, in seconds.

minmessage

Sets the minimum length of a voicemail message, in seconds.

maxgreet

Sets the maximum length of voicemail greetings, in seconds.

skipms

Specifies how many milliseconds to skip forward/back when the user skips forward or backward during message playback.

maxsilence

Indicates how many seconds of silence to allow before ending the recording.

silencethreshold

Sets the silence threshold (what we consider "silence"—the lower the threshold is, the more sensitive it is).

maxlogins

Sets the maximum allowed number of failed login attempts.

externnotify

Supplies the full path and filename of an external program to be executed when a voicemail is left or delivered, or when a mailbox is checked.

externpass

Supplies the full path and filename of an external program to be executed whenever a voicemail password is changed.

directoryintro

If set, overrides the default introduction to the dial-by-name directory.

charset

Defines the character set for voicemail messages.

adsifdn

Specifies the ADSI feature descriptor number to download to.

adsisec

Sets the ADSI security lock code.

adsiver

Indicates the ADSI voicemail application version number.

pbxskip

Causes Asterisk not to add the string [PBX]: to the beginning of the subject line of a voicemail notification email.

fromstring:

Changes the From: string of voicemail notification email messages.

usedirectory

Permits a mailbox owner to select entries from the dial-by-name directory for forwarding and/or composing new voicemail messages.

pagerfromstring

Changes the From: string of voicemail notification pager messages.

emailsubject

Specifies the email subject of voicemail notification email messages.

emailbody

Supplies the email body of voicemail notification email messages.

 Please note that both the emailsubject and emailbody settings can use the following variables to provide more in-depth information about the voicemail:

- VM_NAME
- VM_DUR
- VM_MSGNUM
- VM_MAILBOX
- VM_CALLERID
- VM_CIDNUM
- VM_CIDNAME
- VM_DATE

mailcmd

Supplies the full path and filename of the program Asterisk should use to send notification emails. This option is useful if you want to override the default email program.

Voicemail Zones

As voicemail users may be located in different geographical locations, Asterisk provides a way to configure the time zone and the way the time is announced for different callers. Each unique combination is known as a *voicemail zone*. You configure your voicemail zones in the [zonemessages] section of *voicemail.conf*. Later, you can assign your voicemail boxes to use the settings for one of these zones.

Each voicemail zone definition consists of a line with the following syntax:

    zonename=timezone | time_format

The *zonename* is an arbitrary name used to identify the zone. The *timezone* argument is the name of a system time zone, as found in */usr/share/zoneinfo*. The *time_format* argument specifies how times should be announced by the voicemail system. The *time_format* argument is made up of the following elements:

'*filename*'

> The filename of a sound file to play (single quotes around the filename are required)

${*VAR*}

> Variable substitution

A *or* a

> The day of the week (Saturday, Sunday, etc.)

B *or* b *or* h

> The name of the month (January, February, etc.)

d *or* e

> The numeric day of the month (first, second... thirty-first)

Y

> The year

I *or* l

> The hour, in 12-hour format

H

> The hour, in 24-hour format—single-digit hours are preceded by "oh"

k

> The hour, in 24-hour format—single-digit hours are *not* preceded by "oh"

M

> The minute

P *or* p

> A.M. or .P.M.

Q

> "today", "yesterday," or ABdY (note: not standard strftime value)

q

> "" (for today), "yesterday", weekday, or ABdY (note: not standard strftime value)

R

> 24-hour time, including minutes

For example, the following example sets up two different voicemail zones, one for the Central time zone in 12-hour format, and a second in the Mountain time zone, in 24-hour format:

```
[zonemessages]
central=America/Chicago|'vm-received' Q 'digits/at' IMp
mountain24=America/Denver|'vm-received' q 'digits/at' H 'digits/hundred' M 'hours'
```

Defining Voicemail Contexts and Mailboxes

Now that the system-wide settings and voicemail zones have been set, you can define your voicemail contexts and individual mailboxes.

Voicemail contexts are used to separate out different groups of voicemail users. For example, if you are using Asterisk to host voicemail for more than one company, you should place each company's mailboxes in different voicemail contexts, to keep them separate. You might also use voicemail contexts to create per-department dial-by-name directories.

To define a new voicemail context, simply put the context name inside of square brackets, like this:

```
[default]
```

Inside a voicemail context, each mailbox definition takes the following syntax:

```
mailbox=password,name[,email[,pager_email[,options]]]
```

The *mailbox* argument is the mailbox number.

The *password* argument is the code the mailbox owner must enter to access his voicemail. If the password is preceded by a minus sign (-), the password may not be changed by the mailbox owner.

The *email* and *pager_email* arguments are email addresses where voicemail notifications will be sent. These may be left blank if you don't want to send voicemail notifications via email.

The *options* argument is a pipe-separated list of voicemail options that may be specified for the mailbox. (These options may also be set globally by placing them in the [general] section.) Valid voicemail options include:

tz
> Sets the voicemail zone from the [zonemessages] section above. This option is irrelevant if envelope is set to no.

attach
> Attaches the voicemail to the notification email (but *not* to the pager email). May be set to either yes or no.

saycid
> Says the Caller ID information before the message.

cidinternalcontexts
> Sets the internal context for name playback instead of extension digits when saying the Caller ID information.

sayduration
> Turns on/off the duration information before the message. Defaults to on.

saydurationm
> Specifies the minimum duration to say when sayduration is on. Default is 2 minutes.

dialout
> Specifies the context to dial out from (by choosing option 4 from the advanced menu). If not specified, dialing out from the voicemail system will not be permitted.

sendvoicemail

Specifies the context to send voicemail from (by choosing option 5 from the advanced menu). If not specified, sending messages from within the voicemail system will not be permitted.

callback

Specifies the context to call back from. If not specified, calling the sender back from within the voicemail system will not be permitted.

review

Allows senders to review/rerecord their messages before saving them. Defaults to off.

operator

Allows senders to hit 0 before, after, or while leaving a voicemail message to reach an operator. Defaults to off.

envelope

Turns on/off envelope playback before message playback. Defaults to on. This does not affect option 3,3 from the advanced options menu.

delete

Deletes voicemails from the server after notification is sent. This option may be set only on a per-mailbox basis; it is intended for use with users who wish to receive their voicemail messages *only* by email.

nextaftercmd

Skips to the next message after the user hits 7 or 9 to delete or save the current message. This can be set only globally at this time, not on a per-mailbox basis.

forcename

Forces new users to record their names. A new user is determined by the password being the same as the mailbox number. Defaults to no.

forcegreetings

Forces new users to record greetings. A new user is determined by the password being the same as the mailbox number. Defaults to no.

hidefromdir

Hides the mailbox from the dial-by-name directory. Defaults to no.

You can specify multiple options by separating them with the pipe character, as shown in the definitions for mailboxes 9855 and 6522 below.

Here are some sample mailbox definitions:

```
[default]
; regular mailbox with email notification
101 => 4242,Example Mailbox,somebody@asteriskdocs.org

; more advanced mailbox with email and pager notification and a couple of
; special options
102 => 9855,Another User,another@asteriskdocs.org,pager@asteriskdocs.org,
attach=no|tz=central
```

```
; a mailbox with no email notification and lots of extra options
103 => 6522,John Q. Public,,,tz=central|attach=yes|saycid=yes|
dialout=fromvm|callback=fromvm|review=yes
```

vpb.conf

This file is used to configure Voicetronix cards with Asterisk.

zapata.conf

The *zapata.conf* file is used to define the relationship between Asterisk and the Zaptel driver. Because *zapata.conf* is specific to Asterisk, it is located with the other Asterisk configuration files in */etc/asterisk/*. As with *zaptel.conf*, the *zapata.conf* file contains a multitude of choices reflecting the multitude of hardware it supports, and we won't try to list all of the options here. In this book we've covered only the analog interfaces to the Zaptel driver, as described in Chapter 3.

zaptel.conf

The *zaptel.conf* file is not located with the other Asterisk *.conf* files—the Zaptel driver is available to any application that can make use of it, so it makes more sense to store it in a non-Asterisk-specific directory (*/etc/*). *zaptel.conf* is parsed by the *ztcfg* program to configure the TDM hardware elements in your system. You configure three main elements in the *zaptel.conf* file:

- A way of identifying the interfaces on the card within the dialplan
- The type of signaling the interface requires
- The tone language associated with a particular interface, as found in *zonedata.c*

 Be very careful not to plug your FXS module into a telephone line. The voltage associated with the phone line, especially during an incoming call, will be much too high for the module to handle and may permanently damage it, rendering it useless!

Within the *zaptel.conf* file, we define the type of signaling that the channel is going to use. We also define which channels to load. The options in the configuration file are the information that will be used to configure the channels with the ztcfg command.

The actual parameters available in the *zaptel.conf* file are quite extensive, as a wide variety of PSTN interfaces make use of the Zaptel telephony engine. Also, as this technology is rapidly evolving, anything we write now may not be accurate by the time you read it. Consequently, we won't try to list all of the options here.

In this book, we have focused on the Zaptel analog interfaces as provided by the Digium TDM400P card (see Chapter 3).

Asterisk Command-Line Interface Reference

To access the Asterisk command-line interface (CLI), pass the -c or -r argument to the Asterisk executable. In other words, type this from your shell prompt:

```
# asterisk -r
```

If you want the system to provide you with more information about what it is doing (an excellent idea, especially when you're new to Asterisk), you can add the argument -v, as many times as you'd like:

```
# asterisk -vvvvvvvvr
```

The more vs you include, the more vvvvvvvvverbose the output will be.

The CLI allows you to interact with a running Asterisk server, and it will be very useful to you for troubleshooting and monitoring.

Since the CLI employs tabbed name completion, you can press the Tab key to see a list of possible commands. This makes the CLI very easy to use. Let's take a look at the commands.

!

!command

Executes a given shell command. If followed immediately by a carriage return, Asterisk starts an interactive shell. You can return to the Asterisk CLI by executing an exit command.

abort halt

`abort halt`

Cancels a requested Asterisk shutdown (betcha never get the chance, though!). This command is only for the very fast-fingered.

add

The add command contains many subcommands that allow you to add functionality to your Asterisk PBX without directly editing the configuration files.

When you add a new line to the dialplan your changes immediately become active, but changes made to the dialplan from the command line are not permanent until you save them (see save dialplan). All comments are stripped from the *extensions. conf* file upon a save dialplan. The add commands are useful for making temporary changes and for ad hoc testing, but we recommend that permanent changes to the dialplan be made directly to *extensions.conf* in */etc/asterisk/*.

add extension

add extension *exten,priority,app,app_data* into *context* [replace]

Adds a new extension into the specified context. If an extension with the same priority exists, and the optional replace argument is given, replaces the existing extension.

    add extension 500,1,Dial,IAX2/guest@misery.digium.com/s@default into local

add ignorepat

add ignorepat *pattern* into *context*

Adds a new ignore pattern into the specified context.

    add ignorepat 9 into local

add queue member

add queue member *channel* to *queue* [penalty *penalty*]

Allows you to add a channel to a specified queue, optionally specifying a penalty with the penalty option.

    add queue member SIP/1000-d448 to customer_service penalty 10

agi

When you're running an AGI program, you can turn debugging on and off with the use of agi debug and agi no debug, respectively.

agi debug

agi debug

Turns on AGI debugging.

agi no debug

agi no debug

Turns off AGI debugging.

answer

answer

Answers an incoming call on the CONSOLE (OSS) channel. The OSS channel must be configured in *oss.conf* before the answer command is available.

database

The Asterisk database is a simple implementation based on Version 1 of the Berkeley database. You can add entries to the database, remove entries from the database, and view entries in the database with the following commands.*

database del

database del *family key*

Deletes an entry in the Asterisk database for a given family and key.

    database del phones 1000/username

database deltree

database deltree *family* [*keytree*]

Deletes a family or a specific keytree within a family in the Asterisk database.

    database deltree phones

---

* For more about the Asterisk database, see Chapter 6.

database get

database get *family key*

Retrieves an entry in the Asterisk database for a given family and key.

    database get phones 1000/username

database put

database put *family key value*

Adds or updates an entry in the Asterisk database for a given family, key, and value.

    database put phones 1000/username bob

database show

database show [*family* [*key*]]

Shows contents of database, or specific families, keys, and values.

    database show phones

debug channel

debug channel *channel_name*

Allows a debug of a specific active channel. See also show channels.

    debug channel SIP/1000-e54f

dial

dial [*extension*[*@context*]]

Dials a given extension (optionally, in the context specified) through the CONSOLE channel. This command is available only if *chan_oss.so* or *chan_alsa.so* is loaded in the *modules.conf* file.

    dial 1000@phones

dont include

dont include *context_to_be_removed* in *context*

Removes a specified include from a context.

    dont include local-extensions in incoming

dump agihtml

dump agihtml *filename*

Dumps a list of AGI commands in HTML format to the given filename. The file will be saved to the */tmp/* directory by default, but a full path may be specified.

exit

exit

Closes the command-line interface, if you connected to the Asterisk console via the –r flag. You cannot use the quit and exit commands to shut down the PBX (as would be the case if the Asterisk were running in the foreground). To shut down the PBX rather than exiting the console, see the stop and restart commands.

extensions reload

extensions reload

Reloads the dialplan configuration from the *extensions.conf* file. In other words, it reloads only your dialplan; nothing else. This command is safe to run even when calls are active. Any new channels being created will be based on the newly reloaded dialplan.

hangup

hangup

Hangs up any currently active calls placed using the CONSOLE channel. This command is only available if *chan_oss.so* or *chan_alsa.so* is loaded in the *modules.conf* file.

help

help [*command* [*subcommand*]...]

Displays help for commands and command-line usage. A single question mark or tab will do the same.

    help show applications

iax2

Subsets of this command allow you to manage your IAX connections.

iax2 debug

`iax2 debug`

Enables IAX debugging.

iax2 no debug

`iax2 no debug`

Disables IAX debugging.

iax2 provision

`iax2 provision host template [forced]`

Used to configure an IAX device, such as Digium's IAXy. Provisions the given peer or IP address using a template matching either `template` or *' (if the template is not found). Templates are configured in the *iaxprov.conf* file, usually located in */etc/asterisk/*. If `forced` is specified, empty provisioning fields will be provisioned as empty fields.

```
iax2 provision 192.168.1.100 default
```

iax2 show cache

`iax2 show cache`

Displays currently cached IAX dialplan results. Related to the `switch => ` statement for remote dialplans. Remote dialplans are cached for a period of time (600 seconds); they then expire and must be requeried if used again.

iax2 show channels

`iax2 show channels`

Displays detailed information about active IAX channels.

iax2 show firmware

`iax2 show firmware`

Shows available IAX firmware.

iax2 show peer

`iax2 show peer peer_name`

Shows details on a specific IAX peer.

```
iax2 show peer iaxfwd
```

iax2 show peers

```
iax2 show peers [registered] [like pattern]
```

Lists all known IAX2 peers. The optional registered argument causes only peers with known addresses to be listed. The optional regular expression pattern is used to filter the peer list.

```
iax2 show peers registered like iax*
```

iax2 show provisioning

```
iax2 show provisioning [template]
```

Lists all known IAX provisioning templates, or the details of a specific template.

iax2 show registry

```
iax2 show registry
```

Lists details and status of all registration requests.

iax2 show stats

```
iax2 show stats
```

Displays statistics for the IAX channel driver.

iax2 show users

```
iax2 show users [like pattern]
```

Lists all known IAX2 users. The optional regular expression pattern is used to filter the user list.

```
iax2 show users like iax*
```

iax2 trunk debug

```
iax2 trunk debug
```

Requests the current status of IAX trunking. Trunking is enabled for a peer with trunk=yes in *iax.conf*.

include context

```
include context in context
```

Includes the specified context in another context.

```
include local-users in incoming
```

indication

The loadzone option in a channel configuration file configures the tone zone to use for a channel. A *tone zone* is a set of indications, as configured in *indications.conf*, that contains information about all the various sounds that are common to telephones in a particular country—dial tone, ringing cycles, busy tones, and so on. A loaded tone zone is applied to a Zaptel channel, which will behave according to the definition for its tone zone. The idea is to deliver familiar telephone sounds, wherever in the world the users might be. Individual channels can have different indication sets configured, which means that a single Asterisk system can provide familiar telephony behavior to people from different countries. The defaultzone is used if nothing is specified for the channel.

indication add

indication add *country indication* "*tonelist*"

Adds the given indication to the country. See also show indications.

    indication add us dial "350+440"

indication remove

indication remove *country indication*

Removes the given indication from the country. See also show indications.

    indication remove us dial

init keys

init keys

Initializes private RSA keys using the passcode specified by the user. Keys are generated with the use of the *astgenkey* script. Keys generated with the use of a passcode must be initialized with the –i flag when starting Asterisk, or with the init keys command from the CLI.

load

load *module_name*

Loads the specified module into Asterisk.

    load chan_oss.so

local show channels

`local show channels`

Shows the status of Local channels.

logger

In the *logger.conf* file, you can specify the various levels of detail the system will record in its logs. The following commands allow you to reload and rotate those files. Logs are typically stored in the */var/log/asterisk/* directory.

logger reload

`logger reload`

Reloads the log files. Required after making a change to the *logger.conf* configuration file.

logger rotate

`logger rotate`

Rotates and reopens the log files. When rotating, the old file is renamed to include a .n, where n is the highest numbered *logfile.n* + 1. If *logfile.n* does not exist, the file is renamed to *logfile.0*.

meetme

The meetme command can be used for a variety of purposes, including listing all active conferences, the number of parties in a conference, the number of marked users, the active length of a conference, and whether a conference was created dynamically or statically.

A timing interface must be loaded in order for this command to be available.

The following meetme subcommands can be used from the console to control active conferences.

meetme kick

`meetme kick` *confno* `[`*user_number* `| all]`

Kicks (i.e., removes) one or all participants from an active conference.

    meetme kick 100 all

meetme list

`meetme list` *confno*

Lists the associated channel names of conference participants and monitors status.

    meetme list 100

meetme lock

`meetme lock` *confno*

Locks a conference from allowing any joins.

    meetme lock 100

> As the number of users in a conference grows, so does the load on the CPU, as it has to mix all of the incoming streams into one, and then transmit the result back out to all the participants. If you have advertised a public conference and it suddenly becomes too popular, you may want to lock out any further participants in order to preserve sound quality.

meetme mute

`meetme mute` *confno user_number*

Mutes a user in the conference.

    meetme mute 100 1

meetme unlock

`meetme unlock` *confno*

Unlocks a conference, allowing channels to join the active conference.

    meetme unlock 100

meetme unmute

```
meetme unmute confno user_number
```

Unmutes a user in the conference who is muted.

```
meetme unmute 100 1
```

pri

If you are running the ISDN-PRI protocol on any of your T1 spans, the following commands will help you with troubleshooting.

pri debug

```
pri debug
```

Turns on PRI debugging.

pri intense debug span

```
pri intense debug span span
```

Enables very verbose debugging information for the D-channel of your PRI. This information is invaluable when troubleshooting PRI connections to non-Asterisk systems (such as the PSTN).

```
pri intense debug span 1
```

pri no debug

```
pri no debug
```

Turns off PRI debugging.

pri show debug

```
pri show debug [span]
```

Displays the status of PRI debugging and intense debugging for all spans or, optionally, a single defined span.

pri show span

`pri show span` *span*

Displays extended information about a PRI span.

    pri show span 1

quit

See exit.

reload

`reload [`*module* `...]`

Reloads configuration files for all listed modules that support reloading (or for all supported modules, if none are specified).

    reload res_crypto.so

remove

The `remove` command contains many subcommands that allow you to remove functionality from your Asterisk PBX without directly editing the configuration files.

This function can be used for ad hoc testing, but if you want to make the changes permanent, it is recommended that you edit the various configuration files directly, from */etc/asterisk/*.

remove extension

`remove extension` *exten@context* `[`*priority*`]`

Removes a whole extension from a context. If the priority is specified, removes that priority only within the given extension. Subsequent priorities within the extension will be renumbered if you use the n priority-naming scheme.[*]

    remove extension 500@default 3

---

[*] If you have explicitly numbered your priorities, you will create a gap in your extension. This can easily be corrected by adding a NoOp( ) command in the removed priority (e.g., add extension 500,3,Noop into default).

remove ignorepat

remove ignorepat *pattern* from *context*

Removes the ignore pattern from the given context.

    remove ignorepat 9 from local

remove queue member

remove queue member *channel* from *queue*

Drops the active channel from the given queue. Queue members are the active channels within a queue.

    remove queue member SIP/1000-d448 from customer_service

restart

When a restart is performed, all channels are cleared (i.e., hung up) and all modules are reloaded. You can also instruct Asterisk to restart only when there no longer any active channels, thus preventing calls from being dropped.

restart gracefully

restart gracefully

Causes Asterisk to stop accepting new calls and perform a cold restart when all active calls have ended.

restart now

restart now

Causes Asterisk to immediately hang up all calls and perform a cold restart.

restart when convenient

restart when convenient

Causes Asterisk to perform a cold restart when all active calls have ended. New calls are accepted, and only when all calls have completed is the restart performed. Use this command very carefully, as you have no way of knowing when the conditions for the restart will be met. On a busy system, the restart might not occur until well after you've forgotten you requested it. The best practice on a busy system is to execute restarts manually.

save dialplan

save dialplan

Saves the current dialplan from the command line to the *extensions.conf* file. It is important to remember that all comments are stripped from the dialplan upon saving. It is recommended that permanent changes to the dialplan be made directly in the *extensions. conf* file and then reloaded (see extensions reload) to preserve comments.

set

The set command is used to control the amount of debugging information on the console. If connecting to a remote Asterisk console, be aware that changes made to the level of debugging have global scope—that is, they affect all consoles. Also be sure to lower the debugging level before exiting if you are logging to a text file (see logger).

set debug

set debug *level*

Sets the level of core debug messages to be displayed. 0 means no messages are displayed. Equivalent to -d[d[d...]] on startup.

    set debug 10

set verbose

set verbose *level*

Sets the verbosity level on the console. A setting of 0 means that no information on calling activity will be displayed. If you request 10, you'll be seeing a lot of activity indeed (especially on a busy system). This command has the exact same effect as the -v[v[v...]] flags you provide on startup.

    set verbose 10

show

The show subcommands are used to display all kinds of information about your system.

show agents

```
show agents
```

Provides summary information about agents configured in *agents.conf*.

show agi

```
show agi [topic]
```

Displays usage information on the given command, when called with a topic as an argument. If called without a topic, provides a list of AGI commands.

```
show agi channel status
```

show application

```
show application application [application [application [...]]]
```

Displays extended information about one (or, optionally, more than one) given application.

```
show application dial
```

show applications

```
show applications
```

Lists brief explanations of all currently available applications.

show channel

```
show channel channel
```

Displays extended information about the given channel.

```
show channel SIP/1000-3d43
```

show channels

```
show channels [concise]
```

Lists the currently defined channels and some information about them. If concise is specified, the format is abridged and presented in a more easily machine-parsable format.

show dialplan

show dialplan [*context*]

Shows the current state of the dialplan as loaded into memory. If a context name is appended to the end of the command, only that context will be shown. The show dialplan command is useful for verifying the order of pattern matching as well.

show dialplan incoming

> If you type show dialplan and then press the Tab key a few times, you'll be presented with a list of all the contexts in your dialplan. On the Asterisk CLI, the Tab key can yield all kinds of neat information. If in doubt, press Tab.

show indications

show indications [*country* [...]]

Displays a condensed list of countries, or optionally a detailed list of indications for one or more countries. See also indications add and indications remove.

show indications us

show keys

show keys

Lists the encryption keys on your system. Keys are stored in */var/lib/asterisk/keys/* and are loaded with the *res_crypto.so* module.

show manager command

show manager command *command*

Shows extended information about a Manager command. See also show manager commands.

show manager command setvar

show manager commands

show manager commands

Lists all available Manager commands and their privilege levels, and gives a brief synopsis of each.

show manager connected

```
show manager connected
```

Lists all currently connected Manager agents. Manager agents are configured in *manager.conf*.

show modules

```
show modules
```

Lists currently loaded modules, gives a brief description of each, and shows the module use count.

show parkedcalls

```
show parkedcalls
```

Lists currently parked calls.

show queue

```
show queue queue
```

Provides extended information about a particular queue.

```
show queue customer_service
```

show queues

```
show queues
```

Provides extended information about all queues.

show translation

```
show translation
```

Displays a table of all codecs and their relative translation times between formats (provided in milliseconds). The higher the number, the more work is required to transcode between those formats. If the formats are native (i.e., the same), no transcoding is required—Asterisk simply routes the packets, which requires very little processing time.

show uptime

```
show uptime
```

Displays Asterisk's total uptime and the time since the last reload.

show version

```
show version
```

Displays the currently installed version of Asterisk. The version is controlled through the *.version* file in the Asterisk sources. When updating the Asterisk source code, be sure to perform a make update to update this value. The correct version is required when submitting a bug report to the bug tracker (located at *http://bugs.digium.com*—be sure to read the bug submission guidelines *before* submitting bugs!).

show voicemail users

```
show voicemail users [for vm_context]
```

Displays the voicemail context, mailbox number, voicemail zone, and number of new messages for all voicemail users configured in *voicemail.conf*. Optionally, displays information for a specific voicemail context.

```
    show voicemail users for default
```

show voicemail zones

```
show voicemail zones
```

Displays the currently configured voicemail zones and their associated time zones and message formats.

sip

The subsets of the sip command allow you to manage your SIP connections.

sip debug

```
sip debug
```

Turns on SIP debugging. This will be very verbose.

sip debug ip

```
sip [no] debug ip dotted_ip_notation
```

Debugs (or disables debugging of) SIP messages from a specific IP address. This is useful when trying to debug messages coming from a peer who is not yet registered with you or is not configured in *sip.conf*.

```
    sip debug ip 192.168.1.100
```

sip debug peer

`sip [no] debug peer peer_name`

Debugs (or disables debugging of) SIP messages from an individual peer, referenced by the peer name configured in *sip.conf*. Debugging information can be displayed for a dynamic host only if that host is registered with you. If you are trying to debug a registration issue, see `sip debug ip`.

```
sip debug peer john
```

sip history

`sip [no] history`

Enables or disables SIP history recording. See also `sip show history`.

sip no debug

`sip no debug`

Turns off SIP debugging.

sip reload

`sip reload`

Reloads the SIP channel module. This is the equivalent of performing a `reload chan_sip.so`. Reloading the SIP channel is required to load changes to *sip.conf* and *sip_notify.conf* into memory. Active SIP channels are not dropped during a `sip reload`.

sip show channel

`sip show channel channel`

Displays extended information about an active SIP channel. See also `sip show channels`.

```
sip show channel 00036bdd-39
```

sip show channels

`sip show channels`

Displays a list of all active SIP channels. The value in the Call ID column is used by the `sip show channel` command to display extended information about an individual channel. See also `sip show channel`.

sip show history

`sip show history channel`

Provides a detailed log history for a given SIP channel.

`sip show history 00036bdd-39`

sip show peer

`sip show peer peer_name`

Displays detailed information about a peer configured in *sip.conf*.

`sip show peer john`

sip show peers

`sip show peers`

Lists and displays the status of all SIP peers.

sip show registry

`sip show registry`

Lists and displays the status of all peers with whom you are registered.

sip show user

`sip show user user_name`

Displays detailed information about a user in *sip.conf*.

`sip show user 1000`

sip show users

`sip show users`

Displays a listing of all users configured in *sip.conf*.

soft hangup

`soft hangup channel`

Requests a hangup on a given channel.

`soft hangup SIP/1000-4248`

stop

Asterisk has various ways of controlling how and when it stops the system. The options are similar to the restart commands. You can instruct Asterisk to stop only when there no longer any active channels, thus preventing calls from being dropped.

stop gracefully

stop gracefully

Stops the system when all currently active calls have completed, and does not accept new calls.

stop now

stop now

Stops immediately, terminating all active calls.

stop when convenient

stop when convenient

Stops the system when all currently active calls have completed. New calls are accepted, and the system will stop only when there are no longer any active calls. Using this command is not a good idea, since you have no real way of knowing when the necessary condition for stopping the system will occur.

unload

unload [-f | -h] *module_name*

Unloads the specified module from Asterisk. The −f option causes the module to be unloaded even if it is in use (which may cause a crash), and the -h option causes the module to be unloaded even if the module says it cannot be, which will almost always cause a crash.

    unload app_math.so

zap

The Zaptel interfaces allow Asterisk to interact via a physical medium, either analog or digital. This may include telephones, analog PSTN connections, or digital circuits such as T-1/E-1 circuits.

zap destroy channel

`zap destroy channel` *channel_number*

Immediately removes a channel, whether or not it is in use.

    zap destroy channel 1

zap show cadences

`zap show cadences`

Displays the configuration of the various ring cadences (ring tones) Asterisk has configured for an analog circuit (FXS).

zap show channel

`zap show channel` *channel_number*

Displays extended information about a particular Zaptel channel.

    zap show channel 1

zap show channels

`zap show channels`

Lists all Zaptel channels and their associated extensions, languages, and default Music on Hold classes.

Index

Symbols

!command, 337
(hash), comment marker, 36
$ (dollar sign)
   $[ ], enclosing expressions, 99
   ${ }
      referencing function values, 102
      referencing variable value, 90
${EXTEN} channel variable, 95
& (ampersand), concatenating destinations
      for Dial( ), 86
* (asterisk), wildcard character, 58
, (comma), separator for application
      arguments, 81
. (period), wildcard matches, 93
/* */ comment tags in zconfig.h, 38
/tmp/ directory, 56
[ ] (square brackets)
   context names in, 78
   enclosing macro definitions, 110
^ (caret), beginning of line matching in
      regular expressions, 101
_ (underscore), beginning patterns, 93
| (pipe character)
   separating mailbox option/value
      pairs, 108
   separator between application
      arguments, 81

Numbers

0V logic reference, 18
1.544-Mbps bit stream (DS-1), 132
23B+D (ISDN-PRI), 135

30B+D (ISDN-PRI), 135
64-kbps channel (DS-0), 132

A

A/D (analog-to-digital) converter, 129
abort halt command, 337
AbsoluteTimeout( ), 229
AbsoluteTimeout( ) application, 103
adapters, telephony, 29
Adaptive Differential Pulse-Code Modulation
      (ADPCM), 146
add command, 338
add extension command, 338
add ignorepat command, 338
add queue member command, 338
add-ons, 46
AddQueueMember( ) application, 229
adsi.conf file, 313
ADSIProg( ) application, 230
adtranvofr.conf file, 313
AgentCallbackLogin( ) application, 230
AgentLogin( ) application, 230
AgentMonitorOutgoing( ) application, 231
agents.conf file, 313
aggressive residual echo suppression, 39
AGI (Asterisk Gateway Interface), 156–174
   debugging, 172
   fundamentals of
      communication, 156–158
      calling AGI script from dialplan, 158
      standard pattern of
         communication, 157
      STDIN, STDOUT, and STDERR, 157

We'd like to hear your suggestions for improving our indexes. Send email to *index@oreilly.com*.

AGI (Asterisk Gateway Interface) (*continued*)
 Perl AGI library, 163
 PHP AGI library, 168
 Python AGI library, 172
 reference, 292–300
 writing scripts in Perl, 159–163
 writing scripts in PHP, 163–168
  important steps to remember, 168
 writing scripts in Python, 169–172
  important steps to remember, 172
agi debug command, 173, 339
agi no debug command, 339
AGI( ) application, 158, 231
agi-bin/ directory, 55
agi-test.agi script, 159
alarm system for your home,
  controlling, 199
AlarmReceiver( ) application, 232
alarmreceiver.conf file, 315
A-law companding algorithm, 132
 use with G.711 codec, 145
aliasing, 129
alsa.conf file, 315
alternating current (AC) voltage, analog
  phone ringer, 120
AMD CPUs
 IRQ latency and, 15
 power FPUs, 14
analog circuits, 131
Analog Display Services Interface
  (ADSI), 313
analog ground start lines, 62
analog interface cards, 23
analog interfaces, 58
analog telephones, 26
analog telephony, 119–122
 echo, why it occurs, 151
 parts of analog telephone, 120–122
 Tip and Ring wires, 122
Analog Terminal Adaptor (ATA), 29
analog waveform, digitally encoding, 123
analog waveform, digitization of, 12
analog-to-digital (A/D) converter, 129
answer command, 292, 339
Answer( ) application, 80, 82, 233
 Festival( ) application and, 177
APIC-enabled motherboards, IRQ control, 16
AppendCDRUserField( ) application, 233
applications
 AGI, 158
 Background( ), 84

Congestion( ), 96
Dial( ), 86–88
extensions, 80
 s (start) extension, 82
Goto( ), 84
modules.conf file, 303
reference, 229–291
SayDigits( ), 95
Zapateller( ), 115
arguments (application), 81
 Dial( ), 86
 Goto( ), 85
arguments, using in macros, 111
AstDB (Asterisk database), 112–115
 deleting data, 113
 retrieving data from, 113
 rotating key for DUNDI information, 186
 storing data, 113
 using in the dialplan, 114
astdb file, 55
Asterisk
 acceptance of, 5
 challenges to, 205
 community for development and
  support, 5
 compiling, 41–46
 configuring for Festival, 176
 directories used by, 54–57
 Documentation Project, 7
 future of, 200–208
 Internet Relay Chat (IRC) channels, 7
 loading, 52
  CLI commands, 52
 mailing lists, 6
 Manager, 180–182
 passing call control to an external
  program, 11
 sizes of systems, 14
 source code, obtaining, 32–34
 things now possible, 196–200
 versions, xv
 VoIP, 152–155
 Wiki, 7
Asterisk Gateway Interface (see AGI)
asterisk package, 31
asterisk program
 -c (console) flag, 75
 −r (remote) flag, 75
 running with -h switch, 52
Asterisk program (see asterisk package)
asterisk.conf file, 316
Asterisk::AGI Perl module, 163

asterisk-addons package, 46
asterisk-sounds package, 32, 56
    installing, 46
Asterisk-users list, 195
ATA (Analog Terminal Adaptor), 29
audio
    built-in components on motherboards, 17
    digital, 122
    quality problems on inadequate
        systems, 9
audio formats, translation costs, 83
audio streams, packetization for transport
        over IP networks, 138
Authenticate( ) application, 233
authentication
    IAS FWD incoming calls, 73
    IAX connections, 72
    IAX protocol, 140
    inbound and outbound, Asterisk
        scheme, 152
    with Manager, 182
    secret (password) in SIP, 69
    SIP, 141
    SIP client, 71
auto-attendants, 84
Automated Attendant (AA), IVR vs., 198
Automatic Ringdown circuit, 66

B

B- and D-channels, PRI circuits, 135
Background( ) application, 84, 234
    sound files, specifying, 180
BackgroundDetect( ) application, 235
bandwidth
    analog telephony, 120
    network, future of, 206
    PCM-encoded telephone circuit, 132
Basic Rate Interface (BRI) ISDN circuits, 25,
        27
B-channels (bearer channels), 134
Berkeley DB Version 1 database, 112
best effort method, QoS and, 151
binary one (1), voltage and, 19
binary zero (0), relation to 0-volt signal, 19
BIOS
    control over IRQ assignment, 16
    USB activation, 50
bison parser, 32
    not found, error caused by, 47
bit resolution, analog wave samples
    increasing resolution, 126

bitrates
    ADPCM (Adaptive Differential
        PCM), 146
    Variable Bitrate (VBR) codecs, 147
bit-resolution, analog wave samples, 123
    effects on quality of digitally encoded
        waveform, 125
blacklist, looking up numbers on, 115
Boolean operators, 100
BOOST_RINGER option (zconfig.h file), 38
boostringer parameter, activating for
        Zaptel, 41
BRA (Basic Rate Access) ISDN circuits, 134
BRI (Basic Rate Interface) ISDN circuits, 27,
        134
bridged calls, echo cancellation, 152
buffering output
    flushing after every write in Python AGI
        script, 172
    turning off in AGI Perl scripts, 159, 163
    turning off in AGI PHP script, 164, 168
bug fixes, 33
built-in audio and video components on
        motherboards, 17
business case for Asterisk, 7
    flexibility for growth, 10
busy destination
    Dial( ) application handling of, 87
busy message, voicemail, 108
busy signal, fast, 96
Busy( ) application, 235

C

CAC (Carrier Access Corporation) ground
        start signaling, 40
Call Detail Records (see CDRs)
call files, 182–184
call parking, 115
call transfer, configuring on FXO
        channel, 64
call waiting
    configuring for FXO channel, 64
    on analog line, 121
Caller ID
    configuring for FXO channel, 64
    LookupCIDName( ), 258
    SetCallerID( ) application, 277
CALLERIDNUM channel variable, 104
CallingPres( ) application, 235
Carrier Access Corporation (CAC) ground
        start signaling, 40

CAS (Channel Associated Signaling), 133
CBR (constant bitrate) encoding (MP3
    files), 55
cdr.conf file, 316
cdr_manager.conf file, 317
cdr_odbc.conf file, 318
cdr_pgsql.conf file, 318
cdr_tds.conf file, 318
CDRs (Call Detail Records), 46
    challenges to obtaining, 179
    recording, 178
    storage directory, 57
    storing in a database, 179
Celeron processors, Asterisk lab systems
        on, 14
cellular telephone networks, end of, 203
central office (CO), signaling incoming call
        on analog phone, 120
Cepstral text-to-speech engine, 177
chan_h323.so module, 142
chan_iax2.so module, 139
chan_mgcp.so module, 144
chan_oh323.so module, 142
chan_sip.so module, 140
chan_zap (channel module), 35
ChangeMonitor( ) application, 237
ChanIsAvail( ) application, 237
Channel Associated Signaling (CAS), 133
channel banks, 24
    CAC, FXS ground start signaling, 40
CHANNEL STATUS command, 292
channel variables, 91
    ${EXTEN}, 95
    CALLERIDNUM, 104
channels
    configuration, 58
    FXO and FXS, 60–67
        configuring FXS, 65–67
        FXO configuration, 61–65
    hardware, signaling methods and
            options, 64
    separation of B- and D- channels in
            ISDN, 134
    SIP, configuring, 67–72
    STDIN, STDOUT, and STDERR, 156
    timeouts, setting, 103
    VoIP (Voice over IP) channels, 209–228
CheckGroup( ) application, 238
checkresult function, 168, 170
checkresult subroutine, 161
chkconfig --add asterisk command, 44
chkconfig command, 37

circuits
    electrical (see electrical circuits)
    OC (optical carrier), 132
    types in PSTN, 131
        DS-0, 131
        T-carrier, 132
circuit-switched telephone networks, 26, 130
    BRI ISDN, 27
Cisco VoIP proprietary protocol (SCCP), 144
CLI (command-line interface), 52
    reference, 337–358
client, configuration in SIP, 70
clocking mechanism (ztdummy), 25
closed thinking in telecommunications
        industry, 191
codecs, 12, 144–148
    compressed, DSP load on system, 11
    configuring for IAX, 73
    G.711, 145
    G.723.1, 146
    G.726, 146
    G.729 codec, 46
    G.729A, 146
    GSM, 147
    IAX outbound connections, 74
    iLBC (Internet Low Bitrate Codec), 147
    MP3, 148
    quick reference, 145
    Speex, 147
    VoIP, 137
codecs.conf file, 319
coder/decoder (see codecs)
command line, storing values from in
        AstDB, 113
comment tags (/* */) in zconfig.h file, 38
commoditization of telephony hardware and
        software, 193
communications technologies, integration
        of, 208
communications terminals, 29
companding, 128
    A-law, in E-1 circuits, 132
    G.711 codec, 145
    m-law algorithm, on T-1 circuits, 132
compiler, GCC, 32
compiling
    Asterisk, 41–46
        alternative make arguments, 42–44
        common issues, 47
        Makefile options, 44
        precompiled binaries, using, 45
    libpri, 41

Zapata drivers, telephony, 36
Zaptel, common problems, 48–50
Zaptel drivers, 35–41
    zconfig.h file, 38–40
    ztdummy, 36
complexity of open systems, 207
COmpression/DECompression (see codecs)
compression/decompression (see codecs)
computer power supplies, 17
concatenating destinations for Dial( ), 86
Concurrent Versioning System (see CVS)
conditional branching, 103–106
    GotoIf( ) application, 103
    time-based, with GotoIfTime( ), 105
conference rooms, 199
conferencing
    MeetMe( ) application, 117
    MeetMeCount( ) application, 117
    system requirements and, 11
    timing source, 25
    video-conferencing, 202
CONFIG_CALC_XLAW, 38
configuration
    initial, of Asterisk, 58–76
        debugging, 75
        FXO and FXS channels, 60–67
        IAX connections, inbound, 72–73
        IAX connections, outbound, 74
        interface configuration files, 59
        SIP channel, 67–72
configuration files, 301–336
    /etc/asterisk/ directory, 54
    Asterisk, disabling overwrites, 45
    default, installing for Asterisk, 42
    voicemail.conf, 107
Congestion( ) application, 96, 238
Conjugate-Structure Algebraic-Code-Excited
        Linear Prediction (CS-ACELP), 146
connections, maximum number to be
        supported by the system, 11
console
    connecting to Asterisk console, 75
    Linux, specifying for Asterisk CLI
        output, 54
    remote console on TTY9, 157
constant bitrate (CBR) encoding (MP3
        files), 55
contexts
    [globals], 91
    adding to dialplan for internal
        calls, 88–90

    adding to dialplan for long-distance
        calls, 97
    calls entering without specific destination
        extension, 82
    dialplan, 78
        [general], 79
        defining extensions, 79–81
    DUNDi, creating and mapping to dialplan
        contexts, 186
    FXS channel, 66
    IAX FWD incoming calls, 73
    incoming calls on FXO interface, 65
    internal, FXS channel, 67
    outbound dialing, adding to dialplan, 96
    peer connections and, 153
    SIP channel, 70
    using within another context, via
        includes, 97–98
    voicemail, 107, 108
ControlPlayback( ) application, 239
cords, 129
core file, dumping after Asterisk crash, 53
    safe_asterisk script, 53
countries, phone system sounds for, 62
CPUs
    choosing for Asterisk system, 13–15
    performance effects on Asterisk
        system, 12
    performance information, web sites, 14
crash notifications, 54
CRLF, terminating lines in commands to
        Manager, 182
cryptographic library, Asterisk, requirement
        of OpenSSL, 32
CSV (Comma Separated Values) file, CDR
        details, 178
CSV format (CDRs), 57
Curl( ) application, 239
Cut( ) application, 239
CVS
    obtaining Asterisk source code from, 34
    stable CVS branch vs. releases, 33

D

D/A (digital-to-analog) converter, 125
data groupings in AstDB (families), 113
database del command, 292, 339
database deltree command, 293, 339
database get command, 293, 340
database put command, 293, 340
database show command, 340

database, Asterisk (AstDB), 112–115
  deleting data, 113
  retrieving data from, 113
  storing data, 113
  using in the dialplan, 114
DateTime( ) application, 240
DBdel( ) application, 113, 240
DBdeltree( ) application, 113, 240
DBget( ) application, 114, 241
DBput( ) application, 241
D-channels, 134
  PRI circuits, 135
DeadAGI( ) application, 158, 241
debug channel command, 340
debug file, output to, 53
DEBUG output on the console, 76
debug profiling information, 45
debugging, 75
  AGI Perl script output, writing to Asterisk
    console, 163
  AGI programs, 172
  connecting to Asterisk console, 75
  enabling, with verbosity, 75
  Festival server, starting, 176
Denial of Service (see DoS attacks)
depmod errors during compilation, 50
destination argument, Dial( ) application, 86
/dev/ directory, dynamic population with
    udevd, 50
device drivers, 35
dial tone
  configuring on FXS channel, 66
  FXO and FXS channels, 60
Dial( ) application, 86–88, 242
  DIALSTATUS variable, indicating success
    of call, 112
  t and/or T options, call parking and, 116
  voicemail with busy or unavailable
    message, 108
Dial( ) statement, 73
dial-by-name directory, 109
dialing 9 before calling an outside number, 96
dialpad (analog phones), 120
dialplan, 77–98
  adding logic, 84–98
    Background( ) and Goto( )
      applications, 84
    context for internal calls, 88–90
    Dial( ) application, 86–88
    enabling outbound dialing, 95
    handling invalid entries and
      timeouts, 85

  includes, 97–98
  pattern matching, 92–95
  variables, 90–92
AstDB, using, 114
call parking, 115
calling AGI script from, 158
calling Festival from, 176
conditional branching, 103–106
conferencing with MeetMe( ), 117
configuration for IAX incoming calls, 73
configuration on FXO channel, 65
configuring for dundi local context, 189
configuring for FXS channel, 67
configuring for IAX FWD outgoing
    connection, 74
configuring for SIP channel, 71
creating simple, 81–84
  "Hello World!" example, 83
  s (start) extension, 82
expressions and variable
    manipulation, 99–102
functions, 102
macros, 110–112
  calling from dialplan, 111
scripting logic, 11
sound recordings, creating, 181
syntax, 77
  contexts, 78
  extensions, 79–81
voicemail, 106–109
  adding to dialplan, 108
Zapateller( ) Application, 115
DIALSTATUS variable, 112
DID (Direct Inward Dialing), 197
DiffServ (differentiated service), 150
digital circuits, 58, 131
digital circuit-switched telephone
    network, 130–135
  circuit types, 131
  digital signaling protocols, 133
digital interface cards, 23
Digital Signal Processing (DSP), 2
digital signaling protocols, 133
  Channel Associated Signaling (CAS), 133
  ISDN (Integrated Services Digital
    Network), 134
  SS7 (Signaling System 7), 135
digital signals
  advantages of, 26
  conversion to analog with telephony
    adaptors, 29
digital telephones, 26

digital telephony, 122–130
   Pulse-Code Modulation (PCM), 123–130
digital-to-analog (D/A) converter, 125
DigitTimeout( ) application, 103, 246
Digium cards
   analog interface card for Asterisk, 23
   IRQ latency and, 13
Digium Dev-Lite kit, 58
   with FXO and FXS interface, 59
direct current (DC) voltage, powering analog
     phones, 120
directories
   sounds directory, 82
   specifying where to install Asterisk, 45
   staging directory, changing, 45
   used by Asterisk, 54–57
Directory( ) application, 109, 246
DISA( ) application, 247
disconnects, far-end, 62
distributed IVR, 198
Distributed Universal Number Discovery
     (see DUNDi)
Dixon, Jim, 2
dmesg command, checking USB controller
     type, 50
dnsmgr.conf file, 319
Documentation Project, Asterisk, 7
Domain Name System (DNS), mapping
     E.163 numbers into, 204
Domain Name System Service records (DNS
     SRV records), 69
domain/realm (SIP X-Lite client), 71
dont include command, 340
DoS (Denial of Service) attacks on VoIP
     communications, 141
DPDISCOVER query, 186
drivers, unloading from memory with
     rmmod, 63
DS-0 (64-kbps channel), 132
DS-1 (1.544-Mbps bit stream), 132
DSP (Digital Signal Processing), 2
   system requirements for, 11
DTMF (Dual-Tone Multi Frequency), 120
   WAIT FOR DIGIT command, 162
Dual-Tone Multi Frequency (see DTMF)
dump agihtml command, 341
DumpChan( ) application, 248
DUNDi (Distributed Universal Number
     Discovery), 184, 204
   configuring Asterisk for use
     with, 185–189

dundi.conf file, 319
   contexts, mapping to dialplan
     contexts, 186
   defining DUNDi peers, 187
   general configuration, 185
DUNDiLookup( ) application, 184, 248
dust in equipment rooms, 22
dynamic IP addresses, 154

E

E&M (Ear & Mouth or recEive & transMit)
     signaling, 133
E.164 ITU standard for phone number
     assignment, 28
E.164 numbering specification, 204
E-1 lines, 58, 132
e164.org, 204
EAGI( ) (enhanced AGI) application, 158
EAGI( ) application, 248
Ear & Mouth signaling (see E&M signaling)
echo, 151
   managing, 152
   why it occurs, 151
echo cancellation, 11
   choosing method, 38
   disabling, 39
   removing echo on analog lines, 64
echo suppression, aggressive, 39
echo training, 64, 152
Echo( ) application, 249
   testing X-Lite soft phone, 72
   verifying bidirectional communication on
     FXS channel, 67
   verifying bidirectional communications
     for FXO channel, 65
electrical circuits, 21
electrical regulations
   power quality and, 21
   safety of users, 20
electrical signals, translation of sound waves
     to, 26
Electronic Numbering (ENUM) system, 319
emergency calls, outbound, 96
   enabling on FXO line, 64
encoding audio digitally, 123–130
EndWhile( ) application, 249
entity ID (dundi.conf), 186
ENUM (Telephone Number Mapping)
     group, 204
enum.conf file, 319
ENUMLookup( ) application, 249

environment variables, 92
environment, system, 18–22
  electrical circuits, 21
  equipment room, 21
  grounding, 19
  power conditioning and UPSs, 18
equipment room for systems, 21
ERROR messages, 75
error reporting, turning off HTML messages
    in PHP, 163, 168
/etc/asterisk/ directory, 54
/etc/asterisk/manager.conf file, 180
/etc/rc.d/init.d/ or /etc/init.d/ directories, 37
  automatically executing Asterisk at
      startup, 42
  Red Hat-style initialization scripts, 44
Eval( ) application, 250
EXEC command, 293
Exec( ) application, 250
ExecIf( ) application, 251
exit command, 341
expressions, 99–102
  basic, 99
  operators, 100
extconfig.conf file, 320
extensions
  added to [internal] context, 89
  defining in dialplan contexts, 79–81
  dialplan contexts, 78
  invalid entries and timeouts, 85
  priorities, 80
    unnumbered, 81
  s (start) extension, 82
  timeout argument for Dial( ), 87
extensions reload command, 341
extensions.conf file, 59, 320
  context name for SIP channel, 70
  dialplan specification, 77
  dundi local context, 188, 189
  global variables, 91
  IAX call to FWD echo test application, 74
  IAX FWD incoming calls, 73
  instructions to perform inside context on
      FXO line, 65
  parsing of expressions with bison, 47
  sample file, 78
external program, passing call control to, 11

F
families (data groupings in AstDB), 113
far-end disconnects, 62

fast busy signal, 96
FastAGI( ) application, 158, 251
fax machines, 2100-Hz tone during
    negotiation, 39
fear campaign to undermine telephony
    revolution, 205
features.conf file, 320
  call parking, 116
Festival, 175–177, 200
  calling from dialplan, 176
  configuring Asterisk for, 176
  setting up to use with Asterisk, 175
  starting the Festival server, 176
  using with Asterisk, text2wave
      utility, 177
Festival( ) application, 176, 252
festival.conf file, 320
festival.scm file, altering for use with
    Asterisk, 175
fflush command (PHP), 167
fflush function (PHP), 168
fgets command, 165
fgets function, 168
fiber optic circuits, 131
  SONET and OC, 133
file pointers, 156
filename, specifying for Playback( ), 82
find-me-follow-me, 197
firmware/ directory, 55
flash (electronic analog phones), 121
Flash Operator Panel (FOP), 182
Flash( ) application, 252
Floating Point Unit (see FPU)
Foreign eXchange Station (see FXS)
ForkCDR( ) application, 252
FPU (Floating Point Unit), 12
  processor selection and, 13
friend connections (SIP), 69
friends, 152, 153
full-duplex audio communications, 28
fully qualified domain name (FQDN), 69
  SIP X-Lite client configuration, 71
functions, dialplan, 102
FWD (Free World Dialup) account via
    IAX, 72–74
  configuring outbound connections, 74
  dialplan configuration for incoming
      calls, 73
  iax.conf file, 72
fwrite function, 168

FXO (Foreign eXchange Office)
    channel configuration, 58, 61–65
        dialplan, 65
        testing by dialing in, 65
        Zapata hardware, 63–65
        Zaptel hardware, 61–63
    channels, 60
    connection to analog phone line, 59
    determining port on TDM400P card, 60
FXS (Foreign eXchange Service)
    channel configuration, 65–67
        Zapata hardware, 66
        Zaptel hardware, 65
    channels, 60
    connection to analog phone, 59
    determining port on TDM400P card, 60
    dialplan configuration, 67
FXS (Foreign eXchange Station)
    channel configuration, 58
    ports provided by TDM400P cards, 23

G

G.711 codec, 145
G.723.1 codec, 146
G.726 codec, 146
G.729 codec, 46
G.729A codec, 146
gatekeeper (H.323), 143
gateways, telephony, 29
GCC compiler, 32
gcc compiler
    Asterisk compilation using make
        program, 42
    attempting to build Zaptel without, 48
    installing, with dependencies, 47
[general] context, 79
General Peering Agreement (GPA), 185
GET DATA command, 171, 294
GET FULL VARIABLE command, 294
GET OPTION command, 294
GET VARIABLE command, 294
GetCPEID( ) application, 253
GetGroupCount( ) application, 253
GetGroupMatchCount( ) application, 253
getnumber function, 171
glare, 62
global variables, 91
    channel for outbound calls, 95
GNU make program, 42

GNU tar application, extracting compressed
        source code, 34
GNU/Linux, 30
Golovich, James, 163
Goto( ) application, 84, 254
    repeating greeting after playing back
        number dialed, 85
GotoIf( ) application, 103, 254
GotoIfTime( ) application, 105, 255
GPA (General Peering Agreement), 185
ground start (gs), 62
ground start signaling (CAC), 40
ground, defined, 19
grounding, 19
    0V logic reference, 20
    power supplies, 18
gs (see ground start)
GSM codec, 147
    converting WAV files to, 180
GSM codec optimizations, 44

H

H.323 protocol, 142
    NAT and, 143
    security, 143
hacker's PBX, 5
handset (analog phone), 122
HANGUP command, 295
Hangup( ) application, 80, 83, 256
hardware
    Zapata
        configuring for FXO channel, 63–65
        configuring for FXS channel, 66
    Zaptel
        configuring for FXO channel, 61–63
        configuring for FXS channel, 65
hardware drivers, Zaptel, 32
    compiling, 35–41
hardware selection, 10–18
    choosing a motherboard, 15–17
    choosing a processor, 13–15
    performance issues, 11–13
    power supply requirements, 17
hardware, telephony, 22–25
harmonic noise on electrical circuits, 21
HasNewVoicemail( ) application, 256
HasVoicemail( ) application, 256
HDLC functionality in Zaptel drivers, 39
head Asterisk, 33
help command, 341

high-fidelity voice, 201
hobby systems
    hardware selection, 14
    system requirement guidelines, 10
home automation, 199
hook switch, 121
HTML error messages, turning off in AGI
        PHP script, 163
HTML error messages, turning off in
        PHP, 168
humidity (equipment rooms), 21
hybrid transformer, 121

I

IAX (Inter-Asterisk eXchange protocol), 139,
        209–219
    channel configuration, 58
    channel definitions, 215–219
    configuring inbound connections, 72–73
        dialplan, 73
        iax.conf file, 72
    configuring outbound connections, 74
    future of, 139
    general settings, 209–214
    NAT (Network Address Translation)
        and, 140
    register statements, 214
    retrieving dialplan information from
        remote Asterisk box, 214
    security, 140
iax.conf file, 59, 321
    configuration to accept calls from FWD
        users, 72
    configuring to place call on FWD
        network, 74
iax.conf.sample configuration file, 73
iax2 debug command, 342
iax2 no debug command, 342
IAX2 protocol, RSA key checks, 48
iax2 provision command, 342
iax2 show cache command, 342
iax2 show channels command, 342
iax2 show firmware command, 342
iax2 show peer command, 342
iax2 show peers command, 343
iax2 show provisioning command, 343
iax2 show registry command, 74, 155, 343
iax2 show stats command, 343
iax2 show users command, 343
iax2 trunk debug command, 343
IAX2Provision( ) application, 257

images/ directory, 55
ImportVar( ) application, 257
inbound IAX connections,
        configuring, 72–73
include context command, 344
includes, 97–98
incoming calls
    context for, 78
    users, 153
incoming context (FXO interface), 65
indication add command, 344
indication remove command, 344
indications.conf, 344
indications.conf file, 321
init keys command, 344
initialization scripts, installing with make
        config, 44
initialization scripts, Red Hat-style, 53
installing Asterisk, specifying directory, 45
Integrated Services Digital Network (see
        ISDN)
integration of communications
        technologies, 208
Intel CPUs
    IRQ latency and, 15
    powerful FPUs, 14
Interactive Voice Response (IVR), 198
interfaces
    configuration files, 59
    PSTN, 22–25
    types to which Asterisk can connect, 58
interference
    with analog signals, 122
internal calls, context for, 88–90
internal context
    FXS channel, 67
    FXS port, 66
    SIP channel, 70
International Telecommunication Union
        (ITU)
    closed thinking, 191
    H.323 protocol, 142
Internet transport protocols, real-time media
        streaming and, 138
Internet connectivity with ISDN-BRI
        circuit, 134
Internet gateways to telephony services,
        locating, 184
Internet Low Bitrate Codec (iLBC), 147
Internet Relay Chat (IRC) channels, 7
Interrupt Requests (see IRQs)

---

invalid entries, handling, 85
INVITE requests in DoS attacks, 141
invites (in SIP), 69
IP (Internet Protocol)
    evolving your old PBX to, 196
    transport mechanism for
        video-conferencing (H.323), 142
    (see also VoIP)
IP addresses
    domain/realm, SIP X-Lite client, 71
    register statement for FWD IAX
        server, 73
    SIP endpoints, 69
IP telephones, 27
    connecting to Asterisk, 59
    definition of a phone call, 29
    echo in, 151
IRQs (Interrupt Requests)
    latency, 13
    motherboard selection, latency and, 15
ISD, BRI, 134
ISDN, 134
    BRA, 134
    BRI, 25
    limited standards compliance, 192
    PRI, 134
        commands, 347
        libpri library, 32
    PRI/PRA, 134
ISDN telephones, 27
"It Still Does Nothing" (ISDN), 134
IVR (Interactive Voice Response), 198

J

jitter buffering (IAX), 73

K

kernel modules, ztdummy, 25
kernels
    layers of interaction with Asterisk, 35
    optimizations for Asterisk, 12
    Version 2.6, support for Asterisk, 13
    ztdummy driver and, 36
kewlstart (ks), 62
    on FXS channel, 66
Key Telephone Systems (KTSs), 26
keys and key families, deleting from
        AstDB, 113
keys/ directory, 55
ks (see kewlstart)
KTSs (Key Telephone Systems), 26

L

label argument, GotoIfTime( ), 105
Label Switched Path (LSP), 150
large Asterisk installations, requirements, 10
large systems
    processor selection, 15
    system requirement guidelines, 10
Last Mile (PSTN), 122
latency
    IAX outbound calls, 74
    monitoring between Asterisk server and
        SIP phone, 69
laxprov.conf file, 321
legacy telecommunications equipment
    hardware for Asterisk connections, 22–25
LEN( ) function, 103
libnewt libraries, 32
    zttool program and, 37
libpri library, loading, 52
libpri package, 31
    compiling, 41
    installing before asterisk, 32
Link button, analog phones, 121
Linux
    directories used by Asterisk, 54–57
    GNU utilities, selecting and
        configuring, 30
    kernel Version 2.6, support for
        Asterisk, 13
    layers of interaction between Asterisk and
        kernel, 35
    precompiled Asterisk binaries, 45
    problems with IRQs, 13
    Red Hat-based distributions, xv
    systems running udevd, allowing Zaptel
        access to, 50
    Zaptel drivers, 35
    ztdummy kernel module, 25
load command, 345
loadzone, 62
local show channels command, 345
locking down your phones, 199
logarithmic companding, 128
logger reload command, 345
logger rotate command, 345
logger.conf file, 57, 321, 345
    enabling DEBUG output to console, 76
logic ground, 18, 20
long-distance calls, context for, 97
LookupBlacklist( ) application, 115, 257
LookupCIDName( ) application, 258

loop start (ls), 62
low barrier to entry in open source
    telephony, 208
low-pass filter, 129
ls (see loop start)
lsmod command, 50
    verifying loading of zaptel module, 51
    verifying loading of ztdummy and its use
        by zaptel, 51
LSP (Label Switched Path), 150

M

Macro( ) application, 111, 258
macros, 110–112
    calling from the dialplan, 111
    defining, 110
    using arguments, 111
mail command, crash notifications, 54
mailboxes, creating, 107
MailboxExists( ) application, 259
mailing list, Asterisk users, 195
mailing lists, 6
make clean command, 37
make config command, 37
    installing Red Hat-style initialization
        scripts, 44
make install command, 47
make mpg123 command, 44
make program
    alternative compile-time
        arguments, 42–44
    Asterisk compilation with gcc, 42
    compiling Zapata telephony drivers, 36
make rpm command, 48
make samples command, 42
    disabling configuration file overwrites, 45
make update command, 46
make upgrade command, 47
Makefile
    optimizing Asterisk compilation, 44
    ztdummy, creating, 36
Manager interface, 180–182
    commands, 181
    Flash Operator Panel (FOP), 182
manager.conf file, 180, 323
MARK2 echo canceller, 39, 152
    aggressive suppression, 39
math program IVR, 198
Math( ) application, 259
mathematical operators, 101
MD5 hash, use in SIP authentication, 141

media
    redirection in SIP, 69
    transmission on SIP channel, 67
Media Access Control (MAC) address, 186
    identifying DUNDi peers by, 187
medium systems
    processor selection, 15
    system requirement guidelines, 10
meetme command, 345
meetme kick command, 346
meetme list command, 346
meetme lock command, 346
meetme mute command, 346
meetme unlock command, 347
meetme unmute command, 347
MeetMe( ) application, 117, 259
meetme.conf file, 117, 324
MeetMeAdmin( ) application, 261
MeetMeCount( ) application, 117, 261
messaging, uniting text with voice, 203
MGCP (Media Gateway Control
    Protocol), 143
mgcp.conf file, 144, 324
Milliwatt( ) application, 262
m-law companding, 129
m-law companding algorithm
    use with G.711 codec, 145
    voice encoding on T-1 circuit, 132
m-law/A-law precompilation, 38
MMX instructions, 44
MMX optimization, 38
modem.conf file, 324
modems
    2100-Hz tone during negotiation, 39
    external vs. internal, for Asterisk, 17
modinfo command, 41
modprobe command, 41
    loading zaptel with, 51
    loading ztdummy, 51
modules
    Asterisk, directory for loadable
        modules, 54
    monitoring loading by starting Asterisk
        with -c flag, 75
    Zaptel, loading, 50–52
modules.conf file, 54, 302–312
mohmp3/ directory, 55
Molex connector on the TDM400P for FXS
    modules, 61
Monitor( ) application, 262
monitoring children, 199

motherboards
    choosing for Asterisk system, 15–17
    PCI Version 2.2 support, 49
    USB controller for ztdummy, 25
    VIA-based, compiling Asterisk, 45
Moving Picture Experts Group Audio Layer 3
        Encoding Standard (see MP3)
MP3
    as codec, 148
    files in mohmp3/ directory, 55
    playing natively with Asterisk add-on, 46
MP3Player( ) application, 263
mpg123 program, 44
MPLS (Multiprotocol Label Switching), 150
multi-line phones, SIP, 71
multiple procesors, use of, 16
Multiprotocol Label Switching (see MPLS)
multitasking processes, Asterisk and, 12
music on hold
    licensing and, 148
    mohmp3/ directory, 55
    MP3, use with VoIP systems, 148
    SIP channel, 69
    streaming MP3s, 44
    timing source, 25
MusicOnHold( ) application, 263
musiconhold.conf file, 325

N

named extensions, 90
naming
    contexts, 78
    extensions, 79
NANP (North American Number Plan), toll
        fraud and, 94
NAT (Network Address Translation)
    H.323 protocol and, 143
    IAX and, 140
    SIP and, 141
    SIP extensons and, 69
NBScat( ) application, 263
Network Address Translation (see NAT)
network bandwidth, 206
Network Interface Card (NIC), 17
networking, built-in on motherboards, 17
NIC (Network Interface Card), 17
NoCDR( ) application, 264
noise on electrical circuits, 21
nonce, 141

nonlinear processor (NLP), making stronger
        with aggressive echo
        suppressor, 39
NOOP command, 295
NoOp( ) application, 264
Nortel
    Nortel Business Communications
        Manager, 1
    proprietary VoIP protocol,
        UNISTIM, 144
North American Number Plan (NANP), tool
        fraud and, 94
NOTICE messsages, 75
numbering extensions, 89
Nyquist's Theorem, 128

O

ob_implicit_flush function, 164
ob_implicit_flush(false) command, 168
OC (optical character) circuits, 132
OC-1 circuit, 133
off-hook (analog circuit), 121
on-hook (analog circuit), 121
Open Settlement Protocol (OSP), 325
open source telephony, 193–200
    fear campaign against, 205
    open architecture, 194
    opportunities, 207
    passionate community, 195
    rapid response to new technologies, 195
    responding to customer needs, 194
    standards compliance, 195
    things now possible, 196–200
        conference rooms, 199
        home automation, 199
        legacy PBX migration gateway, 196
        low-barrier IVR, 198
OpenH323 Gatekeeper, 143
OpenSSL, 32
    development library, installing, 48
operating system, debugging AGI scripts
        from, 173
operators, 100
opermode parameter, 41
optical carrier (OC) circuits, 132
option string argument, Dial( )
        application, 87
osp.conf file, 325
oss.conf file, 325
outbound calls
    enabling in dialplan, 95
    making from internal context, 97

outbound IAX connections, configuring, 74
outgoing connections, peers, 153
outgoing/ directory, 56
output buffering
    flushing after every write in Python AGI
        script, 172
    turning off in AGI Perl scripts, 159, 163
    turning off in AGI PHP script, 164, 168

P

package managers, 45
    RPM, 48
packages, 31
    requirements, 32
packet-based
    connections, 136
    telephone network, exclusive Asterisk
        connection to, 25
packetization of audio streams for transport
        over IP networks, 138
packet-switched networks, 135
Park( ) application, 264
ParkAndAnnounce( ) application, 265
ParkedCall( ) application, 265
parking calls, 115
parser generator program (bison), 32
passwords
    DUNDi user, 189
    secret (SIP), 69
    voicemail mailbox, 107
pattern matching, 92–95
    DUNDi dialplan, 189
    examples, 94
    syntax, 93
PauseQueueMember( ) application, 265
pbx_dundi module, 186
pbx_dundi.so module, 188
PBXs (Private Branch eXchanges), 1
    dedicated electrical circuit for, 20
    digital, 26
        terminals, 29
    electrical circuits, 21
PBXs (Public Branch eXchanges), legacy,
        Asterisk as migration gateway, 196
PC platforms, Asterisk on, 17
PCI hardware
    access by Zaptel and other device
        drivers, 51
    zaptel module, using with, 51
PCI hardware for timing, 36
PCI ID (TDM400P Revision H card), 40

PCI slots, server vs. workstation
        motherboards, 16
PCI Version 2.2, 49
PCM (Pulse-Code Modulation), 123–130
    Adaptive Differential PCM
        (ADPCM), 146
    aliasing, 129
    bandwidth of PCM-encoded telephone
        circuit, 131
    digitally encoding analog waveform, 123
    encoded analog waveform, 124
    encoding method in E-1 circuits, 132
    G.711 codec, 145
    increasing sampling resolution and
        rate, 126
    Nyquist's Theorem, 128
PDAs, connecting to voicemail to retrieve
        messages, 29
peering, 203
peers, 152
    connections defined as, 153
    DUNDi, defining, 187
    IAX outbound calls, authentication, 72
    IAX, placing outbound calls, 74
    SIP connections, 69
performance
    issues for hardware selection, 11–13
    shortcomings on inadequate systems, 9
peripherals
    connection to electrical receptacle, 20
    IRQ latency and, 13
Perl
    AGI library, 163
    writing AGI scripts in, 159–163
        summary of important steps, 163
phone trees, 84
phone.conf file, 325
phones (see telephones)
PHP
    AGI library, 168
    writing AGI scripts in, 163–168
        important steps to remember, 168
        invoking PHP with –q switch, 163,
            168
physical telephones, 25–28
pid (process id) information, 57
PLAR (Private Line Automatic Ringdown)
        circuit, 66
platform selection process, 9
Playback( ) application, 82, 266
    sound files, specifying, 180

Playtones( ) application, 266
ports
    RTP (Real-time Transport Protocol), 68
    SIP, 71
power supplies, 17
    computer, 17
    power quality issues and, 20
    redundant, 18
power-conditioning, UPSs, 19
PRA (Primary Rate Access) ISDN circuits, 134
precompiled Asterisk binaries, 45
Prefix( ) application, 267
PRI (Primary Rate Interface) ISDN
            circuits, 134
pri debug command, 347
pri intense debug span command, 347
PRI libraries (see libpri package)
pri no debug command, 347
pri show debug command, 348
pri show span command, 348
Primary Rate Interface (PRI) ISDN, 27
print command, 163
print STDERR command, 163
priorities (extension), 80
    Dial( ) application and, 87
    s extension, 82
    unnumbered, 81
privacy.conf file, 325
PrivacyManager( ) application, 267
Private Branch eXchanges (see PBXs)
Private Line Automatic Ringdown (PLAR)
            circuit, 66
process ID (pid) information, 57
processes, running concurrently on the
            system, Asterisk and, 12
processors
    choosing for Asterisk system, 13–15
    multiple, use of, 16
professional's PBX, 5
profiling information (debug), 45
Progress( ) application, 268
prompts
    asterisk-sounds package, 46
    customizing system prompts, 179
    recording from the dialplan, 181
proprietary digital telephones, 26
proxy servers, SIP, 71
PSTN (Public Switched Telephone
            Network), 3
    circuit types, 131
    connecting Asterisk to, 22–25

echo cancellation, 11
G.711 codec, 145
Last Mile, 122
Public Switched Telephone Network (see
            PSTN), 3
public/private key system, 55
Pulse-Code Modulation (see PCM)
Putland, Karl, 172
Pyst module, 172
Python
    AGI library, 172
    writing AGI scripts in, 169–172
        important steps to remember, 172

Q
qcall/ directory, 56
QoS (Quality of Service), 148–151
    best effort method and, 151
    challenges to, 206
    differentiated service (DiffServ), 150
    guaranteed service, 150
    TCP, UDP, and SCTP, 148–149
Quality of Service (see QoS)
quantization, 123
    companding and, 129
Queue( ) application, 268
queues.conf file, 326–328

R
Random( ) application, 269
RAS (Remote Access Server), turning Asterisk
            into, 39
Read( ) application, 269
RealTime application, 270
Realtime Transport Protocol (see RTP)
RealTimeUpdate( ) application, 270
recEive & transMit signaling (see E&M
            signaling)
RECEIVE CHAR command, 295
RECORD FILE command, 162, 295
Record( ) application, 270
recordings, creating from the dialplan, 181
Red Hat Linux, 31
    initialization scripts, 53
    initialization scripts, installing with make
            config, 44
Red Hat Package Manager (RPM), 48
redundant power supplies, 18
referencing variables, 90
    Unix environment variables, 92
register statement (iax.conf.sample), 73

register statements, 154
  IAX channels, 214
regular expressions
  operator, 101
  Perl-compatible, in AGI PHP script, 165
regulatory wars, 205
reinvite (in SIP), 69
  situations where it won't be issued, 70
releases, stable CVS branch vs., 33
reload command, 348
reloading/restarting after editing
     configuration files, 59
Remote Access Server (RAS), turning Asterisk
     into, 39
remote Asterisk console, connecting to, 75
remove command, 348
RemoveQueueMember( ) application, 271
REN (Ringer Equivalence Number), 120
res_crypto.so module, 188
  for RSA key checks, 48
Reservation protocol (RSVP), 150
ResetCDR( ) application, 271
resource requirements for Asterisk, 9
resources, 302
ResponseTimeout( ) application, 103, 272
restart command, 349
RetryDial( ) application, 272
Ring wire (analog phones), 122
ringer (analog phone), 120
ringing (telephone), boosting voltage for, 38
ringing tone, generating with Dial( ), r
     option, 87
Ringing( ) application, 273
rmmod (remove module) command, 63
root access for writing to /usr/src/
     directory, 33
RPM (Red Hat Package Manager), 48
rpt.conf file, 329
RSA key checks, module for, 48
RSA public/private key pair, authentication
     for IAS FWD incoming calls, 73
RSVP (Reservation protocol), 150
RTP (Realtime Transport Protocol), 67
  IAX and, 139
  use with H.323, 143
rtp.conf file, 329

S

s (start) extension, 82
safe_asterisk script, 52, 53
  remote console on TTY9, 157

sampling analog waveform for digital
     encoding, 123
  increasing resolution and rate, 126
  quality effects of sampling rate, 125
sampling frequency for digitally encoding
     analog signal, 128
Sangoma cards, 13
Sangoma Technologies, SS7 support, 135
save dialplan command, 350
SAY ALPHA command, 296
SAY DATE command, 296
SAY DATETIME command, 296
SAY DIGITS command, 297
SAY NUMBER command, 162, 167, 297
SAY PHONETIC command, 297
SAY TIME command, 297
SayAlpha( ) application, 273
SayDigits( ) application, 95, 273
sayit function, 170
saynumber function, 170
SayNumber( ) application, 273
SayPhonetic( ) application, 274
SayUnixTime( ) application, 275
scalability of Asterisk, 10
SCCP (Skinny Client Control Protocol), 144
scripts, AGI
  writing in Perl, 159–163
  writing in PHP, 163–168
  writing in Python, 169–172
SCTP (Stream Control Transmission
     Protocol), 149
secret (SIP authentication password), 69
security
  enforcement with dialplan contexts, 79
  H.323 protocol, 143
  IAX protocol, 140
  remote console on TTY9, disabling, 157
  SIP, 141
  system environment, 22
SECURITY file, 79
SEND IMAGE command, 161, 298
SEND TEXT command, 161, 298
SendDTMF( ) application, 275
SendImage( ) application, 275
SendText( ) application, 276
SendURL( ) application, 276
server-class motherboards, PCI slots, 16
servers
  electrical circuits, 21
  hardware selection, 10–18
    choosing a motherboard, 15–17
    choosing a processor, 13–15

performance issues, 11–13
  power supply requirements, 17
large Asterisk systems, 10
SET AUTOHANGUP command, 298
SET CALLERID command, 298
set command, 350
SET CONTEXT command, 299
SET EXTENSION command, 299
SET MUSIC ON command, 299
SET PRIORITY command, 299
SET VARIABLE command, 299
Set( ) application, 91, 276
  retrieving data from AstDB, 113
  storing data in AstDB, 113
SetAccount( ) application, 277
SetAMAFlags( ) application, 277
SetCallerID( ) application, 277
SetCallerPres( ) application, 278
SetCDRUserField( ) application, 278
SetCIDName( ) application, 279
SetCIDNum( ) application, 279
SetGlobalVar( ) application, 91, 279
SetGroup( ) application, 279
sethdlc utility, 39
SetLanguage( ) application, 280
SetMusicOnHold( ) application, 280
SetRDNIS( ) application, 280
SetVar( ) application, 281
shortcomings of traditional telephone
        systems, 3
show functions command, 103
show subcommands, 351–354
sidetone, 121, 151
signaling information (D-channels), 134
signaling methods
  for analog circuits, 62
  hardware channels, 64
  kewlstart, on FXS channel, 66
signaling protocols
  digital, 133
    Channel Associated Signaling
        (CAS), 133
    ISDN, 134
    SS7, 135
Signaling System 7 (SS7), 135
sinusoidal (sine) wave, 124
SIP (Session Initiation Protocol), 140,
        219–228
  channel configuration, 58
  channel definitions, 222–228

configuring, 67–72
  client, 70
  dialplan, 71
  sip.conf file (example), 68
conversion of proprietary digital signals
        to, 27
future of, 141
general parameters, 219–222
NAT and, 141
security, 141
sip commands, 354–356
SIP Interoperability Test (SIPIT), 195
sip show registry command, 155
sip.conf file, 59, 330
  context name, 70
  enabling DNS SRV record lookups, 69
  example, 68
sip_notify.conf file, 330
SIPAddHeader( ) application, 281
SIPDtmfMode( ) application, 281
SIPGetHeader( ) application, 282
Skinny Client Control Protocol (SCCP), 144
skinny.conf file, 330
small systems
  choosing CPU, 14
  system requirement guidelines, 10
Smith, Allison, 180
soft hangup command, 356
soft phones, 28
  connecting to Asterisk, 59
  X-Lite SIP client, 70
SoftHangup( ) application, 282
SOHO systems, system requirement
        guidelines, 10
SONET (Synchronous Optical
        Network), 133
sound files (pre-recorded)
  playing for unanswered or busy
        destinations, 87
  playing over a channel, 82
  playing with Background( ), 84
sound prompts (asterisk-sounds
        package), 32
sound recordings, creating from the
        dialplan, 181
sound waves, translation to electrical
        signals, 26
sounds (phone system), for particular
        countries, 62
sounds directory, 82

sounds.txt file, 56
sounds/ directory, 56
sounds-extra.txt file, 56
source code (Asterisk)
    obtaining, 32–34
        extracting source code, 33
        from CVS, 34
    updating, 46
sox application, 180
spam, 205
speech processing, 200
speech recognition, 201
Speex codec, 147
Spencer, Mark, ix, 195
spool directory, 56
SRV records, DNS, 69
SS7 (Signaling System 7), 135
stable Asterisk releases, 33
staging directory, 45
start (s) extension, 82
stations, 29
STDERR, 157
    ensuring open file handles in AGI PHP
        script, 164
    summary of AGI script tests, 163
    using fflush function after writing to, 168
    writing to with fwrite function, 168
STDIN, 157
    ensuring open file handles in AGI PHP
        script, 164
    reading variables from using the fgets
        function, 168
STDOUT, 157
    AGI script sending commands to
        Asterisk, 158
    ensuring open file handles in AGI PHP
        script, 164
    using fflush command after writing
        to, 167
    using fflush function after writing to, 168
    writing to with fwrite function, 168
steps, 129
stop commands, 357
StopMonitor( ) application, 282
StopPlaytones( ) application, 282
Stream Control Transmission Protocol
        (see SCTP)
STREAM FILE command, 160, 163, 167,
        300
strict language checking, AGI scripts in
        Perl, 159, 163

string length of a variable, calculating, 103
StripLSD( ) application, 283
StripMSD( ) application, 283
STS-1 circuit, 133
SubString( ) application, 284
Suffix( ) application, 284
switch hook, 121
symbolic link to Linux kernel sources
        depmod errors during compilation, 50
        Linux 2.4 kernel, 36
Synchronous Optical Network (see SONET)
sys.stdin.readline command, 172
sys.stdout.flush command, 172
sys.stdout.write command, 172
system prompts, customizing, 179
system requirement guidelines, 10
System( ) application, 285

T

T-1 lines, 58, 132
    CAS (Channel Associated Signaling), 133
tailor-made private telecommunications
        networks, 207
tar application, extracting compressed source
        code, 34
T-carrier circuits, 132
TCP transport-layer protocol
    SIP and, 67
    VoIP Quality of Service and, 148
TDD MODE command, 300
TDM (Time Division Multiplexing)
        hardware, 41
TDM bridged calls, echo cancellation, 152
TDM400P card, 23
    determining FXO and FXS ports, 60
    requirement for PCI Version 2.2, 49
    with one FXO module, verifying hardware
        and ports, 63
TDM400P Revision H card, PCI ID, 40
technology (or transport), Dial( )
        application, 86
teenagers' calls, managing, 200
telecommunications networks (private),
        tailor-made, 207
telecommunications systems
    hardware, 22–25
    power supply requirements, 17
telephone extensions, 79
Telephone Number Mapping (ENUM)
        group, 204
telephone numbering plan, maintaining, 204

telephones, 25–29
    analog, parts of, 120–122
    analog interfaces, 58
    boosting ringing voltage, 38
    echo generated by low-quality
        phones, 151
    multiline, SIP, 71
    physical, 25–28
        analog phones, 26
        digital telephones, 26
        IP telephones, 27
        ISDN telephones, 27
    soft phones, 28
telephony
    analog, 119–122
    digital, 122–130
        Pulse-Code Modulation
            (PCM), 123–130
    digital circuit-switched network, 130–135
    fear campaign by traditional industry
        players, 205
    open source, promise of, 193–200
        open architecture, 194
        passionate community, 195
        rapid response to new
            technologies, 195
        responding to customer needs, 194
        standards compliance, 195
        things now possible, 196–200
    opportunities for open source, 207
    packet-switched networks, 135
    paradigm shift, 193
    traditional, problems with, 190
        clinging to the past, 193
        closed thinking, 191
        limited standard compliance, 192
        slow release cycles, 192
telephony adaptors, 29
telephony drivers, Zapata, compiling, 36
temperature (equipment rooms), 21
terminals, 29
text2wave utility (Festival), 177
text-messaging systems, uniting with
        voice-messaging, 203
text-to-speech engines, Cepstral, 177
three-way calling
    on analog line, 121
    on FXO channel, 64
Time Division Multiplexing (TDM)
        hardware, 41
time limit, turning off for PHP invocations
        from command line, 168

Time To Live (ttl) field (dundi.conf), 186
time zone (tx) option, mailboxes, 108
time-based conditional branching, 105
TIMEOUT( ) function, 103
timeouts
    Dial( ) application argument, 86
    handling for user input, 85
timing device (ztdummy), 36
timing source, systems without hardware
        mechanism, 25
Tip and Ring wires (analog phones), 122
TLS (Transport Layer Security), 141
/tmp/ directory, prompt recordings in, 181
toll fraud
    NANP (North American Number Plan)
        and, 94
    securing your system from, 79
tone zone, 344
    setting in zconfig.h, 40
touch-tone dialing, 120
Transfer( ) application, 285
translation costs between audio formats, 83
Transmission Control Protocol (see TCP)
Transport Layer Security (TLS), 141
transport protocols, Internet, real-time media
        streaming and, 138
transport-layer protocols for SIP, 67
trunking
    IAX protocol, 139
    iax2 trunk debug command, 343
    use of ISDN, 134
trunking provided by BRI ISDN, 27
TrySystem( ) application, 285
TXTCIDName( ) application, 286

U

udev daemon, 51
UDP transport-layer protocol
    use with SIP, 67
    VoIP Quality of Service and, 149
UHCI USB controller, 36
    not accessible on Linux 2.4 kernels, error
        caused by, 49
    verifying, 50
UHCI-type USB controller, 25
unanswered calls, Dial( ) application
        handling of, 87
unavailable message for voicemail, 108
Unified Messaging, 203
Uninterruptible Power Supplies (see UPSs)
UNISTIM (Nortel VoIP protocol), 144

Unix environment variables, accessing from Asterisk, 92
unload command, 357
UnpauseQueueMember( ) application, 286
UPSs (Uninterruptible Power Supplies)
    power conditioning and, 18
    redundant power supplies and, 18
URL argument, Dial( ) application, 88
USB controllers, for ztdummy, 25
usbcore module, verifying loading of, 50
usb-uhci module, 36
    use by ztdummy, shown in lsmod output, 52
    verifying loading of, 50
use strict command, 159, 163
User Datagram Protocol (see UDP)
UserEvent( ) application, 286
users, 152
    connections defined as, 153
    DUNDi, 188
    IAX, 72
    IAX FWD incoming calls, 73
    SIP connections, 69
/usr/lib/asterisk/modules/ directory, 54
/usr/src/ directory
    extraction and compilation of Asterisk source, 33
    symbolic link to kernel source, 36

V

/var/lib/asterisk/ directory, 55
    subdirectories, 55
/var/lib/asterisk/sounds/ directory, 82
    custom system prompts, 180
/var/log/asterisk/ directory, 57
/var/log/asterisk/cdr-csv directory, 57
/var/run/ directory, 57
/var/spool/asterisk/ directory, 56
Variable Bitrate (VBR) codecs, 147
variable manipulation, 99–102
variables
    AstDB, setting, 113
    ImportVar( ) application, 257
    Macro( ) application, 111
    reading from STDIN with fgets function, 168
    sent by Asterisk to AGI script at startup, 157, 159
    sent by Asterisk to AGI script on startup, 165

variables, using in dialplan, 90–92
    adding variables to dialplan, 92
    channel variables, 91
    environment variables, 92
    global variables, 91
VERBOSE command, 300
Verbose( ) application, 287
verbosity, setting level for debugging messages, 75
versions, xv
VIA-based motherboards, compiling Asterisk on, 45
video, 202
video camera connected to PC, live chats through, 29
video components (built-in) on motherboards, 17
video-conferencing, 202
    IP transport mechanism (H.323), 142
Virtual Private Network (VPN), use with IAX, 140
VMAuthenticate( ) application, 287
voice menus, creation with Background( ), 84
Voice of Asterisk, 180
voice prompts for Asterisk, sounds/ directory, 56
voice, high-fidelity, 201
voice-compression algorithms, 144
voicemail, 106–109
    accessing, 109
    adding to dialplan, 108
    applications, 256
    creating mailboxes, 107
    dial-by-name directory, 109
VoiceMail( ) application, 108, 287
voicemail.conf file, 107, 330–336
voicemail/ directory, 56
VoiceMailMain( ) application, 109, 288
voice-messaging, uniting with text messaging, 203
Voicetronix, analog cards, 23
VoIP (Voice over IP), 2, 137–155
    access to, for legacy PBX, 197
    alternative route to an extension number or PSTN telephone number, 184
    Asterisk and, 152–155
        register statements, 154
        users, peers, and friends, 152
    bottleneck engineering, 205
    codecs, 144–148
    echo, 151

Free World Dialup (FWD) service
    provider, 72
gateways providing access to PSTN
    circuits, 24
mobility with Wi-Fi, 203
need for protocols, 138
packet-switched networks, 135
proprietary protocols, 144
protocols, 139–144
    H.323, 142
    IAX, 139
    MGCP, 143
protocols in use today, 68
Quality of Service (QoS), 148–151
    best effort method and, 151
    DiffServ, 150
    guaranteed service, 150
    TCP, UDP, and SCTP, 148–149
Quality of Service, challenges to, 206
Session Initiation Protocol (SIP), 140
SIP and IAX protocols, 58
spam, 205
Zapata Telephony Project, 2
(see also SIP)
voltage, 19
    AC and DC, in analog phones, 120
    boosting for telephone during ringing, 38
VPN (Virtual Private Network), use with
    H.323 protocol, 143

W

WAIT FOR DIGIT command, 162, 300
Wait( ) application, 288
WaitExten( ) application, 288
WaitForRing( ) application, 289
WaitForSilence( ) application, 289
WaitMusicOnHold( ) application, 289
WARNING messages, 75
watchdog (Zaptel), 40
WAV files, converting to GSM format, 180
wcfxo driver, error encountered in
    loading, 49
wctdm driver
    errors encountered in loading, 49
    loading with modprobe, 62
    passing module parameters to configure
        Zaptel, 41
weather reporting IVR, 198
web page for this book, xvi
wget command, obtaining latest stable source
    code via, 33

while (<STDIN>) loop, 163
While( ) application, 289
Wi-Fi, 203
Wiki (Asterisk), 7
wildcard matches, 93
Wi-MAX, 203
Win-modems, avoiding, 17
wireless, 202
workstation-class motherboards, 16

X

X-Lite client, 70
    testing, 71

Y

yum install bison command, 48
yum install gcc command, 47
yum install rpmbuild command, 48

Z

zap commands, 357
Zapata
    channel module (chan_zap), 35
    compiling, telephony drivers, 36
    hardware configuration for FXO
        channel, 63–65
    hardware configuration for FXS
        channel, 66
Zapata telephony drivers (see zaptel package)
Zapata Telephony Project, 2
zapata.conf file, 59, 336
    configuration for FXS channel, 66
    echo cancellation, enabling, 152
    hardware configuration for FXO
        channel, 63–65
Zapateller( ) application, 115, 290
ZapBarge( ) application, 290
ZapRAS program, 39
ZapRAS( ) application, 290
ZapScan( ) application, 291
Zaptel
    compiling, 35–41
        common problems, 48–50
        passing module parameters to
            configure, 41
        zconfig.h file, 38–40
        ztcfg and zttool programs, 37
    hardware configuration for FXO
        channel, 61–63

Zaptel (*continued*)
   hardware configuration for FXS
      channel, 65
   loading modules, 50–52
      zaptel, 51
      ztdummy, 51
Zaptel cards, motherboard selection and, 16
zaptel module, loading, 51
zaptel package, 31
zaptel.conf file, 51, 59, 336
   configuring FXO channel with FXS
      signaling, 61
   storage in /etc/ directory, 54
zconfig.h file, 38–40
   BOOST_RINGER option, 38
   disabling echo cancellation, 39
   disabling m-law/A-law
      precompilation, 38
   echo cancellation method, choosing, 38
   echo canceller algorithms, 152
   enabling aggressive echo suppression, 39

   enabling CAC ground start signaling, 40
   enabling MMX optimization, 38
   enabling ZapRAS, 39
   enabling Zaptel watchdog, 40
   setting default tone zone, 40
   TDM400P Revision H card PCI ID, 40
zonedata.c file
   phone system sounds for particular
      countries, 62
   tones (dial, busy, ring, stutter), 40
ztcfg program, 37
   verifying loading and configuration of
      Zaptel hardware and ports, 63
ztdummy driver
   compiling, 36
   unresolved symbol link when loading, 49
ztdummy module, 25
   loading, 51
zttool program, 32, 37
   determining state of your hardware, 63

About the Authors

Jim Van Meggelen is President and CTO of Core Telecom Innovations, a Canadian-based provider of open source telephony solutions. He has over 15 years of enterprise telecom experience, for such companies as Nortel, Williams, and Telus, and has extensive knowledge of both legacy and VoIP equipment from manufacturers such as Nortel, Cisco, and Avaya.

Jim was the architect of two of the world's largest managed enterprise voice networks, each solution serving roughly 20,000 users in more than 1,000 communities across Canada and providing telecommunications in 5 different languages through 6 time zones, administered completely from a central location. These networks pioneered the use of extensive automation and database control in a branch voice network—functionalities not generally available in proprietary telecommunications systems. Jim has now moved on from the world of proprietary telecom, and is commited to open source telephony.

Jim is one of the principal contributors to the Asterisk Documentation Project. He enjoys teaching, public speaking, improvisational acting, and writing.

Jared Smith is one of those rare individuals whose beloved hobby is the same as his profession. The son of a computer store owner, Jared wrote his first computer program at the age of 7 on his Commodore 64. The obvious choice of major for this geek-in-embryo was Computer Engineering, and Jared received his Bachelor of Science degree with a minor in Computer Science from Utah State University. He now has over a decade of professional systems administration and programming experience in the simulation, market research, and web analytics industries. As a key architect of one of the world's largest Asterisk installations, Jared has a wealth of hands-on telephony and VoIP knowledge, which he shares through users groups and various public speaking engagements. He is an active member of the Asterisk community and a co-founder of the Asterisk Documentation Project.

Jared is active in his community, donating Asterisk services to local schools and serving in his church. The greatest joy in Jared's life comes from spending time with his children, Caleb and Sydney Jo, and his wife, Jenny.

Leif Madsen is a graduate of the Telecommuncations Technology program from the Sheridan Institute of Technology and CEO of LeifMadsen Enterprise, Incorporated, a documentation and consulting firm specializing in Asterisk. He was one of the first Digium Certified Asterisk Professionals (dCAP), and assists with the Astricon conferences and trainings organized by IPsando, LLC.

Leif first took an interest in Asterisk while attempting to find a voice conferencing solution for himself and his friends. After someone suggested trying Asterisk, the obsession began. Wanting to contribute and be involved with the community, and noticing the lack of Asterisk documentation, he co-founded the Asterisk Documentation Project.

Colophon

Our look is the result of reader comments, our own experimentation, and feedback from distribution channels. Distinctive covers complement our distinctive approach to technical topics, breathing personality and life into potentially dry subjects.

The animals on the cover of *Asterisk: The Future of Telephony* are starfish. Starfish are classified as *Asteroidea*. They are a group of echinoderms, spiny-skinned invertebrates found only in the sea. Most starfish have five-fold symmetry (arms or rays in multiples of five), though some species can have four or nine arms. But all starfish are radially symmetrical: they have arms or rays branching out from a central body disc. There are over 1,500 species of starfish.

Starfish live on the floor of the sea and in tidal pools, clinging to rocks and moving (slowly) using a water-based vascular system to manipulate their hundreds of tiny, tube-like legs, called *podia*. A small bulb or *ampulla* at the top of the tube contracts, expelling water and expanding the starfish's leg. The ampulla relaxes, and the leg retracts. Starfish use muscles to bend their legs, but it is the flow of water pressure that keeps the feet moving. At the tip of each leg, starfish have suction cups that allow them to pry open clam, oyster, or mussel shells. Many starfish can push their stomachs out through their mouths in order to digest their prey in its shell. Starfish are carnivores; they eat coral, fish, and snails, as well as bivalves.

Starfish can flex and rearrange their arms to fit into small places as they move over the ocean floor. At the end of each arm, they have eyespots, primitive sensors that detect light and help the starfish determine direction. Starfish also have the ability to regenerate a missing limb. Some species can even regrow a complete, new starfish from a severed arm.

Colleen Gorman was the production editor, and Rachel Wheeler was the copyeditor for *Asterisk: The Future of Telephony*. Ann Schirmer proofread the book. Colleen Gorman and Marlowe Shaeffer provided quality control. Ellen Troutman wrote the index.

Ellie Volckhausen designed the cover of this book, based on a series design by Edie Freedman. The cover image is a 19th-century engraving from the Dover Pictorial Archive. Karen Montgomery produced the cover layout with Adobe InDesign CS using Adobe's ITC Garamond font.

David Futato designed the interior layout. This book was converted by Keith Fahlgren to FrameMaker 5.5.6 with a format conversion tool created by Erik Ray, Jason McIntosh, Neil Walls, and Mike Sierra that uses Perl and XML technologies. The text font is Linotype Birka; the heading font is Adobe Myriad Condensed; and the code font is LucasFont's TheSans Mono Condensed. The illustrations that appear in the book were produced by Robert Romano, Jessamyn Read, and Lesley Borash using Macromedia FreeHand MX and Adobe Photoshop CS. The tip and warning icons were drawn by Christopher Bing. This colophon was written by Colleen Gorman.

Better than e-books

Buy *Asterisk: The Future of Telephony* and access the digital edition FREE on Safari for 45 days.

Go to www.oreilly.com/go/safarienabled
and type in coupon code V2IJ-6NFH-DHWU-FTHM-TGYE

Search
thousands of
top tech books

Download
whole chapters

Cut and Paste
code examples

Find
answers fast

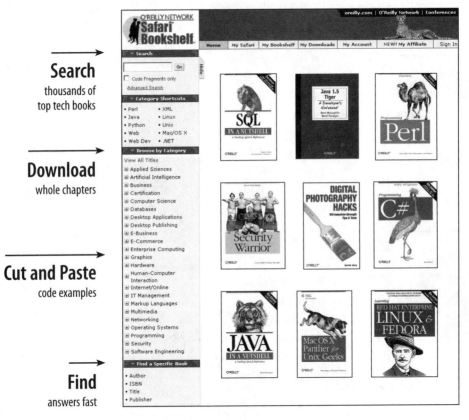

Search Safari! The premier electronic reference
library for programmers and IT professionals.

Related Titles from O'Reilly

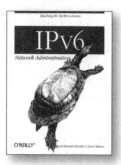

Networking

802.11 Security

802.11 Wireless Networks: The Definitive Guide, *2nd Edition*

Asterisk: The Future of Telephony

BGP

Building Wireless Community Networks, *2nd Edition*

Cisco Cookbook

Cisco IOS Access Lists

Cisco IOS in a Nutshell, *2nd Edition*

DNS & BIND Cookbook

DNS & BIND, 4th Edition

Essential SNMP, *2nd Edition*

IP Routing

IPv6 Essentials

IPv6 Network Administration

LDAP System Administration

Managing NFS and NIS, *2nd Edition*

Network Troubleshooting Tools

RADIUS

sendmail 8.13 Companion

sendmail, *3rd Edition*

sendmail Cookbook

SpamAssassin

Switching to VOIP

TCP/IP Network Administration, *3rd Edition*

Unix Backup and Recovery

Using Samba, *2nd Edition*

Using SANs and NAS

Windows Server 2003 Network Administration

Our books are available at most retail and online bookstores.

To order direct: 1-800-998-9938 • *order@oreilly.com* • *www.oreilly.com*

Online editions of most O'Reilly titles are available by subscription at *safari.oreilly.com*

Keep in touch with O'Reilly

Download examples from our books

To find example files from a book, go to: *www.oreilly.com/catalog* select the book, and follow the "Examples" link.

Register your O'Reilly books

Register your book at *register.oreilly.com* Why register your books? Once you've registered your O'Reilly books you can:

- Win O'Reilly books, T-shirts or discount coupons in our monthly drawing.
- Get special offers available only to registered O'Reilly customers.
- Get catalogs announcing new books (US and UK only).
- Get email notification of new editions of the O'Reilly books you own.

Join our email lists

Sign up to get topic-specific email announcements of new books and conferences, special offers, and O'Reilly Network technology newsletters at:

elists.oreilly.com

It's easy to customize your free elists subscription so you'll get exactly the O'Reilly news you want.

Get the latest news, tips, and tools

www.oreilly.com

- "Top 100 Sites on the Web"—PC Magazine
- CIO Magazine's Web Business 50 Awards

Our web site contains a library of comprehensive product information (including book excerpts and tables of contents), downloadable software, background articles, interviews with technology leaders, links to relevant sites, book cover art, and more.

Work for O'Reilly

Check out our web site for current employment opportunities:

jobs.oreilly.com

Contact us

O'Reilly Media, Inc.
1005 Gravenstein Hwy North
Sebastopol, CA 95472 USA
Tel: 707-827-7000 or 800-998-9938
    (6am to 5pm PST)
Fax: 707-829-0104

Contact us by email

For answers to problems regarding your order or our products:
order@oreilly.com

To request a copy of our latest catalog:
catalog@oreilly.com

For book content technical questions or corrections: **booktech@oreilly.com**

For educational, library, government, and corporate sales: **corporate@oreilly.com**

To submit new book proposals to our editors and product managers:
proposals@oreilly.com

For information about our international distributors or translation queries:
international@oreilly.com

For information about academic use of O'Reilly books:
adoption@oreilly.com
or visit:
academic.oreilly.com

For a list of our distributors outside of North America check out:
international.oreilly.com/distributors.html

Order a book online

www.oreilly.com/order_new

 ®

Our books are available at most retail and online bookstores.
To order direct: 1-800-998-9938 • *order@oreilly.com* • *www.oreilly.com*
Online editions of most O'Reilly titles are available by subscription at *safari.oreilly.com*